BUILDING
EARLY AMERICA

Carpenters' Hall, the home of the Carpenters' Company of the City and County of Philadelphia, heads a narrow courtyard off Chestnut Street. It was built in 1770–73 from the designs of Robert Smith, a member of the Company. The First Continental Congress met here in 1774, and the Provincial Convention of Pennsylvania in 1776. The Hall was restored in 1857 and has been open to the public without charge ever since.

BUILDING EARLY AMERICA

Contributions toward the History of a Great Industry

THE CARPENTERS' COMPANY
OF THE
CITY AND COUNTY OF PHILADELPHIA

CHARLES E. PETERSON, *Editor*

CHILTON BOOK COMPANY
RADNOR, PENNSYLVANIA

Copyright © 1976 by
The Carpenters' Company of the City and County of Philadelphia
First Edition   All Rights Reserved
Published in Radnor, Pennsylvania, by Chilton Book Company
and simultaneously in Don Mills, Ontario, Canada
by Thomas Nelson & Sons, Ltd.
Manufactured in the United States of America

Library of Congress Cataloging in Publication Data
    Building Early America.
        Proceedings of a symposium held on the occasion of the
250th anniversary of the Carpenters' Company of the City
and County of Philadelphia, Mar. 27–29, 1974, in Philadelphia.
        Includes bibliographical references and index.
    1. Building – United States – History – Congresses.
    2. Carpenters' Company of the City and County of Phila-
delphia.   I. Peterson, Charles Emil, 1906–   II. Carpenters' Company of the City and County of Philadelphia.
TH23.B73 1975   690'.0973   75-26532
ISBN 0-8019-6294-3
ISBN 0-8019-6295-1 pbk.

For
SIGFRIED GIEDION
and
TURPIN C. BANNISTER
*who would have enjoyed
working with us*

Contents

Acknowledgments

Preparations for the symposium BUILDING EARLY AMERICA and the book of proceedings presented herewith have been in train for over five years. It would be hardly possible to list all the friends on both sides of the Atlantic whose help and encouragement brought them to pass.

But we must thank here the speakers who came to Philadelphia from near and far to share with us their knowledge of buildings and building. Afterward many went through laborious conversions to prepare chapters for this book. To their names should be added those fellow experts who presided so effectively at the symposium sessions: Professor Anthony N. B. Garvan, Dr. Walter J. Heacock, Dr. Brooke Hindle, Mr. Russell V. Keune and Mr. Robert M. Vogel.

For the financial underpinning of the symposium and book we are indebted to the generosity of the Annenberg School of Communications; M. A. Bruder & Sons, Incorporated; the Certain-teed Products Corporation; Mrs. Lammot duPont Copeland; the English-Speaking Union (Philadelphia Branch); the General Building Contractors Association of Philadelphia; the Daniel J. Keating Company; the Kilgore Company; Mr. Robert L. McNeil, Jr.; John McShain Charities; John J. Manley, Incorporated; the National Heritage Corporation; the National Trust for Historic Preservation; the Newcomen Society in North America; the Henry F. Ortlip Company; the Pennsylvania Society of Sons of the Revolution; the Rittenhouse Foundation; the Smithsonian Institution; the United States National Park Service; the Victorian Society in America; the Warner Company; Williard, Inc. and to the many individual subscribers.

In addition there were many members of the Carpenters' Company who liberally contributed. Further, the last three Company Presidents—Paul C. Harbeson (1973), John H. Ball (1974) and Richard W. Schwertner (1975) gave of their time and energy.

The able assistance of the wives of Company members who gave so abundantly of their talents in support of the symposium is especially appreciated.

Lastly, for bringing this book to press, we want to thank Editor Paul J. Driscoll of the Chilton Book Company for his patience, Mrs. Rita R. Robison for her expert services as assistant editor and Secretaries Hilda I. Guadalupe and June M. Bukowski for their endurance and industry from the first to the last. The detailed index at the end of this volume we owe to Thomas C. Guider.

CHARLES E. PETERSON

is a Fellow of the American Institute of Architects,
Past President of the Society of Architectural Historians
and of the Association for Preservation Technology.
In Britain he is a member of the Association for the Study
of Conservation of Historic Buildings, Honorary Correspondent of the
Committee on Training of Architects for Conservation,
and a Benjamin Franklin Fellow of the Royal Society of Arts.

Preface

THE ANNIVERSARY OF 1974

When a fraternity of builders comes to celebrate its anniversary marking a quarter of a millennium there is very little precedent to be consulted. Few other American organizations have lived as long as the Carpenters' Company of the City and County of Philadelphia. When the Revolutionary War overturned all local and provincial governments on the seaboard, we were already 50 years old. As associations go, only the most ancient of our colonial churches can claim seniority.

Back in 1924, on the occasion of our *200th* birthday, the festivities were mingled with the celebration of the *150th* anniversary of the First Continental Congress, a great political event which, because of the delegates' unwillingness to meet on public property, had taken place in our Hall. There was a great deal of oratory by public figures, including President Calvin Coolidge at the Academy of Music, much marching by military units and a big convocation at the Valley Forge encampment site. Planning for a celebration in 1974, however, required something quite different, as the public mood had changed and military flourishes were no longer fashionable.

The Carpenters' Company today is not an historical society, in spite of its evident respect for history and a stubborn devotion in caring for its old Hall. But our membership, approaching the 250th anniversary, felt that some review of the past—some recognition of the builder's part in the history of the new nation—was called for. When the American Revolution broke upon the colonies, Philadelphia was not yet a century old. But it was already the metropolis of the Atlantic Seaboard, its energetic and ambitious population having taken advantage of a liberal government, a fine site on salt water near waterpower in abundance and a rich, agricultural hinterland. The builders of the city had already produced a grand State House, churches in variety, a theater, an outstanding hotel, many inns and boarding houses and a hospital. In the center was Carpenters' Hall, a meeting place made available to the Continental Congress by a friendly society of mechanics in close league with Benjamin Franklin and his associates. Conveniently located between Boston and Savannah, Philadelphia was the logical convention city of its time.

Historians have a tendency to remember battle dates and elections while forgetting the people who actually created a civilization. Our master builders, largely unknown today, were not quoted in newspapers and their portraits were not painted for posterity. Even the business arrangements for erecting the finest structures are often missing from the manuscript records of their owners. Yet we know that the Carpenters' Company controlled the construction business of the city, as medieval guilds had done earlier in Europe. Members protected the mysteries of their craft so well that today it is not easy to recreate the workings of the early construction industry.

Thus, for 1974, the history of the building business seemed an appropriate and worthwhile subject for examination and we were fortunate to get the ready cooperation of an unusual group of experts, each an authority on some aspect of the subject. Proof that the symposium was an original and pioneering effort lies in the fact that few of the speakers knew each other before they assembled at *Building Early America*, March 27–29, 1974. They were meeting their colleagues for the first time, just as had the delegates to that First Continental Congress.

The outline of the proceedings was partly shaped by our cordial relations with the Carpenters' Company of London, one of the ancient livery companies still active in that great city. When we learned that a substantial delegation would be crossing the Atlantic for the occasion, we extended our program to acknowledge the American debt to the builders of Britain. The five speakers from England helped round out our presentation in every way.

The program of events operated on several levels, but the three day celebration came together at a great dinner held at the Union League Club in honor of our guests from overseas. The round of toasts and other expressions of good will were continued when, six weeks later, a large part of our Company flew over to enjoy the bountiful hospitality of our British friends.

A Symposium in Two Parts

It is a happy characteristic of our times that Early American buildings are more generally appreciated today than ever before. And more effort and money are being devoted to preserving them. Across the country personnel are being assembled and trained; the need for informed and reliable literature about buildings is becoming more urgent. The Carpenters' Company, which had taken the lead in 1857 by restoring its Hall and opening it free to the public, took the initiative again in 1974 with *Building Early America*.

The most likely audience for our symposium seemed to be the people who are entrusted with the care of old buildings: a large number came. So the first two days were devoted to the history of building and especially to materials and construction techniques. It seemed logical to start with the beginning, reaching back to British and Continental foundations and then forward until about 1860. The third day was spent in examining current problems in preserving and restoring these early American buildings, including organization and training personnel for such preservation.

Speakers could not be found for every important subject and a review of the entire history of building was too vast to attempt in three days. Certain areas, such as the evolution of the construction business as a trade, were hardly touched upon. But, even with the gaps in subject matter, the symposium focused attention on many aspects of a major industry whose beginnings are little understood.

This Book

The volume presented herewith corresponds to the agenda of the symposium. In converting lectures to chapters of a book, a special effort was made to incorporate material of permanent value for reference; no attempt was made to standardize the treatments by various authors. They are individuals, pursuing individual topics.

We have, we believe, managed to suggest some of the problems encountered by our predecessors as the handicraft age of building evolved through mechanical invention into the Age of Steam. We hope that the American public, including academia, will become more aware of our subject: the history of a great and useful industry. If future writers criticize us for having left out important subjects, our efforts will not have been in vain. We hope that other books will join this volume. Nothing less than a whole shelf full can do justice to the subject. As a contribution, we present these essays with pride.

CHARLES E. PETERSON

Introduction

BROOKE HINDLE

Dr. Hindle is director of the National Museum of History and Technology. He is the author of *The Pursuit of Science in Revolutionary America* (1956), *David Rittenhouse* (1964), *Technology in Early America* (1965), and editor of *America's Wooden Age* (1975)

This volume represents a new direction of achievement in the study of building. The papers published here were initially offered at a conference participated in by some of the most active people in this field. They concentrate attention upon the realities of building in England and early America in a manner that has long been called for but never seriously attempted. This effort reflects throughout the energy and insight of Charles E. Peterson, who conceived it.

Nearly a decade ago, Charles Peterson stopped by to visit me in my New York hospital room where I had been for a couple of months. He had one topic on his mind, building. He was convinced that historians had denied building the attention it deserved. Although he knew of my interest in carpenters and craftsmen, he insisted that building was a more central dimension of history than I recognized. We talked and talked—paying no attention to the fact that all the lights had gone out. Ultimately he found his way out in the dark, for the lights never came on. It was the night of the Great Blackout of 1965.

This book and the conference from which it derives show that Charles Peterson was right about the central character of building and also about the need to examine it in its own terms. The problem was not a failure of historians to recognize building as one element of their study, for many have done that. It was really a question of the frame of reference within which the subject was viewed, for all efforts at historical synthesis are distortions and each distorts some elements more than others. They relate discrete elements to a particular frame of inquiry; they always highlight those elements which relate most easily and directly to their fundamental interpretation. Subjects which may be important in their own right are given slight attention when their relationships are peripheral. So it has been with the study of building within all those subdivisions of history which have given it attention.

One of the earliest and potentially most important subfields to encompass building was the history of science. Building was regarded as a subdivision of the history of technology, and the history of technology has always been considered within the scope of the history of science. However, it began as a poor, poor sister. George Sarton's famous remark, that he was interested in "thinkers, not tinkers," described well the focus upon scientific ideas and the great minds which had conceived them.[1] More recently, science in its social dimensions has received more attention from historians of science—but the history of technology remains with them a minor concern, and the history of building of exceedingly limited interest.

Lately the history of technology has attained a fairly well defined identity of its own outside the history of science. This is an area of inquiry which seems more hospitable to the history of building, but even here achievements have not measured up to expectations or potentialities. Building is characteristically subsumed under civil engineering, taking a role that is different but nonetheless secondary. Some historians of technology are primarily concerned with internal developments, others with social and cultural dimensions. Although building may profitably be studied within each of these frameworks, so far little has been done either way.

Some American economic historians specifically include building within their area of concern, and most of them admit it along with other economic activities. Much of the importance of building in American society can be perceived in terms of its role in the economy. Yet, once again, this is an approach which does not give the primary emphasis to building and which does not find its questions within the internal determinants of building.

Perhaps the history of architecture has offered the most congenial home for the study of building. Here, however, the focus is upon art and aesthetics, and the view of the builder and engineer as a kind of inferior partner of the architect has tended to leave building in a position that is less than ideal.

What has been needed can be easily enough stated, and this volume gives hope that it may be possible to achieve. The need is not to cut the study of building free from the history of science, technology, economics, and architecture—but to launch building-centered inquiries in addition to the others. The objective is not antiquarianism, not an effort to discover how the work of a lefthanded hammerer differs from a righthanded one. Building needs to be studied in terms of its own values and questions—and within the largest possible framework. Building must be viewed as a central element in social history.

The beginning is to ask the right questions. In the nineteenth century a visiting English scholar questioned an American acquaintance about building practices, "Your wooden houses, I can't understand. Why don't you put up something in stone and brick that will be solid at the end of three hundred years, as we do in England?"

The American, not in the least abashed, responded, "It is because we don't want that kind of a house. Changes, improvements, new comforts of all sorts come so fast, that we don't want a house to last too long. This house is what I want, but not what my children will want. Even I want to make some structural change every ten years. I can now do it without being ruined, as I could not in one of your three-century dwellings."[2]

The answer is certainly insufficient and may even be wrong, but the question is exactly right. The questions that need to be asked now are questions about how American building reflected the characteristics of American society and how it, in turn, influenced that society. Many of the right questions are asked on the following pages; some of them are brilliantly answered. This volume points the direction in which further historical inquiries about building should move.

NOTES

1. This is a widely quoted remark which represents very clearly George Sarton's view of the history of science. It has not been possible to find a specific written source for it, and it may not have appeared in writing.
2. Quoted in Marvin Fisher, *Workshops in the Wilderness: The European Response to American Industrialization, 1830–1860* (New York, 1967) p. 70.

PART ONE

Building History

CHAPTER 1

The King's Work:
A Thousand Years of British Building

J. MORDAUNT CROOK

DR. CROOK is Reader in Architectural History at Bedford College, the University of London. He is the author of *Victorian Architecture* (1971), *The Greek Revival* (1972), *The British Museum* (1972), and coauthor of *The History of the King's Works*, Vols. V and VI (1973–1976).

The title of this chapter is neither exaggerated nor irrelevant. Royal buildings are as old as royalty. And though Americans have seldom cared for kings, their architecture grew out of a building tradition nourished over centuries by the patronage of the British monarchy. The King's Works began with the Saxons and survive today as part of Britain's Department of the Environment. In 1951 the British government decided to sponsor a mammoth *History of the King's Works,* which would tell the story of royal building from King Alfred to Queen Victoria, from the timber stockades of the Saxons to the date in Victoria's reign — 1851 — when the King's Works became a modern department of state.

Now, after a quarter of a century, that histo-

ry has at last been written. Ten scholars, led by Howard Colvin, have compiled six formidable volumes[1] containing hundreds of plates and plans, thousands of pages and millions of words. These books are not topographical catalogues. They are not limited to architectural description. They are not specifically concerned with the economics of building, nor with its technology, nor with aesthetic judgment and the theory of design — although all those factors are inevitably involved. They are concerned with architecture as an expression of government. And they are chiefly based on the gigantic collection of governmental records housed in London's Public Record Office. They deal with the buildings — civil and religious, military and domestic, public and private — sponsored by the British monarchy. And they deal with the economic, administrative and human machinery which made those buildings possible. In other words, *The History of the King's Works* is a massive study in architectural politics. Too massive, certainly, for easy reading. Hence the justification of this essay: a prospectus rather than a summary, indicating the scope and scale of the enterprise.

Our earliest royal buildings date from the seventh, eighth and ninth centuries. The Saxon court was peripatetic. The monarchs of old England wandered the country fighting, praying, governing and carousing. They left behind them timber encampments like King Aethelfrith's at Yeavering in Northumberland, or King Alfred's at Cheddar in Somerset. Impressive earthworks like Offa's Dyke, stretching a hundred miles from the Severn to the Dee and constructed in the eighth century to separate the Mercians from the Welsh, marked the limits of their conquests. In Christian churches built by royal workmen, St. Augustine prayed for King Aethelbert of Kent and St. Paulinus for King Edwin of Northumbria. King Alfred's Carolingian-style church at Athelney was famous in the ninth century; King Edgar's abbey at Abingdon, designed by the saintly mason Aethelwold, was famous in the tenth century. But the only Saxon royal church of international significance was Edward the Confessor's monastery at Westminster, consecrated in 1065. The earliest example of Romanesque architecture in England, probably designed by Norman master masons and executed by Anglo-Saxon craftsmen, this first Westminster Abbey survives only as an image in the Bayeux Tapestry.

The Norman Conquest of 1066 was imposed by two instruments: the mounted knight and the moated castle. The organisation of feudalism took on formidable architectural expression. Colchester Castle and London's White Tower were the first stone keeps to be built in England, perhaps deriving from the fortress-palaces of Carolingian France. No less than thirty-nine castles, mostly of the motte and bailey type, were built by William the Conqueror or his chief supporters, men like Gundulf Bishop of Rochester and the sadistic engineer Robert of Bellême. The later Norman Kings were rather less prolific. William II and Henry I built mostly in Normandy, but the earthworks of 1139 at Corfe Castle in Dorset still testify to the troubled reign of Stephen. These castles were administrative centres as well as military strongholds. The royal court followed the King, fighting, hunting, governing. Great feudal feasts, when the King entertained Archbishops, Bishops, Abbots, Earls, Thanes and Knights, occurred in Winchester at Easter, in Westminster at Whit-

suntide and in Gloucester at Christmas. The walls of William Rufus's great hall at Westminster, where some notable feasting took place, still survive.

Under Henry II, Richard the Lionheart and John, castles remained the principal area of royal building and by far the largest item in the national budget. Between 1155 and 1215, £46,000 were spent: an annual average of £780 at a time when the basic revenues of the Crown were less than £10,000 per annum. Then there were the royal houses: King John's Hunting Lodge at Writtle in Essex, for example, and the great aisled halls of Woodstock or Clarendon. But castles dominate the records. These castles — King John, for instance, was concerned with about one hundred — were the visible embodiment of royal power. As one chronicler put it, they were "the bones of the kingdom." Henry II was responsible for two particularly remarkable works: Dover (1180–90), the most elaborate rectangular keep in England, and Orford (1166–72), the first polygonal keep in England, marking the transitional phase between rectangular and circular (Fig. 1.1).

It is under the Angevin Kings that a clear picture of the organisation of royal building first begins to emerge. The Saxons had relied largely on slave labour and military conscription. The Normans rationalised this tradition of requisitioned labour and preempted materials by absorbing it into their structure of feudal duties and obligations. Craftsmen were directed by their local sheriffs, who in turn received the King's writ or order and accounted for expenditure to the Exchequer each Michaelmas. But toward the end of the twelfth century local sheriffs began to be replaced by local "keepers of the works," assisted by "clerks of the works." In general the King remained very much master of his own works, inspecting them as he progressed about the country. There was still no central office, but an elite corps of artificers and technicians, sometimes paid out of the Royal Wardrobe or private treasury, can be identified. Men like Ailnoth the Engineer; Richard the Engineer, who is described as "a skilful craftsman and an

Fig. 1.1   Orford Castle, Suffolk (1166–72), constructed during the reign of Henry II. The Norman Conquest of England was imposed by two instruments: the mounted knight and the moated castle. These castles, which symbolized the new feudalism, were administrative centers as well as military strongholds. *(All photos in this chapter courtesy of the Royal Commission Historical Monuments, National Monuments Record, Crown Copyright, unless otherwise stated)*

experienced architect"; Master Elyas of Oxford, variously called "Elyas the Carpenter," "Elyas the Engineer" and "Elyas the Mason"; and below them, people like "Ralph the King's Mason" and "Master Nicholas the King's Carpenter." These are the names which first emerge from the manuscripts.

Under Henry III these tendencies were accelerated. A connoisseur King whose writs read "like an architect's specification," here was a monarch who spent lavishly and well. Westminster is his monument. His was the Painted Chamber; his the ark-like shrine of Edward the Confessor, close by his own tomb. On his twenty royal houses he spent £500 a year, twice as much as King John; on twice as many castles he spent three times as much again. In all, between 1216 and 1272 he spent £113,000 on houses and castles, one tenth of his government's total income. On Westminster Abbey he spent a further £40–50,000, the equivalent of £8–10,000,000 today. He

encouraged the emergence of professional clerks of works, notably Edward of Westminster, his righthand man in matters artistic. In 1256 Henry even anticipated the foundation of a central Office of Works by giving wide powers of control to two particular craftsmen: Master John of Gloucester, the King's Mason, and Master Alexander, the King's Carpenter. And in Westminster Abbey, begun in 1246 by his French-trained mason, Master Henry, he created not just the mausoleum of the Plantagenets but the spiritual home of the English monarchy, a rival to Rheims or St. Denis.

With Edward I we enter the heroic age of the King's Works. Three military campaigns against the Welsh—in 1277, in 1282–83 and in 1294–5—produced a spate of castle building at fourteen strategic points. In 30 years about £100,000 were spent, the equivalent of at least £20,000,000 today. The great Edwardian castles represent the labour of a generation, but their overall conception was the

Fig. 1.2   Conway Castle (a)

Fig. 1.2   Caernarvon Castle (b). These castles, both in Wales, are two of the great fortresses built during Edward I's campaigns against the Welsh, to the designs of Master James of St. George.

work of one man: Master James of St. George. It was he who developed to its logical conclusion the keepless, Crusader-style castle plan, first demonstrated in Britain at Caerphilly. The harbour fortress of Flint, the symmetrical drumtowers of Rhuddlan, the geometrical double circuits of Harlech and Beaumaris, and the asymmetrical bastions of Conway and Caernarvon (Fig. 1.2) all testify to the genius of Master James and to the logistical machinery of the King's Works. Men, money and building materials were drafted to Wales from almost every town and county in England. And there was more to come. James of St. George and his chief assistant, Walter of Hereford, followed Edward I into Scotland and rebuilt Linlithgow and Stirling. Robert of Beverley vastly expanded the Tower of London (Fig. 1.3) between 1275 and 1285 at a cost

of £21,000. Apart from the maintenance of a score of royal houses, Edward I's extraordinary building programme also included Walter of Hereford's Vale Royal (larger than Fountains Abbey) and Michael of Canterbury's Chapel of St. Stephen at Westminster. It was rounded off with William Torel's beautiful tomb of Eleanor of Castille and the twelve Eleanor Crosses marking the Queen's funeral route between Lincoln and London.

By comparison, the King's Works of Edward II and even Edward III seem comparatively small. Edward II continued building in Wales, but spent more on the castles of his favourites: Piers Gaveston's Knaresborough and Hugh le Despencer's Hanley. Edward III's military campaigns against the Scots resulted in castles at Roxburgh and Berwick; his invasion of France produced impressive forti-

7

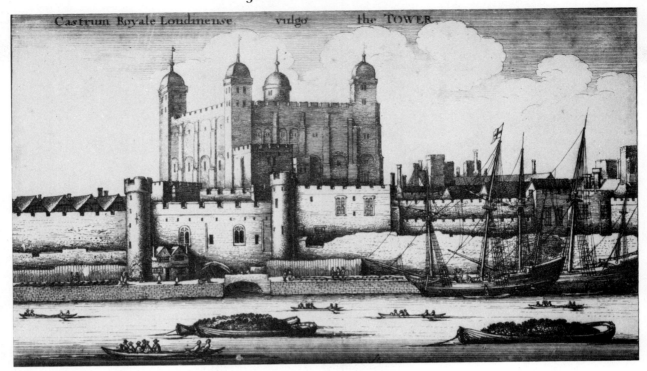

Fig. 1.3   The Tower of London, from a drawing by Wenceslas Höllar. The White Tower at the heart of this military complex, the first stone keep to be built in England, was constructed during the reign of William the Conqueror.

fications at Guines and Calais. At Queenborough on the Isle of Sheppey he spent £25,000 on the construction of a castle to guard the Thames estuary—the climax of Edwardian concentric castle planning. Work at Westminster continued. Carisbrooke Castle received an imposing new gateway. Sheen and Langley were well maintained. But the only project of Edward III which matched those of Edward I was the aggrandisement of Windsor Castle (Fig. 1.4). Between 1350 and 1377 £50,000 were spent—the costliest King's Work of the Middle Ages—in creating the grandest of all fortified palaces and a seat suitable for the newly created Order of the Garter.

It was under Edward III and his successor Richard II that the Office of Works crystallised as a formal administrative unit. During the reconstruction of Windsor, William of Wykeham emerged as a prototype Clerk of the King's Works: his first step toward a famous career as King's Secretary, Chancellor and Bishop of Winchester. And while the coordinating functions of the Clerk developed,

so did the accountancy functions of the Comptroller and the logistical functions of the Purveyor. At the same time the various master craftsmen began to take on new dignity and security as Patent Artisans. The first royal patents to craftsmen were issued in 1336: to William of Ramsey, master mason; to William Hurley, master carpenter; and to Walter the Smith, master smith.

Under Edward III there was still no central Office of Works controlling all royal building. Individual projects were still organised in an empirical manner from the Exchequer, or else from the Wardrobe and the Chamber. Not until 1378 are the King's Works centralised. In that year royal treasurer Thomas Brantingham advised Richard II to set up what was in effect the first Board of Works. It consisted of six men: John Blake, Clerk of Works; William Hannay, Comptroller; Richard Swift, Master Carpenter; Henry Yvele, Master Mason; John of Brampton, Master Glazier. They proved to be the first of a long line of names stretching continuously into the eighteenth

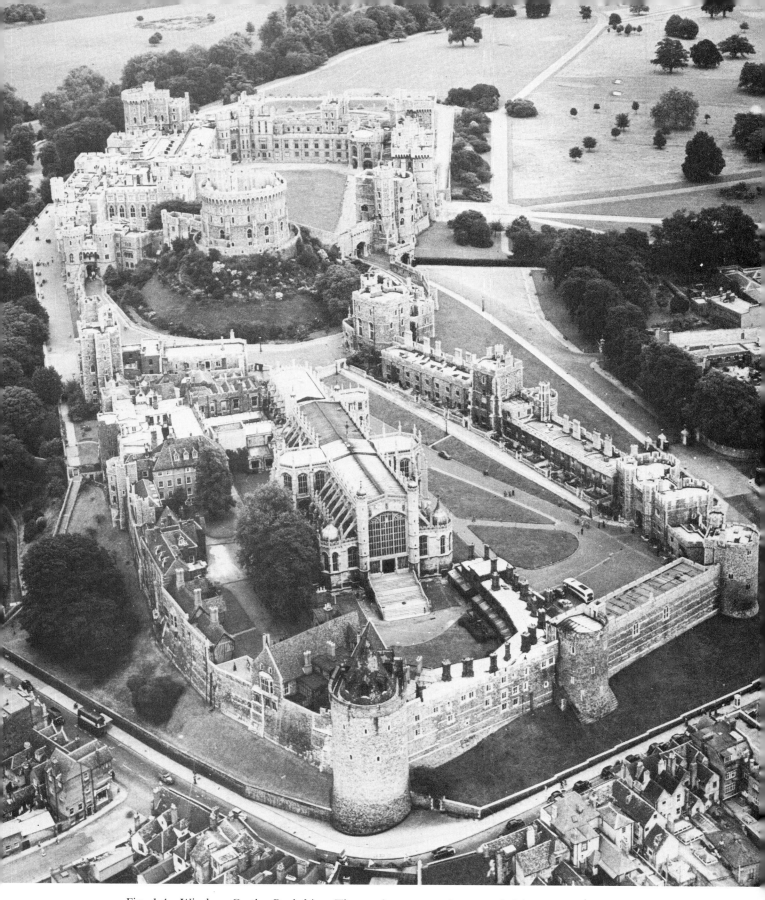

Fig. 1.4 Windsor Castle, Berkshire. The castle was greatly expanded by Edward III during the fourteenth century, as a fortified royal palace and a symbolic center for the newly created Order of the Garter. During the early nineteenth century, George IV employed Sir Jeffry Wyatville to give the castle a picturesque silhouette. (*British Information Service photo*).

century. During the late medieval period the most celebrated Clerk of Works—though for quite extraneous reasons—was Geoffrey Chaucer. He held office from 1389–91, earning the usual stipend of 2s. per day, plus travelling expenses and a clothing allowance. He was of course not an architect but an administrator. As Clerk of the King's Works, Chaucer and his successors, operating from an office close to Westminister Hall, were ultimately responsible for the whole system: the impressment of workmen, the preemption of materials and the contracting of craftsmen; as well as for planning, surveying and accounting. As yet there was no regular system of estimates, nor any predetermined allocation of money. Chaucer and his immediate successors negotiated pragmatically with the late medieval Exchequer. The key creative posts were held by master craftsmen on a cooperative basis.

In Richard II's soaring Westminster Hall (1394–1401) we see the cooperative process at work: Henry Yvele, masonry; Hugh Herland, timber; William Bourgh, glass (Fig. 1.5). But Gothic was primarily a mason's art and the King's Master Mason enjoyed a natural primacy in the King's Works. Appropriately it was the greatest of royal masons, Henry Yvele, who designed the tomb of the formal founder of the King's Works, Richard II. But even lesser master craftsmen won themselves a niche in history: John Wedemer the joiner, for example, who made Richard II's coffin and the lances used at the battle of Agincourt; John Byrchold, who fixed the weathercock on the 500-foot steeple of Old St. Paul's in 1462; and Master Walter the Smith, maker of the first guns manufactured in England, the "ribalds" used at the battle of Crécy.

The Lancastrian King Henry IV built lavishly only on his home ground at Lancaster Castle. Henry V was preoccupied with the reconquest of Normandy: the only building projects by which he is remembered are the fortifications at Harfleur and Rouen and, posthumously, his own superb chantry chapel at Westminster. Under the saintly Henry VI castle building declines, but ecclesiastical and col-

Fig. 1.5 Westminster Hall, London (1394–1401). Built during Richard II's reign, the famous hammer-beamed roof was the work of Hugh Herland, Master Carpenter.

legiate work is magnificently endowed. The project for Eton College is begun in 1441 and dramatically expanded in 1448, the bulk of the chapel being executed under John Smyth's direction in 1469–75. The foundation stone of the chapel of King's College, Cambridge (Fig. 1.6), was laid in 1446, with Reginald Ely as Master Mason.

We owe St. George's Chapel at Windsor not to a royal saint but to a royal warrior, the Yorkist King Edward IV otherwise remembered architecturally for the great hall at Eltham Palace (Fig. 1.7). The great fan-vaulted interior of St. George's was not completed until 1511 by John Aylmer and William Vertue. Even slower was the construction of that miracle of stone, the fan-vaulting of King's College Chapel, Cambridge. Originally designed as a simpler, tierceron vault, it was eventually completed as a multiple fan-vault under the direction of John Wastell between 1506 and

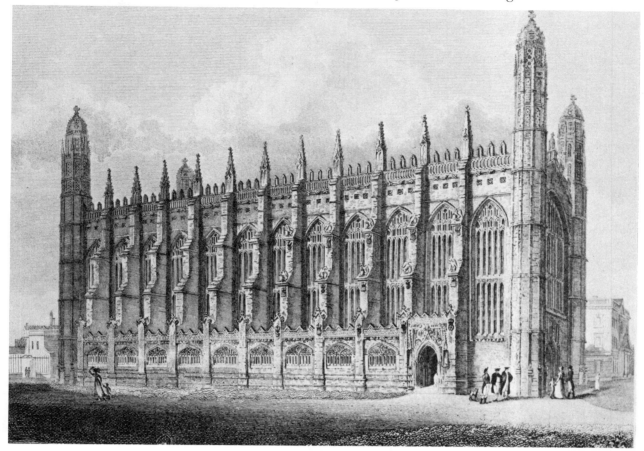

Fig. 1.6 King's College Chapel, Cambridge, from an engraving made in 1805 for John Britton's *Architectural Antiquities of Great Britain*. Begun by Reginald Ely, Master Mason, in 1446, the celebrated fan vault was completed by John Wastall between 1506 and 1515.

1515. Indeed the interior fittings, the stalls, glass and Renaissance screen, were only inserted on the very eve of the Reformation.

Henry VII, seldom regarded as a great patron of the arts, must take the credit for this most lavish of King's Works. He is of course the eponymous patron of Henry VII's Chapel, Westminster, shown in Figure 1.8. Here some £20,000 were spent between 1503 and 1512. Probably designed by Robert Vertue as the shrine of Henry VI, it became instead the setting for Torrigiano's Renaissance tomb of Henry VII. In all, Henry VII spent £28,000 on royal palaces, notably Richmond and Greenwich, and £40,000 on royal churches at Cambridge, Westminster, Richmond and the Savoy.

Henry VIII was no mean builder. Besides continuing work at Windsor, Eton, Cambridge and Westminster, he spent lavishly on his houses at Bridewell in London, New Hall in Essex and Hunsdon in Hertfordshire; he constructed a major new palace at Nonsuch in Surrey; and he expanded the already lavish programmes of Cardinal Wolsey at Whitehall and Hampton Court. At his death he possessed more than 40 royal houses and as many castles. Hampton Court indeed became a symbol of England's new Renaissance monarchy, the Reformation state in action. And on the defense of that state against Catholic Europe, immense sums were expended: fortifications on the Scots border cost over £27,000; at Calais over £120,000; and on the South and East coasts of England more than £181,000 between 1539 and 1553. Tournai and Boul-

Fig. 1.7   Eltham Palace, Kent. The great hall was built during the reign of Edward IV.

Fig. 1.8   Henry VII's Chapel, Westminster Abbey. Built between 1503 and 1512, its design is attributed to Robert Vertue, Master Mason.

ogne were captured and fortified at a cost of more than £160,000. Still, Henry VIII destroyed more than he built: The Dissolution of the Monasteries was an architectural catastrophe.

The Reformation also shifted the balance of architectural patronage away from the King's Works. For the first time in British history the palaces of the aristocracy—Hilaire Belloc's "Reformation profiteers"—were becoming grander than the palaces of the monarchy. This tendency was accelerated by the troubled reigns of Edward VI and Mary, and by the calculated parsimony of Elizabeth I. Elizabeth was a great Queen, but not a great builder. Apart from rare state occasions like the visit of the Duke of Alençon to woo the Virgin Queen, or the Armada Thanksgiving at St. Paul's in 1588, there was no extravagance. Royal castles, except for the Tower of London, were allowed to run down. Royal palaces such as Richmond, Oatlands, Woodstock and Hatfield were maintained on a shoestring budget. The King's Works still retained immense prestige; they were still a magnet for talent; they still set national standards—professional, technical and financial. But under the later Tudors architectural initiative moved elsewhere.

This initiative was only partly recaptured by the Stuarts. James I spent prodigious sums on lavish hunting boxes like Theobalds, Holdenby, Royston and Newmarket, first doubling and then quadrupling Elizabeth's expenditure of £4,000 per annum. But the only substantial works which survive from his reign are two sepulchral monuments at Westminster: Maximilian Colt's tomb of Elizabeth I and William and Cornelius Cure's tomb of Mary Queen of Scots.

In 1615 Inigo Jones became Surveyor of the King's Works. It was under the later Tudors that the Board of Works emerged as a secular, executive body, under the direction of a professional craftsman. James Needham, master carpenter for Henry VIII's hammer-beamed hall at Hampton Court, had become in effect the first Surveyor General in 1532. From 1557 onward, departmental income was regularised

on the basis of monthly issues from the Exchequer. From 1609 onward, departmental expenditure was more closely guarded, with local Clerks of Works—men like the great ichnographer John Thorpe, a clerk from 1591–1601—accounting regularly to the central Board, now situated in Scotland Yard. Charles I never possessed the money to realise his architectural ambitions. Whitehall was never to match the Escorial or the Louvre. And apart from Inigo Jones, his assistant John Webb, and the Master Mason Nicholas Stone, the Caroline Board of Works was not conspicuously talented. Still, it is memorable for two of Jones's finest achievements: The Queen's House at Greenwich (1616–19 and 1630–35) and the Banqueting House, Whitehall (1619–22), shown in Figures 1.9 and 1.10. Such symbols of royalty were lucky to survive the Civil War. Jones himself was expelled from office, fined and very nearly impeached. He fled from London, after burying his treasures in Scotland Yard, and was apparently carried off naked in a blanket by Roundhead soldiers during the siege of Basing House.

On the other hand, John Embree, the Roundhead Serjeant Plumber, did well out of the Civil War, buying up many of the royal paintings and eventually inheriting Inigo Jones's mantle after the interim Surveyorship of Edward Carter. More important, Parliamentary government accelerated the process by which royal houses and palaces were sold off to courtiers and politicians, and royal castles were turned into local administrative centres or else left to decay. After 1649 Office of Works responsibility for royal houses shrank to a mere eight items: Whitehall, Charing Cross Mews, St. James's, Somerset House, Hampton Court, Greenwich, the Tower of London and Windsor Castle.

The reigns of Charles II, James II, William III and Mary, and Queen Anne were in some ways the golden age of the King's Works, the age of English Baroque. Sir John Denham, in Evelyn's words "a better poet than architect," was the first Surveyor Genral after the Restoration. But Denham went mad: "He went to the King, and told him he was the Holy

Fig. 1.9   The Queen's House, Greenwich (1616–19 and 1630–35), designed by Inigo
Jones, Surveyor of Works to Charles I. *(C. J. Bassham photo).*

Ghost." So a Deputy was appointed: Dr. Christopher Wren, the rising star of a new architectural generation. For half a century (1669–
1712) Sir Christopher Wren was Surveyor
General; Hugh May, William Talman and Sir
John Vanbrugh were successive Comptrollers.
Regional Clerks of Works included Nicholas
Hawkesmoor and William Dickinson. Then
there were the master craftsmen: carpenters
like Grinling Gibbons and Emmett; painters
like Thornhill, Verrio and Laguerre; sculptors like Pierce and Cibber; smiths like Tijou;
and gardeners like London, Wise and Bridgeman. But it was also at this time that two factors emerged which were to bedevil the King's
Works for the next century: chronic insolven-
cy and political intrigue. Between 1675 and
1719 the department was never less than
£20–30,000 in debt. The system of preempted materials supplied wholesale by the

Purveyor finally collapsed through lack of
ready cash. Contracting master craftsmen
teetered on the brink of bankruptcy and were
forced to charge exorbitant fees to cover high
interest rates. Labourers were frequently paid
with tickets or promissory notes which they in
turn sold at a discount. In other words, the
King's Works were in the grip of an inflation-
ary spiral. Add to this the corruption and
intrigue of the late Stuart Court and civil
service, and it is a matter for wonder that the
department was still capable of producing
buildings of superlative quality: at Hampton
Court, Wren's Garden Front and Fountain
Court, shown in Figure 1.11 (1689–96,
£130,000); and at Kensington, Vanbrugh's
superb Orangery (1704–5, £6,000).

    This proved to be the last wave of major
palace-building until the Napoleonic period.
Naval and military building for national de-

15

Fig. 1.10   The Banqueting House, Whitehall (1619–22), also designed by Inigo Jones.
Here Charles I was executed in 1649.

fense had now gone over entirely to the Ordnance Office and the Admiralty. The Office of Works was left with royal residences and major public buildings as symbols of royal government. The history of the Stuart and Hanoverian Office of Works is littered with unbuilt palaces. Plans for Whitehall by Inigo Jones, John Webb and Christopher Wren remained largely dreams. Wren planned a palace at Woodstock which was never built and a palace at Winchester which remained a shell. Hugh May did refurbish Windsor Castle between 1675 and 1684 at a cost of £130,000. But Greenwich became a sailor's hospital—by Webb, Wren, Vanbrugh and Hawkesmoor—

and Vanbrugh's schemes for St. James's and Kensington were abandoned by George I. Plans by Lord Burlington for Windsor, and by Edward Lovett Pierce and William Kent for Richmond, were abandoned by George II. Plans for Richmond by Sir William Chambers were abandoned by George III.

The English monarchy did not compete with Versailles, Caserta, Potsdam or St. Petersburg. Even at home, the scale of White Lodge, Richmond or St. James's Palace seemed ludicrously domestic compared with Chatsworth, Blenheim or Castle Howard, or indeed when compared with several dozen noble seats. Hewitt and Kent's Cupola Room at Kensing-

16

Fig. 1.11 Hampton Court Palace. The Fountain Court (1689–96) was added by Sir Christopher Wren to the palace built by Cardinal Wolsey and Henry VIII. Wren was Surveyor General for half a century between 1669 and 1712.

ton (1720–24) or Bridgeman's Serpentine Lake were hardly adequate compensation. Perhaps even more extraordinary is the absence of major royal funerary monuments throughout the eighteenth century. The great age of English monumental sculpture does not include royal tombs.

But then the Hanoverian monarchy was a constitutional one: our German Kings held office by invitation of the great Whig magnates. And it was quite appropriate, therefore, that the eighteenth-century Office of Works should emerge on the one hand as a source of unearned income for political sinecurists, and on the other as the constructor and custodian of governmental buildings. The mid-eighteenth century Office of Works was a stronghold of Palladianism. Roger Morris, Colen Campbell, William Kent, Thomas Ripley, John Vardy, Sir William Chambers, Sir Robert Taylor, James Paine, Henry Flitcroft, Thomas Sandby, Isaac Ware, Stephen Wright: these are all names for architectural historians to conjure with. Palladianism was the governmental style, the hallmark of our Augustan age. The Palladians were in power and the Horse Guards (1750–59) of Whitehall serves as their memorial (Fig 1.12).

Fig. 1.12 The Horse Guards, Whitehall (1750–59). During England's Augustan Age,
the Office of Works became a stronghold of Palladianism.

Above these architects and their contracting tradesmen, however, was a corps of political placemen, nominally members of the Board of Works, sitting in Parliament and drawing unearned civil service salaries in return for party loyalty. Ever since Wren's shameful dismissal by the Whigs in 1718 — and his replacement by a disastrous dilettante, William Benson — the key posts in the King's Works had followed the vagaries of party politics.

Fig. 1.13 Somerset House, London, the first large-scale, purpose-built governmental
complex in the world. Designed by Sir William Chambers, it was begun in 1776 and not
finished until the 1820's.

The department was already financially vulnerable. Now it became politically vulnerable too. Men like George Augustus Selwyn, wit, rake, gambler and multiple sinecurist, or Colonel James Whitshed Keene, a "rather absurd Irishman" obscurely related to Lord North, did nothing but harm to the reputation of the Office of Works. The result was the great purge of 1782—Edmund Burke's Economical Reform Act—which eliminated the Parliamentary politicians and left the architectural politicans in control. Chief among them was Sir William Chambers, Joint Architect with Robert Adam, 1761–67, Comptroller from 1767 to 1782, and Surveyor General from 1782 to 1796. He was a man much respected by his Clerks of Works, men like Kenton Couse for example, who designed the front of 10 Downing Street in 1774–5. Chambers' masterpiece is of course Somerset House on the banks of the Thames, the first large-scale, purpose-built governmental complex in the world (Fig. 1.13). This palace of bureaucracy was begun in 1776 and not finished until the 1820s, at a cost of about half a million pounds. With echoes of Palladio and Piranesi, J.D. Antoine and Sangallo, it quite outshone the Uffizi in Florence.

Chambers was succeeded as Surveyor General to George III by James Wyatt. With Wyatt's arrival in 1796, the Office of Works entered a period of extravagance and confusion. Wyatt supplied the confusion. His accounts are not really corrupt, just wildly in-

Fig. 1.14 Hanover Terrace, Regent's Park, London. One of the many metropolitan improvements, including Regent Street, Trafalgar Square, Piccadilly Circus, the Mall and St. James's Park, designed by John Nash and sponsored by the Prince Regent, later George IV.

competent. George III's eldest son, the future Prince Regent and George IV, happily supplied the extravagance. Parliamentary control of royal building was very difficult to enforce, and the result was four spasms of reform in eight decades. The department was purged in 1782, pruned in 1815, redistributed in 1832 and ministerialised in 1851. But still the expenditure increased. The economic impact of the Napoleonic Wars; the luxury of the Regency royal household; the expansion of government in the Victorian period: all these factors contributed to the dislocation and magnification of the Office of Works. Yet the achievements of the department were considerable. The face of England was transformed by the industrial revolution and the Office of Works transformed the face of London.

Regency London was the London of John Nash (Fig. 1.14). Regent Street, Trafalgar Square, Piccadilly Circus, the Mall, St. James's Park, Regent's Park, Hyde Park Corner — the West End we know today is largely Regency. And the Office of Works, in conjunction with the Crown Estate, played a key part in its development. Not all the dreams came true. Sir John Soane never built his royal palace, though his law courts at Westminster were constructed. For the rebuilding of Windsor Castle, George IV brought in Sir Jeffrey Wyatville over the heads of the three Office of Works architects, Nash, Soane and Smirke. Sir Robert Smirke did achieve the Royal Mint, the General Post Office, the Custom House and the mighty colonnade of the British Museum shown in Figure 1.15 (1823–47), though we

Fig. 1.15   The British Museum, London (1823–47), designed by Sir Robert Smirke. It was the first museum in the world which was public, secular and national.

owe the famous Round Reading Room (1851–57) to his younger brother Sydney. Brighton Pavilion was not part of the Office of Works empire and Carlton House only partly so. For these, Henry Holland and John Nash were paid privately by the King out of the Privy Purse, but Buckingham Palace was paid for out of the Parliamentary grant or Civil List. It was Nash who received this plum commission, turning the Buckingham House of Chambers and Adam into the Buckingham Palace of today. Of course there have been extensive additions since by Edward Blore, Sir James Pennethorne and Sir Aston Webb.

We conclude with the building by which the Victorian Office of Works will always be remembered: the Houses of Parliament. Until 1834 the House of Commons, the House of Lords and several courts of justice continued to meet in the environs of old Westminster Palace. Their quarters were very cramped and stuffy, particularly the Commons meeting in old St. Stephen's Chapel. In 1692 Wren had to expand the Commons by inserting galleries on iron columns—their first structural use in Europe—with wrought iron capitals by Tijou. Many schemes for new Parliament houses failed to materialise: Wren in the 1690s; Benson and Campbell in 1718–19; Hawkesmoor, Burlington and Kent in 1732–33; Kent and Ripley in 1738–39. Then came disaster. The great fire of 1834 destroyed the ancient Palace of Westminster, though Westminster Hall and Westminster Abbey survived. A fierce architectural competition gave the prize to Sir Charles Barry but architectural historians have given much of the credit to A.W.N. Pugin (Fig. 1.16). Though perhaps it is Barry's silhouette which is known throughout the world, it is Pugin's interior decoration which still amazes and delights the visitor: the Queen's Robing Room, the House of Lords and the other dazzling rooms of this gilded labyrinth. By 1860 when Barry died, nearly £2½ million had been spent on this symbol of Parliamentary democracy.

In 1851 the Office of Works emerged as a modern ministry, fully accountable to Parlia-

Fig. 1.16   The Houses of Parliament, London. Rebuilt after the great fire of 1834 which destroyed the ancient Palace of Westminster; designed by Sir Charles Barry and A. W. N. Pugin. *(The Bettmann Archive, Inc.)*

ment. Its responsibilities were prodigious: royal castles and palaces, ancient monuments, public buildings, state occasions, and the multifarious and mundane duties produced by the daily maintenence of civil service offices. Having been for centuries the spearhead of the British building tradition, the Office of Works has now been absorbed into the Department of the Environment, a formidable engine of conservation. The first building in English history described as "an ancient monument" and thus worthy of restoration out of public funds was Tickhill Castle: built by the Normans and preserved on historic grounds in 1562. Today there are some 250,000 "listed buildings" in Britain.

Right at the end of the thousand years covered by the six volumes of *The History of the King's Works* comes one building which any historian must learn to love or hate, the Public Record Office (1850–58) in Chancery Lane, designed by Sir James Pennethorne. Here the records of monarchs and politicians, architects, builders, craftsmen and labourers are all preserved—the records of the King's Works.

NOTE

1. *The History of the King's Works* (Her Majesty's Stationery Office, London): H. M. Colvin, A. J. Taylor and R. Allen Brown, Vols I–II. *The Middle Ages* (1963); H. M. Colvin. Martin Biddle, D. R. Ransome and Sir John Summerson, Vols III–IV, *1485–1660* (in press); H. M. Colvin, J. Mordaunt Crook, Kerry Downes and John Newman. Vol V, *1660–1782* (1976); J. Mordaunt Crook and M. H. Port, Vol VI, *1782–1851* (1973).

CHAPTER
2

America's Wooden Age

ROBERT P. MULTHAUF

DR. MULTHAUF, Senior Historian of Science, the National Museum of History and Technology, is editor of *Isis,* past President of the Society for the History of Technology, and was the convenor of the symposium *Wood in American Culture* (1972).

The immigrant Europeans were confronted in North America with a vast mass of wood, through which they could hardly penetrate, but which was well adapted to the way of life of their soon out-numbered adversaries, the American natives. Wood was thus seen as a menace and its disposal as an urgent "necessity." Mass burnings became common, especially after 1750 when the demand for potash in England created an additonal economic incentive.[1] The new Americans quickly demonstrated their efficiency and the survival of woodlands on the eastern seaboard owes less to the fact that some regions found it desirable to legislate protection of their remaining forests, even in the eighteenth century, than to the continued challenge of apparently inexhaustible forests in the west.

Under these circumstances the Americans found it convenient to depend upon wood for the requirements of daily life to a degree largely forgotten in other parts of the world.[2] And they managed to base on wood a civilization which seems in many ways superior—at least in retrospect—to that which was to succeed it. Rude cabins gave way to elegant houses. Habitually working with wood, the new Americans developed skills which equaled or surpassed those known in the Old World in fabricating such traditionally wooden articles as furniture. Woodworking became a fine art. The islands of Old World elegance, based on brick, stone, and metal, which had been planted in America by the mother country, tended to be replaced by democratic wood as the United States replaced the British colonies of North America. Such was the skill of the craftsman in wood that the affluent found that this entailed little sacrifice of elegance. The growth of an affluent class also owed much to wood, for the forests were the principal source of American exports. Quickly joining the "family of nations," the United States set forth in wooden ships to compete for world trade. By the mid-nineteenth century the American "clippers" were the fastest ships in the world. Merchant princes lived in elegant wooden houses in the north, and in the south even

Fig. 2.1   Virgin spruce forest in Piscataquis County, Maine. This woodland survived long enough to be photographed in 1902. *(Courtesy of U.S. Forest Service)*

more sumptuous houses accommodated a landed gentry whose cotton and tobacco now filled many of the ships.

The early decades of the United States coincided with the coming into fruition of the industrial revolution—the beginning of the iron age. Americans had succeeded in fabricating wood articles such as clocks and scientific instruments, for which metal had become indispensable in Europe. But the large size and small population of their country gave mechanization a particular appeal to the Americans, and they quickly became strenuous participants in the industrial revolution. Through the first quarter of the nineteenth

century they contrived to copy Old World machinery, and to improve on it, while retaining wood as the principal material of construction. But the products of those machines increasingly brought wood into a new relationship to the national economy. The steamboat, railroad, and ironworks were devourers of wood and finally created a demand which was to prove more inexhaustible than the forests of the country.

Not all Americans enjoyed what had become an unplanned and precipitous dash in the direction of "progress," and not all shared the resulting affluence. For these dissenters there was the "unspoiled" West, and they took

Fig. 2.2 "Blackskin" totem, Shakes Island, Wrangell, Alaska. America's wooden age predated the arrival of the Europeans. Photographed in 1940. *(Courtesy of U.S. Forest Service)*

to it by the thousands, riding wooden wagons into the wilderness on a venture scarcely more planned than the civilization left behind. They gained a new appreciation of wood as they left

Fig. 2.3 The "Wayside Inn" made famous by Longfellow. First constructed in 1690 at South Sudbury, Massachusetts. Shown before its destruction by fire in 1955; it has since been reconstructed. *(Courtesy of National Trust for Historic Preservation)*

Fig. 2.4 Room from a house near Essex, Massachusetts, built in 1668 by Seth Story, a carpenter. Now in the Museum of History and Technology, Smithsonian Institution. The oak table-top chair was made in the Connecticut valley in the mid-seventeenth century. *(Courtesy of the Smithsonian Institution)*

Fig. 2.5 This mid-nineteenth century "up and down" sawmill, from Chester Springs, Pennsylvania, was saved from destruction by Charles Peterson. Driven by a water wheel, gearing both drives the saw and moves the log gradually through it. It is now in the Museum of History and Technology. *(Courtesy of the Smithsonian Institution)*

the forests, entered the prairies and then "the great American desert." There a culture existed which appreciated wood but its scarcity limited its use beyond the painstaking assembly of a few logs for house framing. Farther to the west the new colonists encountered an alternative, that of Spanish America, based on clay, but no less sophisticated than the "developed" areas of the United States as it made elegant use of a sparse supply of timber.

That the new colonists continued to drag their wagons westward through these forbidden and forbidding regions was due to the lure of Oregon and California. And they generally found them worth the trip. Wood such as had never been imagined in the East manifested itself on the western coast. The immigrants settled in to enjoy it, for a period which proved to be shorter than they had enjoyed in the East. As they wrested California from

Fig. 2.6   This is a late survivor of the windmills which dotted the eastern seaboard from its first settlement. The mill shown is Spanish and was used for lifting water at Puerto Malino, Puerto Rico. Its present status is unknown. *(Courtesy of the Smithsonian Institution)*

Fig. 2.7 "Mills" were driven by wind or water through wooden gearing such as this, restored from a Pennsylvania flour mill for exhibition in the Museum of History and Technology. (*Courtesy of the Smithsonian Institution*)

Mexico they set out, as had their eastern ancestors, to join "the family of nations" in trade, again in wooden ships and carrying wood as cargo; for the inexhaustible forests of the West were brought to the rescue of the newly wood-poor East. Those most successful created an affluent culture and produced some of the most elegant wooden houses ever known. And finally they reconciled themselves to planting trees as well as cutting them down. But America's wooden age was over.

The country is not treeless. Changing patterns of farming have allowed much open land in the east to revert to forest. With an affection for wood unknown to their ancestors but born of experience, the dwellers of the prairie have through tree planting approximated New England villages on the plains and "shelter belts" aimed to arrest erosion. The surviving relics of our wooden age have become objects of interest, even national treasures; and the survival of a grove of trees planted in the eighteenth century is a thing of wonder. And we have discovered, among the few trees whose remoteness or uselessness enabled them to escape the woodsman's axe, what is now believed to be the oldest living thing.

Fig. 2.8 The decorative use of wood in public structures reached its apex in Philadelphia. Louis Wernwag's "Colossus" bridge over the Schuylkill was built in 1812 and destroyed by fire in 1838. (*Courtesy of the Smithsonian Institution*)

Fig. 2.9   This room, now in the Museum of History and Technology, is from the house of Reuben Bliss in Springfield, Massachusetts. Bliss, who was also a carpenter, built the house in 1754. It shows the advance of luxury since the Story house (Fig. 2.4) of nearly a century earlier. *(Courtesy of the Smithsonian Institution)*

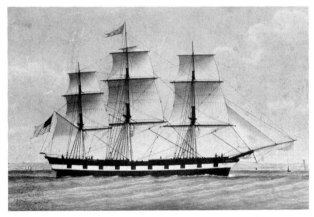

Fig. 2.10   The ship was perhaps the most important and elegant product of the woodworker. This painting of 1858 shows the ship Malabar, of New York, entering the harbor of Marseilles. The artist is unknown. Museum of History and Technology. *(Courtesy of the Smithsonian Institution)*

Fig. 2.11   Complicated though it was, the loom required little more than wood. This example, in the Museum of History and Technology, was made in northern Pennsylvania during the first half of the nineteenth century. *(Courtesy of the Smithsonian Institution)*

Fig. 2.12 Textile manufacture appeared in the 1790s, based on the imitation of English machines. This carding machine was built by Samuel Slater of Pawtucket, Rhode Island, about 1790. Museum of History and Technology. *(Courtesy of the Smithsonian Institution)*

Fig. 2.13 D'Evereux house, constructed at Natchez, Mississippi, in 1840. *(Courtesy of the U.S. Forest Service)*

Fig. 2.14 Travel and settlement in the west were facilitated by military protection. This is Fort Logan, Montana, as photographed in 1939. *(Courtesy of U.S. Forest Service)*

Fig. 2.15 Spanish civilization long predating the United States was encountered in the southwest. Here wood was scarce and judiciously used. This painting of San Antonio, on wood, was made by Pedro Antonio Fresquis (1749–1831), of New Mexico, the first native-born Spanish-American religious artist (Santoro). Museum of History and Technology. *(Courtesy of the Smithsonian Institution)*

NOTES

1. Potash (potassium carbonate) was obtained by leaching wood-ash and was in great demand as a raw material for the manufacture of soap, glass and gunpowder.
2. Russia, Scandinavia, Japan and tropical Africa were also in "wooden ages," and were to remain there longer than did the United States.

Fig. 2.16   That devourer of wood, the railroad, followed close on the heels of the western settlers. This is the Portage bridge over the Genesee River near Buffalo, New York, about 1852 (Erie Railroad). It was 300 feet long and 234 feet high. *(Courtesy of the Smithsonian Institution)*

Fig. 2.17   The mines consumed wood as fuel and in construction at a rate scarcely exceeded by the railroad; many were soon abandoned. These buildings are in the Gallatin National Forest, Montana, photographed in 1961. *(Courtesy of U.S. Forest Service)*

Fig. 2.18 This bristlecone pine in Inyo National Forest, California, is the first tree accepted by scientists to be over 4,000 years old. *(Courtesy of U.S. Forest Service)*

Fig. 2.20 Wood survives as an artistic medium, as shown in *San Miguel y el dragon,* the triumph of good over evil. A recent carving by George López of Cordova, New Mexico. Museum of History and Technology. *(Courtesy of the Smithsonian Institution)*

Fig. 2.19 Wooden house building may have reached its apogee in the Carson house, constructed in 1886 at Eureka, California, by a "lumber baron." It is sometimes called the most photographed house in America. *(Courtesy of the National Trust for Historic Preservation)*

CHAPTER
3

The Origins of The Carpenters' Company of Philadelphia

ROGER W. MOSS, Jr.

Dr. Moss is Secretary and Librarian of The Athenaeum of Philadelphia and author of the forthcoming volume *Master Builders: A History of the Colonial Philadelphia Building Trades.*

The majority of artisans arriving in Philadelphia during the 1680s were British-born and schooled in the London craft tradition. Not all came directly from London, of course, but for generations preceeding the Delaware River migration, English lads had been sent to London to learn their trade. Some stayed on in the metropolis and were absorbed into the labor force; a greater number returned to the provinces with their newly acquired skills. In this manner, the techniques and styles of London were spread throughout the countryside to builders who lacked the advantage of a London appreticeship.

Once in London the craftsman, whether transient apprentice or established master, pursued his trade under the watchful eye of his craft guild, of which there were more than fifty at the beginning of the seventeenth century. Typical of these was the Worshipful Company of Carpenters that had developed from a medieval fraternal society devoted to charitable aid — to succor the sick, to bury the dead, and to comfort the survivors of members who happen to share a common craft. This loose organization was strengthened by the labor shortages following the plague epidemics in the fourteenth century, and by the third quarter of the fifteenth century, the Company had obtained its coat of arms, seal and first charter of incorporation.[2]

As a full-fledged City Livery Company, the Carpenters Company was capable in law to purchase, receive and hold property and to bring suit in the courts. More importantly, the Company was empowered to make such ordinances "to support their businesses and the matters touching the mystery and brotherhood." These powers were spelled out in some detail:

The said master and wardens . . . shall have full power and authority to oversee, search, rule and govern the said commonalty and mystery, and all men occupying the same, their servants, stuffs, works, and merchandizes whatsoever . . . and their defaults to punish and correct, and cause to be corrected, according to their discretions.[2]

From this charter flowed a new set of ordinances that gave scant attention to the original charitable aspects of the association and went

Fig. 3.1 Carpenters' Hall, London, from a sixteenth century map. (A) The Moorfields, (B) Ditch, (C) London Wall, (D) Church of All Hallows, (E) Carpenters' Hall and Garden, (F) Garden of the Drapers' Company and (G) St. Augustine's Priory.

directly to the heart of their intent to control the "mystery and brotherhood" of carpentry. In addition to provisions for searching out violators of building codes and material regulations, the articles ruled that

> every person abiding within the City, its franchieses and liberties, occupying the occupation of Carpentry, or any thing thereto pertaining,

must be a member of the Company; that no person may

> exercise the trade of Carpentry unless he should have first been bound apprentice to the same trade, and served in the same by the space of seven years at the least;

and, that the official carpenters of the City of London (the Common Viewers of the City)

> shall at all tymes herafter be named elected and chosen by the Master & Wardens of the said Fellowshippe of Carpenters.[3]

In part, these ordinances restated powers granted to all guilds by the fourteenth century Royal Charter that had required six men of the craft to support the application of any artisan desiring to be admitted to citizenship or "freedom" in an incorporated town. This virtually made guild membership a prerequi-

site to practicing a trade. Not only could this power be used to exclude unwanted competition from outside the town, but the guilds gradually had become the chief agent for enforcing the seven year apprenticeship established in law by the Elizabethan Statute of Artificers. Thus the guilds could regulate the number of new apprentices entering the "mystery" and the number eventually taken into the "brotherhood."[4]

By the seventeenth century, however, the Company was fighting a rear-guard action to enforce its regulations; the area of London was too large, the nature of the craft too diverse, the resources of the Company too limited. Power over the craft declined as London expanded under Elizabeth and the Stuarts. Then the problem of declining control was made academic by a series of events that began on the morning of September 2, 1666, with a fire in the well-seasoned wood pile of a Pudding Lane baker near the north end of London Bridge. That fire cost the city some £10,730,000 and many of its buildings, and The Carpenters Company lost the last vestige of its traditional control over the brotherhood and mystery of carpentry.[5]

One problem facing the commission charged with rebuilding London was to ensure an adequate supply of skilled labor to replace the 13,200 houses, sundry public buildings and 87 parish churches that had been lost. Over the strident protests of the building trade guilds, nonfreemen were encouraged to flood into the city. These outsiders could, after working for seven years, apply for the same rights as freemen of London without fulfilling the usual requirements of apprenticeship or securing approval from the guilds of their trade. The guilds would never regain this fundamental control. In addition, the Great Fire resulted in the Act for Rebuilding the City of London (1667) that established a sweeping set of building regulations which the city government policed as part of its overall responsibility, thus depriving the guilds of their official control over construction standards. The code was detailed; the most important provisions required that all build-

Fig. 3.2   The Holy Family with Joseph the Carpenter at work. From a sixteenth century mural painting in the former Carpenters' Hall, London. (Jupp, *Historical Account*)

ing should henceforth be structurally standardized and of noncombustible materials. These regulations spread over the country during the next generation and set the major building style of late seventeenth and early eighteenth century British architecture.[6]

The rebuilding of London after the Great Fire did provide temporary prosperity for all the building trades, in spite of the influx of outsiders. Eventually, however, bust followed the boom in the newly reconstructed city of declining population, oversupplied with craftsmen and with guilds shorn of their powers to restrict competition. By the 1680s, a stifling depression had settled over the London builders. In an unrestrained and glutted market, conditions were ripe for young men of frustrated ambition to make their way to the new colony in America, where another city was about to rise and William Penn promised that all sorts of craftsmen would be in demand.[7]

For William Penn and the future of his venture in America, the economic distress of London builders was a happy coincidence. Philadelphia developed rapidly, in part because the town from the first had a relatively large cadre of skilled artisans from all the basic building trades who, by one means or another, found themselves a part of Pennsylvania. Some were sufficiently affluent to purchase land and pay their passage, and among First Purchasers the building trades were well represented. The London carpenter John Day, for example, purchased 1250 acres and transported his wife and children to Philadelphia. Mark Keynton, London carpenter, bought a more modest 125 acres, while the Oxfordshire carpenters John March, Sylvester

Fig. 3.3   Coat of arms of the Worshipful Company of Carpenters, London.

Jourden, and William Cecill between them purchased 2000 acres. Richard Townsend, another London carpenter, who sailed for Philadelphia with Penn on the *Welcome* and left the only known account of that voyage, came under a five-year contract to the Free Society of Traders at a stipulated annual salary of £50. Townsend was a free craftsman, a First Purchaser in his own right, who brought a crew of three indentured servants.[8]

Not all craftsmen drawn to Philadelphia in the 1680s could afford to purchase land or even transport themselves. Those of modest means could exchange four years' labor as an indentured servant for passage. This alternative offered the artisan escape from the glutted British labor market, enriched the province with skilled craftsmen and provided those of sufficient wealth to advance passage money (or to buy an indenture) with a private, noncompetitive labor force. In this manner James Claypoole, Treasurer of the Free Society, sent the brickmaker Edward Cole with instructions to build a house in Philadelphia. It was not uncommon for indentured servants, once free from their contractual obligation, to assume their head right and proceed to establish themselves in the community. An excellent example is James Portues, who is the best known master carpenter of seventeenth century Philadelphia. Indentured to William Wade of Cold Waltham, Sussex, in 1682, Portues was probably brought to Philadelphia to erect a house for the Wade family. The First Purchaser died on board the *Welcome,* and Portues is assumed to have served his remaining time with Wade's nephews and heirs. Several years after becoming free he was commissioned by the wealthy merchant Samuel Carpenter to erect the Slate Roof House—the most famous seventeenth-century house in Philadelphia to survive into the age of photography, making it a familiar touchstone of architectural historians (Fig. 3.4).[9]

James Portues was not the only master house carpenter of importance in seventeenth-century Philadelphia; several others deserve equal attention. John Harrison (d. 1708) has the strongest claim of any known master to first-generation prominence in the history of early colonial Philadelphia building trades. Trained as a carpenter in London where he was presented as an apprentice at Carpenters Hall, he migrated to Philadelphia in the 1690s. Harrison was of the Church of England and one of the founders of Christ Church, Philadelphia, giving him a strong circumstantial claim as master carpenter for the first structure of 1695. One commission may have led to another. When, in 1697, a Swedish Lutheran missionary was sent to America to build a new church at present-day Wilmington, Delaware, he followed a logical pattern that would characterize building in the Lower Counties and the upper Eastern Shore for the next century.

Pastor Eric Bjork turned to Philadelphia for his craftsmen. Joseph Yard and his three sons came as masons, while John Brett and John Smart executed the rough carpentry. When the shell reached completion in the winter of 1698, John Harrison was called in to finish

> all ye Inside work . . . that is all ye pews and Windows, shot work and the Pulpit with a canape over it and a pew of each side of ye communion Table and also with Reals and Banisters about ye church . . . and finish all the Sealing Joice which is to doe and fitt In ye sealing of ye Roof . . . and to doe with plained Boards over ye three doors In ye church where it is left undone.

Harrison so obviously pleased the Wilmington congregation that upon his return to Philadelphia he was commissioned for similar services in the construction of Gloria Dei. Again he worked within Joseph Yard's masonry walls, completing the rough work of Brett and Smart. These two Swedish Lutheran churches, built entirely by the same team of Philadelphia, British-born and trained craftsmen, survive today largely as built. They are the best documented seventeenth century structures in the Delaware River Valley.[10]

That Harrison was a successful builder of no little skill, both his estate of £722 and the surviving work of his hands amply attest. Of even greater importance here is his role as the

Fig. 3.4   The Slate Roof House, Philadelphia, in decline. *(Photograph, The Athenaeum of Philadelphia)*

founder of the most prominent family of Philadelphia master carpenters in the early colonial period. John Harrison's three sons — John, Jr. (d. 1760), Joseph (d. 1734), and Daniel (d. 1768) — are known to be responsible, in part, for work on the State House, Christ Church (Fig. 3.5), and the house for James Logan's mother at "Stenton," while John, Jr. and Joseph are among the earliest known members of the Carpenters' Company of the City and County of Philadelphia.[11]

Carpenters, joiners, brickmakers and bricklayers, glaziers, painters, plasterers, tool makers — by 1700 Philadelphia was amply supplied with a diverse craft community capable of building a worthy town. In a sense all the building trades began on an equal footing in seventeenth-century Philadelphia; no craft guilds or strong government regulations set the style or materials of construction. Providing shelter was a personal matter dependent upon the skill of the settler or his ability to provide or acquire workmen and materials. That these early residents clearly favored brick over wood in construction suggests that the master mason or bricklayer would play the dominant role in planning and erecting buildings. In fact this was not the case throughout most of the colonial period. The master housecarpenter was usually responsible for the entire fabric and subcontracted the brickwork, joinery, glazing, plastering and paint-

Fig. 3.5   Christ Church, Philadelphia. Painted by William Strickland, 1811. *(Historical Society of Pennsylvania)*

ing. The reasons for this are to be found in English craft tradition, where the master carpenter maintained his domination even after urban houses of wood became rare, and the physical nature of house construction. On this subject there is no better authority than the English mathematician Joseph Moxon, who began publishing parts of his *Mechanick Exercises or the Doctrine of Handy-works* on the eve of the founding of Pennsylvania.[12]

Intended to delineate the basic tools and techniques employed in several manual crafts, Moxon's *Mechanick Exercises* provides a close, practical look over the shoulder of the men in joinery, carpentry, and bricklaying, and gives insight into the working relationships of the various building crafts. By tracing the step-by-step method for building a house in the late seventeenth century, he shows clearly the dominant role of the Master Builder ("Master Workman") whom he assumes to be a house-carpenter. Moxon also assumes that the individual commissioning a house would be conversant with books of architecture or have a clear idea of what type of house he wanted. General demands of size and site would be conveyed to the Master Workman who would lay out the ground plan, interior partitions, major structural members, trim and ornaments on a draft rendered to scale from every side. "Unless we shall suppose our Master-Workman to understand *Perspective;* for then he may, on a single piece of Paper, describe the whole Building, as it shall appear to the Eye at any assigned Station." How common such "drafts" were in colonial Philadelphia is difficult to say. The few examples that survive certainly do not suggest the kind of detailing that Moxon assumes. Yet we do know that several master builders owned drawing instruments and were capable of executing fairly skilled plans, even if few understood perspective. "Drawing Designs," The Company rule book stated, "are to be charged by the Carpenter in proportion to the trouble." These probably varied depending on the importance of the structure and the demands of the patron.[13]

"Having drawn the *Draft,*" and secured approval, the carpenter has the cellars excavated. "Then are the Cellar-Walls to be brought up by a Brick-layer" to ground level, or, if the house was to be entirely of brick, to lay all the walls following those regulations established after the Great Fire which Moxon summarizes. These same rules were generally followed in Philadelphia and enforced by the city Regulators. "The Master-Workman [now] appoints his Under-Workmen" to cut the structural members which are set into the brick carcass. With the window frames in place the glazier and plasterers are called in by the Master Workman, interior and exterior trim are completed by the carpenters, or, if the house is to be more embellished, joiners, carvers, painters and marble workers are engaged. This description oversimplified a complex procedure that involves several levels of skill and numerous craftsmen. It does show that the carpenter is the key person from design through delivery. The other crafts were dependent upon him both for their contracts and directions on the job. In short, the carpenter, rather than the bricklayer, whose work spanned only a single phase of construction, was the Master Builder.[14]

The appearance of large public buildings in Philadelphia after 1700 introduced an additional figure who has become a popular myth of colonial architecture. This person represented the commissioning organization. A large vestry or government body could not deal practically with the builders; a supervisor or overseer was appointed to act in their place. Thus the construction of the State House was left in the hands of Andrew Hamilton, Christ Church to John Kearsley, and Library Hall and the President's House to Richard Wells.

When the great hunt for the builders of Philadelphia's colonial buildings began in the late nineteenth century, it was the supervisor whose name was discovered in the archives and vestry minutes as receiving and distributing funds. Consequently these men were dubbed "architect" in endless numbers of popular accounts. We now know that the gentleman amateur architect of the Jefferson type, armed with a library of pattern books from which he pieced together his dream house, was less common than previously thought in colonial America. Of the 187 "Encouragers" to the Philadelphia edition of Swan's *The British Architect,* 62 were Master Builders, 111 housecarpenters, 2 painters, 2 plasterers, 2 cabinetmakers and only 2 merchants and 2 gentlemen (one of whom lived in Maryland). In addition there was a ship joiner, a tanner, a sea captain, and a tallow chandler. In Philadelphia it was the operative builder who supported the first architectural publishing venture in America, not the prospective client. Working up plans—as Edmund Woolley did for the State House—was a regular part of the master's responsibility. The supervisor was doubtless consulted and he probably carried the plans back to his principals for discussion and approval. But the ultimate form and, especially, the final details, were the result of the knowledge and skill of the Master Builder and his crew of workmen.[15]

During the first quarter of the eighteenth century, Philadelphia gradually emerged from under the economic shadow of Boston and carved out its own place in the trade pattern of the Empire. Ship building flourished on the Delaware and a rich hinterland began to fulfill its promise as the staple goods of commerce flowed east. Recovery from the disruptions of life and trade during Queen Anne's War (1702–1713) was rapid. Following the Treaty of Utrecht and the coming to power of Sir Robert Walpole's cautious, peace-loving ministry, Philadelphia began to develop a wealthy merchant class of some size. This benefited the builders, of course, whose own economic well-being was tied to that of their city and individual patrons. But peace and prosperity also brought another flood of immigrants to swell the Pennsylvania population. To craftsmen used to a community where demand exceeded the supply of skilled workers, these new arrivals presaged disquieting competition.[16]

Seventeenth-century Philadelphia had had little need for craft regulations. There were duties levied briefly on sawed boards, shingles and clapboards imported in competition with Pennsylvania sawyers; the quality of leather, the size of bread and cooper's timber were regulated; there were sporadic efforts to keep the cost of goods and services from spiralling upward. But the regulation of competition rather than the control of quality and price did not become a serious problem until after the turn of the century. Then the situation changed. As Philadelphia expanded, artisans, especially journeymen and runaway apprentices of uncertain training, drifted in from Boston and New York to add their numbers to the steady stream of British immigrants. An even more ominous sign must have been the thousands of German-speaking people whom the Proprietor's agents piled onto the Philadelphia docks in greater numbers each year.[17]

Petitioned by tradesmen and artisans for protection against these newly arrived foreign and domestic competitors, the Pennsylvania Assembly and the Philadelphia Common Council faced a difficult decision. How could established, native freemen be shielded without stemming the flow of craftsmen needed to expand the community? In 1712 some master artisans requested that the Assembly allow them to bring in a bill to limit the number of new craftsmen in their trades permitted to enter Pennsylvania. Here they suffered the first of three defeats. "It is the Opinion of this House," the Assembly responded, "that Tradesmen ought rather to be encouraged than discouraged from coming into this Province." By declaring Pennsylvania a great open shop, the Assembly dealt a death blow to officially sanctioned restrictions on competition at the provincial level.[18]

Unable to check the growing number of craftsmen drawn to Pennsylvania from other colonies and abroad, the trades fell back to the city level where the last vestige of medieval control survived in the corporate power to grant or withhold freedom to work in its jurisdiction. As early as 1705, the newly chartered Philadelphia Common Council had drawn an ordinance "for restraining those that are not admitted freemen of this City to keep open Shops, or be master workmen." However well intended, the freedom requirement appears to have been more honored in the breach. Even newly elected Councilmen were occasionally found not to have become free of the city and hastily had to be admitted before they could take their seats![19]

If the Assembly would not exclude foreign competition and the Common Council was powerless to enforce its own freedom laws, the last alternative appeared to be extramunicipal craft companies with the power to regulate their own brotherhoods and mysteries. With this object in mind, "Several Trades of Manufactors within this City" complained to the Common Council that, "notwithstanding their taking out of their Freedoms, Several Strangers are daily Comeing into & Settling therein, not Qualify'd to Exercise their Trades." As a result, "great Damages Accrue not only to Sd Traders themselves, but to the Publick." The Council considered this petition and resolved:

Whereas, Severall of the Sd Manufactors have peticoned to be Incorporated the better to Serve ye Publick in their respective Capacities, It is therefore Ordered that Such of the Trades or Manufactors afsd as desire to be So Incorporated doe with all Expedicon advise with Councill learn'd in the Law, and procure a draught of an Ordinance for that purpose . . .[20]

It is not known if the "Severall Trades or Manufactors" ever brought in their ordinance. The minutes of the Council mention no applications, no debate, no effort to enforce the actions of an incorporated craft. During these years, the Council had attempted to tighten enforcement of the freedom law, but ever-increasing population placed en-

forcement beyond the limited capabilities of the Philadelphia government. Without guilds to police each craft, the concept of municipal freedom was doomed. If the freedom requirement atrophied, then the concept of officially sanctioned craft companies died still-born.[21]

Disregarding the silence of government records on incorporated crafts after 1718, some historians have assumed that the ordinance was brought in, passed, and charters granted to those trades that desired to organize "the better to serve ye Publick." The only example available as proof that such companies existed at all is The Carpenters' Company of the City and County of Philadelphia that has claimed, for nearly 200 years, a founding date of 1724. Unfortunately, the founding and early history of The Carpenters' Compa-

Fig. 3.6   Benjamin Loxley's House in 1828. Water color by W. L. Breton. (*The Athenaeum of Philadelphia*)

ny is obscured by the loss of all Company records prior to the 1760s. No evidence contemporary to the founding has been discovered, and even the 1724 date is uncorroborated. What evidence there is antedating the extant records suggests that the date accepted by The Company and, consequently, most books on American colonial history, is incorrect.[22]

Founding dates were of little concern to the early carpenters. It was enough to say, in 1786, on the occasion of their first published book of prices, that they had "for many years met together, for preserving good order among ourselves, and for the utility and benefit of our fellow-citizens." When The Company applied for its Pennsylvania charter in 1790, however, a more precise date was required. Then The Company laid claim to the 1724 founding, a convenient half century prior to the First Continental Congress meeting in their Hall, that happily provides a dou-

ble excuse for pleasant celebrations every 100 years. To The Company's credit it must be said that an effort was being made in the nineteenth century to document the founding when it was discovered that the required early records were missing. In 1956 a small manuscript booklet entitled "Benjamin Loxley's Memerandum Book, Made and Begun the 6th of June 1768, concerning the Carpenter's Company Interest and Building &cccc" was discovered among a collection of papers purchased by the Historical Society of Pennsylvania. Following an assortment of accounts relating to Company business, was found this cryptic reference: "The Company of Carpenters of Philadelphia 1726/7 in feby Began as pr Book at Mr Foxes." As a modest antiquarian, Loxley may have examined the early records in the 1760s and copied out the reference on the formation of The Company. He might understandably have forgotten his note a

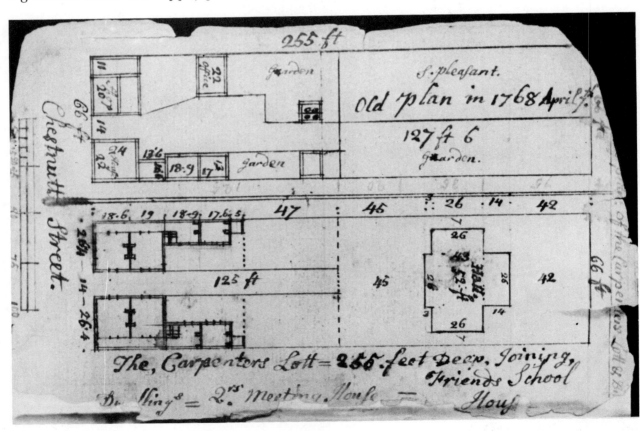

Fig. 3.7 Carpenters' Court plans. Existing layout in 1768 *(above)* proposed Hall and new flanking buildings *(below)*. *(Loxley Memo Book* [MS], Dilks Collection, Historical Society of Pennsylvania)

44

quarter century later when The Company first claimed an earlier founding. The precise date is not especially important, since the difference between the two is so slight; but Loxley's "memerandum" does confirm the existence of a company of carpenters in Philadelphia during the 1720s.[23]

It may be coincidental that on June 26, 1727, just four months after the Loxley note suggests The Carpenters' Company was formed, the Common Council considered the question of city freedoms for the last time:

> An Ordinance for the better Ascertaining & Establishing the Privilidges & Advantages of the Freemen of this City, being read & put to the Vote whether the Same should be Laid Aside, it was the Opinion of this Board that the Examinacon of the same should be at present Suspended.

Had the housecarpenters come together in a final effort to enlist support from the city government? Had that effort, like all those before, failed? The answer may never be known. It seems clear that the policy of the Assembly toward restricting the flow of craftsmen, and the inability of the Common Council to enforce the freedom laws, made it impossible for companies patterned on London guilds to operate effectively in Philadelphia. Without official sanction, such companies would have no control over the "mystery and brotherhood" of their crafts. Unless they were empowered to review all applications for permission to work as a master in the city, and the city was prepared to exclude all craftsmen who failed to gain such approval, these groups would be reduced to impotent fraternal societies. If the tailors, cordwainers, tanners, coopers, and bakers, who regularly petitioned the Assembly and Common Council for relief from outsiders, did form companies immediately following 1718, they failed in their objective and disappeared without leaving a trace.[24]

From its organization there can be little doubt that The Carpenters' Company of Philadelphia was consciously patterned on the Worshipful Company of Carpenters in London, under whose regulations several of the first-generation Philadelphia carpenters had probably entered into apprenticeship. It was born of a similar desire to regulate their craft in the face of rapid, competitive expansion. It also failed, as did its parent in London, directly to control the building trades. As a loosely organized group of master artisans formed, as they later declared, "for the purpose of obtaining instruction in the science of architecture, and assisting such of their members as should by accident be in need of support," The Company appears innocuous enough. But these benevolent and educational objectives are misleading. The influential net of Carpenters' Company was cast over greater numbers of craftsmen than its restricted membership and stated purposes might at first glance suggest. Never numbering more than a minority of the carpenters at work in Philadelphia, and deprived of official sanction to regulate the brotherhood and mystery of house carpentry, The Carpenters' Company found more subtle and informal means to influence the course of Philadelphia history.

How then is the continued existence of The Carpenters' Company to be explained? The answer to that question and the secret of Company power lies in the composition of The Company membership and the nature of their craft. Since the early records are lost, there is only the cumulative list published by The Company in 1786 to provide the names of those members who were no longer alive or active when the surviving minutes begin. While this list is probably not complete for the early years, enough members are known to yield a clear portrait of the founders.

William Coleman is the earliest member of whom we can be certain. His death in 1731 so close to the founding provides additional credibility to The Company's existence in the 1720s. Coleman may have been the son of Alexander Coleman of Fenny Stratford, Buckinghamshire, who was apprenticed to the London master carpenter Edmond Chesmore from 1686 to 1693. His apprenticeship overlapped that of another lad indentured to Chesmore, one William Clark, and a carpenter of that name was also an early member of

The Company. To date there is, unfortunately, no irrefutable proof that these two carpenters with common English names were the same William Coleman and William Clark numbered among the founders of The Carpenters' Company of Philadelphia. Circumstantial evidence does favor the theory that two apprentices of the same London carpenter might have migrated together. Coleman was in Philadelphia by 1701 where he later married Rebeccah Bradford, eldest daughter of the successful first-generation Philadelphia master carpenter Thomas Bradford, who had migrated in 1682. Bradford held 500 acres in Pennsylvania and sat on the Common Council from 1705 to 1720. At his death Coleman and his children were the chief beneficiaries of Bradford's estate. One of these children, William Coleman, Jr., became a prosperous merchant, a Director of the Library Company, a member of Franklin's Junto, and a Director of the Philadelphia Contributionship for the Insurance of Houses from Loss by Fire.[25]

Another of Coleman's children, his daughter Ann, married Edward Warner who, although younger, was also one of the early Company members. Warner became a freeman of Philadelphia in 1717. During the 1720s he lived with the bachelor master builder James Portues, to whom he may have originally been apprenticed. When Portues died in 1736, Warner, together with a former Portues apprentice, Joseph Fox, became executor and chief legatee of his estate. In the 1730s, he held various posts as a Regulator in the Philadelphia city government and, beginning in 1735, sat as an Assemblyman representing Philadelphia County, a position he held until his death in 1754.[26]

James Portues, like William Coleman, Sr., was one of the earliest members of The Company. Understandably, both Warner and Joseph Fox were brought into the brotherhood at an early date; Warner probably at the founding, and Joseph Fox later. About Fox there is more to be said than is possible here, but to continue the development of the early Company, it should be mentioned at this point that he was the son of Elizabeth Yard Fox,

daughter of the first-generation master bricklayer Joseph Yard, Sr., who was a Regulator of streets for the Common Council and, together with Thomas Bradford, one of the earliest members of the city government. Fox, too, became a Regulator and, by 1753, had been elected a member of the Pennsylvania Assembly that he served as Speaker in 1765. His sister married Joseph Rakestraw, carpenter and member of The Company, thereby giving him family ties to one of the largest and most complex carpentry families of colonial Philadelphia. Fox was also linked to the Rakestraws by association with James Portues. His brother-in-law, William Rakestraw, was Portues' apprentice at the time of the elder carpenter's death.[27]

Examples of this type are nearly endless within the building trade crafts. The ancient tradition of sons following the trade of their fathers was still strong in early Philadelphia. Related by birth, marriage or apprenticeship, the carpenters of The Company form a great tangled cousinry that would do justice to the more celebrated Quaker merchant dynasties whose interlocking families have long been viewed as a key to their economic success in the colonial period. Nor do the founders and early members of The Company as a group fit the stereotype of the colonial craftsman. The generation of carpenters that emerged in the early eighteenth century were leaders of the community, men of some wealth and position who, if not of the first rank, were in daily social and political association with city and provincial leaders. Herein lies the key to understanding the importance of The Company as a force in Philadelphia affairs. No other craft was so represented in city government or possessed leaders in such a position to make themselves heard. By standing astride the line dividing those of wealth and power from the middling and lesser sorts, the masters of The Company not only ensured their dominance of building in the city, but provided a necessary link between two elements of colonial society.[28]

In addition to these family ties, The Carpenters' Company had another advantage not available to other crafts. As early as 1698, authority had existed for the Provincial Government of Pennsylvania and, after 1701, the Corporation of Philadelphia, to appoint a number of "Substantial Inhabitants" as "Regulators of the Streets and Water Courses" and as "Regulators of Party Walls and Partition Fences." In effect, these Regulators were inspectors with the responsibility to engage workmen for city projects and to cause the enforcement of those standards governing building in Philadelphia. From the earliest time, a master carpenter was appointed one of the Regulators of Streets and another as one of the Regulators of Party Walls. In the early years, bricklayers shared these positions with carpenters; but, as the century progressed, most of the Regulators were drawn from the leading carpenters who were members of The Company. The first Common Council for which there are records included John Parson, carpenter, and Joseph Yard and John Redman, Sr., bricklayers. By the end of the colonial period, all but one of the Regulators were carpenters. In London these positions had been the prerogative of the Worshipful Company of Carpenters; in Philadelphia there was no such formal understanding, but the result was the same. After the Friendship Carpenters Company was formed, largely of less successful masters, the only carpenters appointed to these posts were members of the two master companies.[29]

It would be difficult to overestimate the direct benefits that might accrue to the holders of the position of Regulator, and indirectly to those craftsmen on favorable terms with them. As officials of the city, posts certainly given only to prominent members of the craft, their ability was vouched for and they could command the most profitable work. In addition, there was the safeguard of avoiding possible violations of city building regulations by employing one of the very men charged with policing these regulations. The Regulators would naturally pass along this advantage to their colleagues, both through favorable association with wealthy and powerful patrons and possible city contracts. Membership in The Company could mean a quick path upward from employee to employer.

While Company carpenters probably benefited from preferential commissions due to their association with government and prominent families, there was an even more practical reason for seeking membership. This centered on the means of determining the price to be charged for all construction in the city. Sir Christopher Wren explained in 1681 that the cost of building was determined by one of three methods: day rates, measured squares of work completed, and by total bid for the job. Of the three, Wren wrote, "I think the best way in this business is to worke by measure: according to the prices in the Estimate or lower if you can, and measure the work at 3 or 4 measurements as it rises. *But you must have an understanding trusty Measurer.*" In Philadelphia measuring was the most popular form of arriving at the final price. For this service the measurer received 3 per cent of the total amount, a tidy sum if the carpenter's bill reached £1128.13.2, as did David Evans's for the John Dickinson House in 1772. But how is one to be sure that he was "an understanding trusty Measurer"?[30]

There are two elements involved in mensuration. First, to determine the amount of work completed the measurer had to be familiar with all types of construction. Second, the price had to be uniform and fair to both the client and the craftsman. In eighteenth century Philadelphia there appears to have been general agreement that both of these requirements were best served by engaging a member of The Carpenters' Company to make the final computation. There were several advantages to this system. Clients would be ensured that they were not being cheated by the craftsman who would, in turn, receive full measure and pay for his work. This is probably what The Company had in mind when it declared itself formed "for preserving good order among ourselves, and for the utility and benefit of our fellow citizens."[31]

United by family and craft, led by their colleagues well placed in government and society, with considerable power over smaller crafts that depended upon the Master Builder for subcontracts, The Company gradually developed a firm grip on building in colonial Philadelphia. Added to this was a virtual monopoly over measuring that provided members with a supplemental income and placed The Company in an ideal position to set prices on various sorts of carpenter's work. As the chief employers of less successful masters, journeymen and day laborers, the Master Builders already could establish the going rates for wages. By setting the prices on which mensuration was based, The Company could enforce uniformity on member and nonmember alike, while maintaining the cost of building against efforts by nonmembers to cut their charges. To protect this monopoly The Company was careful to keep prices secret and, after a blank price book was printed for ease of consultation, a careful record was kept of the number of copies in circulation among the members.[32]

While the masters were unable to restrict the number of carpenters free to work in Philadelphia, they could restrict the number admitted to the privileged ranks of their society. The Company's membership was never large in comparison to the total number of carpenters at work within the city. At the beginning, The Company probably numbered no more than ten or fifteen men. Even during the mid-century building boom, membership did not exceed approximately sixty men; about one for every 400 to 500 persons in the city seems about the usual percentage, with nonmember, free carpenters outnumbering members at a ratio of at least four to one. During the Revolutionary era and up to approximately 1800, the carpenter population appears to have increased until it reached a ratio of perhaps six to one, non-Company to Company.[33]

Except to restrict the taking of slave apprentices by members, The Company made no effort to control apprenticeship as did The Worshipful Company of London. Unlike some other crafts, the carpenters had little difficulty obtaining apprentices, and replacements for Company members who died or became inactive were drawn primarily from the sons and former apprentices of members. The Company Articles provided that "after the decease of any member . . . leaving a son of the same trade, such son . . . may be admitted a member without paying any entrance money." When Ezekiel Worrell died in 1781 his widow asked William Robinson "to propose to the Co. for to Assist her in placeing one of her sons Apprentice to a House Carpenter." James Pearson may have taken the lad, for on January 21, 1788, he proposed Joseph Worrell "Son of the Late Ezekial Worral" as a member of The Company. The Rakestraw family already mentioned is the classic example of three generations elected to membership spanning most of the eighteenth century, and other examples are sufficiently numerous to suggest that The Company's nepotism of the early period continued to be a dominant characteristic throughout the eighteenth century.[34]

At the same time Philadelphia Master Builders were establishing their control over construction in the city, a quiet revolution was taking place in English architecture. In 1715 the first of many design books on English Palladianism appeared. Colin Campbell's *Vitruvius Britannicus*, in three volumes, illustrated the work of Inigo Jones, John Vanbrugh, Christopher Wren and John Webb in the classical style drawn largely from Andrea Palladio's tetralogy *Quattro libri dell' architecttura* that first appeared in Venice in 1570. How quickly these books became available in Philadelphia is difficult to say. Generally there was a lapse of ten or more years between publication and the first appearance here, although as the century progressed this gap narrowed considerably. As early as 1734 Franklin advertised in the *Pennsylvania Gazette* that he had for sale a copy of the "Builders Vademecum." This was probably Batty Langley's *The Builder's Vade-Mecum: or a complete key to the five orders of columns in architecture*. Then, in 1736, the printer reported: "Lent some time since a Book entitled Campbell's *Vitruvius Brittannicos*, the Person who had it is desired to return it to the Printer hereof." This copy predates that ordered by the

Library Company through Peter Collinson in July of 1739 (reordered, August, 1739) and is the earliest known copy of the work in America.[35]

The Library Company, founded by Franklin in 1731 and chartered in 1742, included among its early members Reese Lloyd who was also an early member of The Carpenters' Company. It may have been on Lloyd's suggestion that the first order for Library Company books included "Architect: by And^w Palladio" and "Evelyn's Parallels of ancient and modern architecture." (Several members of The Company were stockholders of the Library Company during the colonial period, and for many years that worthy society kept rooms on the second floor of Carpenters' Hall. When the Library moved to its new hall in 1789 it owned over twenty works on "Civil Architecture".) In 1742 Franklin advertised

Fig. 3.8 Carpenters' Hall, Philadelphia. Front elevation as engraved by Thomas Bedwell for the 1786 Rule Book.

for sale a copy of the "Builder's Dictionary" which was probably *The Builder's Dictionary or Gentleman and Architect's Companion*, 2 vols. (London, 1734), again the first copy of the work known to have been in America. A few months later he had another copy (or the same one?) and a work by Batty Langley too vaguely titled "Langley's Architecture" to be identified. Like Campbell and *The Builder's Dictionary*, this copy of one of Langley's many architectural books is the first known in this country by at least 12 years. From this point the availability of architectural design books rapidly expands. In one order alone in 1760 the bookseller David Hall ordered seventy-eight design books in fifteen different titles![36]

When we turn from the sellers of books to the buyers it is more difficult to determine who owned what. The appraisers of carpenters' estates usually listed books as "1 Bible and a lot of old Books," or, in the case of the carpenter John Songhurst who died in 1735, a tantalizing "Two Folio Books." Specific references to architectural books in the estates of Philadelphia carpenters are rare, even though it is known that over fifty different titles were available in the city prior to the Revolution, and Swan's *The British Architect*, published in Philadelphia in 1774, had nearly two hundred subscribers, most of whom were carpenters. The earliest reference found from reading all carpenters' inventories for the colonial period is to John Harrison, Jr.'s "Four Book of Devinety four of Architecture" in 1760. Given Harrison's role as one of the master carpenters on Christ Church, it is unfortunate that the books are not identified. Edmund Woolley, builder of the State House, provided in his will of 1771 "that all my Carpenters Tools and Books of Architecture be sold." One of these, William Halfpenny's *Practical Architecture* (London, ca. 1730), was purchaseed by his fellow Company member Isaac Coats and eventually was deposited in the Library Company. It was Thomas Coates, a founder of the Friendship Carpenters Company, who owned a library of some twenty-two books and decorated his house with "5 Orders of Architecture in Frames & Glass." Apparently the modern tendency to strip design books for their decorative plates is not so new.

After the Revolution, as the third generation Master Builders began to die, the mention of architectural design books increases. Abraham Jones, Samuel McClure, Richard Armitt, Frazier Kinsley, John Lort, and John King, each owned books of architecture at the time of his death. Most of these are not given by title; however, Joseph Rakestraw's estate included "Pains Architecture" and William Williams owned William Pain's *Practical Builder; easy manner of Drawing and Working the whole or part of a building* (London, 1734).[37]

Architectural historians can expect to find the influence of these books on Philadelphia building from ca. 1740 with a dramatic change following ca. 1750. Previously Philadelphia craftsmen worked in a vernacular English style little altered from that developed in London following the Great Fire. The most advanced builders of the second generation were able to produce structures in the English "Queen Anne" style like Logan's "Stenton," but most construction was probably closer to the unprepossessing Letitia Street House (ca. 1715), the Old Court House (ca. 1707), and the Benezet House (ca. 1700). Great private houses like Isaac Norris's "Fair Hill" and Jonathan Dickinson's plantation are virtually unmentioned because they no longer survive.

The tendency is for architectural historians to jump this second generation in their hurry to get from the caves and cabins of the 1680s to the Palladian houses of the 1750s. As has been shown, this gap of half a century was a metaphase during which prominent seventeenth-century builders—e.g., Bradford, Portues, and the elder John Harrison—combined with the newcomers of the early eighteenth

Fig. 3.9 Carpenters' Company bookplate. Probably 1786. Note coat of arms.

century, like Coleman, Clark and Samuel Powel, to prepare the ground for Warner, Fox and the young Harrisons. As Philadelphia turned toward the mid-eighteenth century her builders were firmly established, organized and armed with design books that would make Philadelphia architecture the finest in America. From this milieu emerged builder-architects who would grace the city with public buildings, churches and private homes worthy of the premier city of America.[38]

NOTES

1. B. W. E. Alford and T. C. Barker, *A History of the Carpenters Company* (London, 1968) pp. 30–33. On the guilds, *see:* Stella Kramer, *The English Craft Guilds* (New York, 1927); George Unwin, *The Guilds and Companies of London* (London, 1925); G. A. Williams, *Medieval London* (London, 1963).

2. The Charter was granted by Edward IV, July 7, 1477, and is reprinted by Edward B. Jupp, *An Historical Account of the Worshipful Company of Carpenters of the City of London* (London, 1848) pp. 12–15.

3. Extracts from the by-laws of the Worshipful Company of Carpenters, reproduced in *ibid.*, pp. 146–152. For a

discussion of the Company and the control of the craft, 1400–1670, *see* Alford and Barker, especially pp. 27–45 and 69–91.

4. On the development of the London Carpenters Company, *see:* Alford and Barker, *A History of the Carpenters Company*, especially chapters I–V; Edward B. Jupp, *An Historical Ac-*

count; H. Westbury Preson, *The Worshipful Company of Carpenters* (London, 1933). The Company records are on deposit at the Guildhall Library, London, but most of the early material has been printed by Bower Marsh (ed.), *Records of the Worshipful Company of Carpenters*, Four Volumes (Oxford, 1913–1916). Williams, *Medieval London*, pp. 157–195.

5. E. S. deBeer (ed.) *The Diary of John Evelyn* (Oxford, 1955), III, September 2–7, 1666; J. P. Malcolm (ed.) *Londinium Redivivum, or, An Ancient and Modern Description of London* (London, 1802–1807), IV, 73–82; *The City Remembrancer: Being Historical Narratives of the Great Plague at London, 1665; Great Fire, 1666; and Great Storm, 1703* (London, 1769), II, 37–44; T. F. Reddaway, *The Rebuilding of London after the Great Fire* (London, 1940); John E. N. Hearsey, *London and the Great Fire* (London, 1965); Summerson, *Architecture in Britain*, pp. 5–6; Frank Jenkins, *Architect and Patron* (London, 1961) pp. 31–39. P. Meadows (ed.) *A Source Book of London History* (London, 1914) reprints the Royal Proclamation of September 13, 1666, that called for the rebuilding of London, pp. 156–158.

6. John Stow, *A Survey of the Cities of London and Westminster* (London, 1720), I, 227. Reddaway, *op. cit.*, p. 27. Summerson, *Architecture in Britain*, pp. 119–122. "One of the major effects of the rebuilding," writes Summerson, "was that a new structural standard of brick domestic architecture was set up for the whole country. The Rebuilding Act was quoted in building agreements far outside the London area." *Ibid.*, p. 122. Alford and Barker, *op. cit.*, chapter VI, 114–117.

7. Alford and Barker, *loc. cit.* Reddaway, *op. cit.*, p. 120. E. H. Phelps Brown and Sheila V. Hopkins, "Seven Centuries of the Prices of Consumables, Compared with Builders' Wage-Rates," *Economica* (1956).

8. "Digested Copy of the Registers of Marriages of the Quarterly Meeting of London and Middlesex, 1657–1719, and of Births, 1644–1719," Genealogical Society of Pennsylvania; *PGSP*, II, 28; IV, 199; *Pennsylvania Archives*, 2nd Series XIX, 205; Walter Lee Sheppard, Jr., *Passengers and Ships Prior to 1684* (Baltimore, 1970), pp. 17–18, 50, 62, 199, 198, 202.

9. Penn advertised that the passage to Philadelphia would cost at most £6 a head. *Some Account of Pennsylvania* (London, 1681), reprinted in Albert Cook Myers (ed.) *Narratives of Early Pennsylvania West New Jersey and Delaware, 1630–1707* (New York, 1912), p. 211. "Extracts from the Letter-Book of James Claypoole," *PMHB*, X,

(1886) 197–202; Sheppard, *op. cit.*, pp. 60, 80. See Richard Townsend's "Testimony," in Robert Proud, *History of Pennsylvania* (Philadelphia, 1797), I, 228. Townsend brought Bartholomew Green, William Smith and Nathaniel Harrison with him on the *Welcome* as indentured servants, Sheppard, *op. cit.*, pp. 11–12, 19, 57. *PMHB*, VIII, 339. James Portues' name is variously spelled. His own spelling as it appears in a petition of February, 1700/1, to William Penn requesting the repair of Front Street at the end of High Street, is used throughout (Logan Manuscripts, Vol. III, 9, HSP). *Pa. Arch.*, 2nd Ser., XIX, 434; Sheppard, *op. cit.*, p. 10. "The request of James Portiss . . . to take up his proportion of head land" of 50 acres (10th 5 mo 1704) in the Taylor Papers, Vol. I, 130, HSP. Scharf and Westcott, I, 147, 159. Later historians have been forced to accept this attribution for the Slate Roof House as no contemporary documentation has been located. Marshall B. Davidson (ed.) *Notable American Houses* (New York, 1971) p. 35. A floor plan of the "Slate Roof House" (ca. 1709) is preserved in the Logan Manuscripts, Vol. XIII, 6, HSP. "Christ Church Journal," Christ Church Archives, Philadelphia, several references to James Portues from April 19, 1711–April 8, 1714. Portues died in 1736, Philadelphia Wills 1736:22.

10. *Records of the Worshipful Company of Carpenters*, I, 67, 185, 198. Philadelphia Wills, 1708:73. In the settlement of Harrison's estate see three payments (March, 1711, to March, 1714) made to the "Carpenters Comp." Harrison was one of thirty-six persons signing a January 18, 1696/7 report that the church was completed. Louis C. Washburn, *Christ Church* (Philadelphia, 1925), p. 97. George Fletcher Bennett, *Early Architecture of Delaware* (New York, 1932), pp. 16–25; Harold Wickliffe Rose, *The Colonial Houses of Worship in America* (New York, 1963), pp. 367–368, 132–133; Harold Donaldson Eberlein and Cortlandt V. D. Hubbard, *Historic Houses and Buildings of Delaware* (Dover, 1963) pp. 209–210. Accounts of the Church at Wikako, Gloria Dei Archives, Philadelphia.

11. Philadelphia Wills, 1760:300; 1734–387; 1768:208. See extensive payments to Daniel Harrison for work done on Christ Church, "Christ Church Journal," April 6, 1726–July 2, 1740. *Pennsylvania Gazette*, January 9, 1734/5. Logan Account Book, 1712–1720, HSP, p. 362. Joseph Jackson, *Early Philadelphia Architects and Engineers* (Philadelphia, 1923) p. 62.

12. Joseph Moxon, *Mechanick Exercises: or*

the doctrine of handy-works, Third Edition (London, 1703). The work first appeared in 1678.

13. *Ibid.*, p. 128. See the drawings reproduced in: Margaret B. Tinkcom, "Cliveden: The Building of a Philadelphia Countryseat, 1763–1767," *PMHB* (January, 1964), LXXXVIII, 3–36. An interesting group of original Philadelphia drawings is preserved as part of the "Fairhill" Norris Collection, HSP. See, also, Charles E. Peterson's introduction to the reprint of the 1786 Carpenters' Company price book (Princeton, 1971) pp. xxii–xxiii.

14. Moxon, *op. cit.*, p. 128–155.

15. Hamilton, a lawyer, was named one of a committee to oversee the Pennsylvania State House; on August 11, 1732, during a debate on the building, he "produced a Draught of the State House, containing the Plan and Elevation of that Building." Hamilton has been credited with the design. In fact, that credit should go to the Master Builder Edmund Woolley. See, Edward M. Riley, "The Independence Hall Group," *Transactions of the American Philosophical Society* (Philadelphia, 1953), XLIII, pp. 7–42. It was the same Dr. John Kearsley who shared membership on the State House committee with Hamilton that was responsible for overseeing the construction of Christ Church. Robert W. Shoemaker, "Christ Church, St. Peter's, and St. Paul's," *ibid.*, 187–198. The literature on both edifices is legion, most of it inaccurate. The fact that Dr. William Thornton did design Library Hall simply adds fuel to the Hamilton-and-Kearsley-as-architects theory. Peterson, "Library Hall," *ibid.*, p. 132.

16. Carl Bridenbaugh, *Cities in the Wilderness*, Chapter V. Clarence L. Ver Steeg, *The Formative Years, 1607–1763*, (New York, 1964) Chapters Six and Seven.

17. "Votes and Proceedings of the House of Representatives of the Province of Pennsylvania," reprinted in *Pennsylvania Archives*, Eighth Series (Harrisburg, 1931–1935) I, 91, 298–299, 411, 554; II, 945, 957, 959, 1054–55, 1298 ff., 858, 1016, 1240.

18. *Pa. Arch.*, Eighth Series, II, 1016, 1020.

19. *Minutes of the Common Council of the City of Philadelphia, 1704 to 1776* (Philadelphia, 1847), p. 34. [Hereafter *MCCCP*.] Scharf and Westcott, *op. cit.*, I, 194n. *MCCCP*, 239. Philadelphia Wills, 1755:206.

20. *MCCCP*, pp. 146–147.

21. *Ibid.*, pp. 115, 117. Several hundred freedoms were redeemed during 1717–1720; the number declined sharply thereafter. *Ibid.*, p. 55.

22. Morris, *op. cit.*, pp. 141–142. Bridenbaugh, *Colonial Craftsman*, p. 144.

23. *Articles of the Carpenters Company of Philadelphia: and their Rules for measuring and valuing house-carpenters work* (Philadelphia, 1786) p. viii. "An Act to incorporate the Carpenters' Company of the City and County of Philadelphia," reprinted in *Reminiscences of Carpenters' Hall* (Philadelphia, 1858) p. 3. Charles E. Peterson, "Benjamin Loxley;" and Louise Hall, "Loxley's Provocative Note," *Journal of the Society of Architectural Historians* (December, 1956) XV (4), 23–27.

24. *MCCCP*, p. 271.

25. *Articles of the Carpenters Company*, iii–iv. *Records of the Worshipful Company of Carpenters*, I, 172, 189, *PMHB* (1892), XVI, 126. Will of Thomas Bradford, Philadelphia Wills, 1721:271. Sheppard, *Passengers and Ships*, p. 43n. *PMHB*, XLIV, 358. *MCCCP*, p. 31. Tolles, pp. 155, 220. Nicholas B. Wainwright, *A Philadelphia Story: The Philadelphia Contributionship* (Philadelphia, 1952), p. 233. William Coleman, Jr. was the owner of "Woodford" plantation (now in Fairmount Park). Of him Benjamin Franklin wrote, he "had the coolest, clearest head, the best heart, and the exactest morals of almost any man I ever met with. He became . . . a merchant of great note, and one of our provincial judges." Eberlein and Hubbard, *op. cit.*, pp. 313–314. Philadelphia Wills, 1731:49 (ad); 1769: 235, 236.

26. *PMHB* (1877), I, 358–359. *MCCCP*, p. 120. *American Weekly Mercury*, July 11, 1723: "There is to be Exposed to sale by way of a Lottery, a new Brick House and Lot, with a good Kitchin, Wash-House, Oven, half of a Well, Necessary-House and a handsome Garden, valued at 230 Pounds, being under the yearly Ground-Rent of Fifty Shillings *per Annum;* . . . Situated on the East side of the Third Street near the Market place in *Philadelphia.*" Edward Warner, carpenter, "living at James Poultis's in the Second Street," was selling 460 tickets at ten shillings each. Will of James Portues, Philadelphia Wills, 1736:22. Portues owned two Negro slaves and two Indian slaves. Warner was collecting money due to Portues for work done at Christ Church as late as 1741. Christ Church Journal, October 23, 1741. *MCCCP*, 326, *Pa. Arch.*, Eighth Series, III, 2293. Philadelphia Wills, 1754:141.

27. Fox was unquestionably an important figure in colonial Philadelphia. Unfortunately all his private papers have been lost and there is little but the public record on which to base a reconstruction of his life. *See* Anne H. Cresson, "Biographical Sketch of Joseph Fox, Esq., of Philadelphia," *PMHB* (1908), XXXII, 175–199. *MCCCP*, p. 500. He was elected a Burgess for Philadelphia City, 1750, *Pa. Arch.*, Eighth Series, IV, 3353, and an Assemblyman for Philadelphia County, 1753, *ibid.*, V, 3621. Philadelphia Wills, 1780:281; 136:22.

28. Philadelphia Wills, 1718:144. Moss, *loc. cit.* Articles of the Carpenters Company, *loc. cit.* Tolles, *op. cit.*, 89–90, 119–121.

29. *Charter to William Penn, and Laws of the Province of Pennsylvania passed between the years 1682 and 1700* (Harrisburg, 1879), pp. 276–277. *MCCCP*, p. 137. The 1774 Regulators of Party Walls, Buildings, and Partition Fences were David Rittenhouse; Robert Smith and William Calladay, members of The Company; and Thomas Morris and Edward Bonsal, members of the Friendship Carpenters Company. *Ibid.*, p. 795. Alford and Barker, *op. cit.*, 42–45.

30. Frank Jenkins, *Architect and Patron* (New York, 1961), pp. 128–129. (Emphasis added.) Tinkcom, *op. cit.*, pp. 27, 31. The Master Builder usually measured the work of his subcontractors, as Jacob Knor did at "Cliveden" and David Evans for the Cadwalader, Dickinson, and Samuel Meredith houses. Nicholas B. Wainwright, *Colonial Grandeur in Philadelphia* (Philadelphia, 1964), pp. 14, 92. Samuel Meredith Papers, University of Delaware Library, 1790.

31. *Articles of the Carpenters Company, loc. cit.*

32. *MCCCP*, 1786 ff. Louise Hall, *loc. cit.* has made a careful study of carpentry price books and discusses at some length the efforts of The Company to protect theirs from being pirated.

33. For documentation *see* Roger W. Moss, Jr., "Master Builders, A History of the Colonial Philadelphia Building Trades" (University of Delaware Ph.D. Dissertation), 1972, pp. 149–150 and Appendix.

34. Articles of the Carpenters Company, *op. cit.*, p. xii. There are several works on colonial apprenticeship, the best to date is Ian M. G. Quimby, "Apprenticeship in Colonial Philadelphia" (Unpublished thesis, University of Delaware, 1963), pp. 15–18, 32.

Quimby found that in carpentry, as in the two other largest crafts (cordwainers and tailors), most apprentices were bound for periods of 4 to 7 years. His findings would also suggest that carpentry had stabilized by the midcentury as the number of apprentices taken by carpenters rose much slower than the number taken by the other large crafts whose products were exportable. *MCCCP*, July 16, 1781, January 21, 1788.

35. Colin Campbell (?–1729) published *Vitruvius Britannicus* from 1715 to 1725 and aroused the interest of the Earl of Burlington in the Palladian style. Colvin, *A Biographical Dictionary of English Architects*, pp. 119–120; Summerson, *Architecture in Britain*, p. 192 ff. The standard work on the introduction of design books to America is Hellen Park, "A List of Architectural Books Available in America Before the Revolution," *JSAH* (October, 1961) XX, 115–130. Of particular interest for Philadelphia is Charles Hummel's unpublished thesis (University of Delaware, 1957), on the influence of English design books on Philadelphia furniture in the period 1760 to 1780. Franklin's advertisements were not discovered by Park or Hummel; consequently the publisher's role in introducing these design books has yet to be credited. *Pennsylvania Gazette*, May 23, 1734, March 3, 1736/7, May 18, 1738. Park did not find Campbell until 1741, and no copy of Langley's *Builder's Vade-Mecum*. The Library Company copy of the latter work (acquired in modern times) bears the date 1735 and Colvin (*op. cit.*, pp. 354–355) does not list it at all. William Salmon's work of a similar title, *The London and Country Builder's Vade Mecum; or the complete and universal Estimator, comprising the London and Country Prices . . .*, did not appear until 1745, and in America is not found until 1755. *Ibid.*, p. 519; Park, *op. cit.*, p. 130. See, *Catalog of the Library Company, 1741* (Philadelphia, 1956); Alfred Coxe Prime (ed.), *The Arts and Crafts in Philadelphia, Maryland and South Carolina, 1721–1785* (New York, 1929); Henry-Russell Hitchcock, *American Architectural Books* (Minneapolis, 1946); Charles E. Peterson (edited with introduction), *The Carpenters' Company of the City and County of Philadelphia 1786 Rule Book* (Princeton, 1971).

36. "Minutes of the Proceedings of the Directors of the Library Company of

Philadelphia," MSS at the Library Company. Peterson, *The Carpenters' Company Rule Book,* p. xiv. *Pennsylvania Gazette,* August 12, 1742, December 12, 1742. Park found the *Builder's Dictionary (n.a.)* in 1750 and the earliest Langley work *(Builder's Treasury)* in 1753, if one does not count his *Practical Geometry* that first appeared in America in 1746. Park, *loc. cit.*

David Hall Letter Book, II, orders of December 22, 1760 and May 10, 1762, American Philosophical Society Library.

37. Philadelphia Wills: (Songhurst) 1735: 419; (Harrison) 1760:300; (Woolley) 1771:103; (Coates) 1774:10; (Jones) 1781:415; (McClure) 1788:84; (Kinsley) 1791:176; (Lort) 1795:83 ad.; (King) 1804:35; (Rakestraw) 1792: 190; (Williams) 1794:237. Chester County Wills (Armitt), ad. 4132. Most of these men were not subscribers to Swan.

38. George B. Tatum, *Penn's Great Town* (Philadelphia, 1961), plates 6,7,8. "Fair Hill" MSS, Norris Collection, HSP.

CHAPTER 4

The Eighteenth-Century
Frame Houses of Tidewater Virginia

PAUL E. BUCHANAN

MR. BUCHANAN, a veteran investigator of Virginia building technology, is Director of Architectural Research, The Colonial Williamsburg Foundation.

The eighteenth-century frame houses of Tidewater Virginia represent a simple form of architecture, based on the English Georgian style as it was brought over in memories of the colonists and in numerous builders' handbooks. This simplicity is probably the fountainhead of their aesthetic appeal; yet, no two buildings are alike. Their distinctive characteristics are directly related to local conditions, for the colonist had to adapt his buildings to unfamiliar factors such as climate and topography. The limited availability of materials, tools, and skilled craftsmen contributed to the simplicity. The facets of frame construction discussed in this paper present conclusions derived from my personal observation and collection of facts and fallacies acquired during 30 years of working with eighteenth-century Virginia architecture.

Both the individuality and the similarity of the houses were the results of owners' varying requirements, budgets, and the circumstances under which these structures were built. In towns and cities distinctive town plans divided the property into regular lots, specified definite street setbacks, and designated minimum building sizes. The individuals who purchased the lots built houses to meet their particular needs. Buildings were controlled by the limitations of obtainable materials, especially those which had to be imported from England.

All window glass was imported and most lights were 8 inches by 10 inches, giving similar proportions to the windows. The quality of imported paint pigments was erratic, making considerable variation in colors. The copious use of whitewash, especially on fences and outbuildings, generally gave one an impression of white, although other exterior paint colors were used.

There was a certain uniformity among locally produced components and in their installation. Brick sizes were different from building to building but they generally averaged 3 inches per course or horizontal row. Weatherboards usually had 6-inch exposures as did the wood roof shingles. Most chimneys extended 5 or 6 feet above the ridge of the roof. A limit of 20 feet for simple timber spans was a structural uniformity. Many such standardized elements resulted in design similarities among colonial buildings.

The weatherboards[1] were fastened with rose head nails.[2] Rake boards were a signifi-

Fig. 4.1 Brush-Everard House, Williamsburg. The framing system shown in Fig. 4.10 was used in this house. *(All photos and drawings in this chapter courtesy of Colonial Williamsburg. Drawings by James F. Waite and the author.)*

cant part of the exterior design. Tapered toward the ridge of the roof and often ending in a decorative profile at the eaves, they offered a subtle detail which is often missed by contemporary designers of "colonial" houses.[3] The corner boards were generally beaded; and if single, they were on the long side of the building, another subtlety often unnoticed today.[4]

Professional architects, as defined today, were unknown in colonial Virginia. The small frame houses were constructed very economically by local carpenters according to what they had seen, learned, and practiced as apprentices. Their "plans" were simply the verbal instructions stipulated by the owners.

They seldom made drawings although details were sometimes roughly sketched on the job. Large houses were generally of brick; their owners relied on amateur architects—gentlemen who had been trained to see and understand buildings as a practical plantation necessity. Their education in mathematics had included architecture. A few builders were also capable of designing these larger houses. Occasionally small-scale models were made but apparently none have survived. Working drawings were rare until after the middle of the century.

By the eighteenth century, a group of native-born Americans had been trained in the

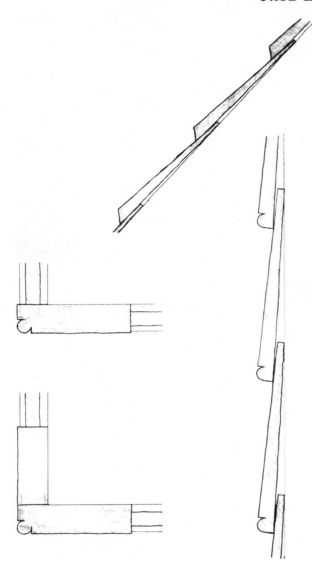

Fig. 4.2 Clapboards, weatherboards and corner boards typically used in eighteeth century construction.

PLAN TYPES

There were only five basic plan types used in Tidewater Virginia. Seldom were any two frame buildings identical, as additions and different combinations of details made them diverse in appearance.

The simplest plan (Fig. 4.3) was a one-room building with an exterior, or perhaps interior, chimney. It usually had only a steep ladder to the attic, which may or may not have had dormer windows. The ground floor could have been brick, wood, or simply clay; the room was usually between 9½ and 10 feet high. There were many such houses, each differing in some detail and each having its own distinct individuality.

The second type of plan featured the addition of a passage on the end opposite the chimney (Fig. 4.4). Although we call this a stair hall today, it was then called a passage. It allowed a complete stair to the second floor and offered much more privacy. There were two variations of this plan type.

The first variation entailed the addition of a chimney on the opposite end with the partition moved nearer the center of the building.

building trades. There was also a constant influx of immigrant craftsmen from England who brought over the latest styles which, controlled by the availability of materials and the climate, were absorbed into the regional style. The work was done by lump sum contract, by day labor, and by measure with standard prices charged for each item based on the current London price books. Some slaves were capable craftsmen and a few became excellent bricklayers and carpenters, but the majority provided only the necessary heavy labor.

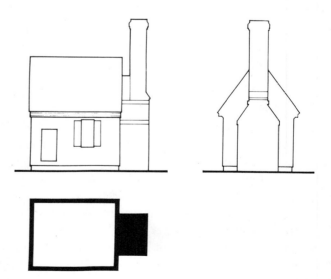

Fig. 4.3 Plan Type I has one room and one chimney, with a second floor reached by a steep ladder-type stair.

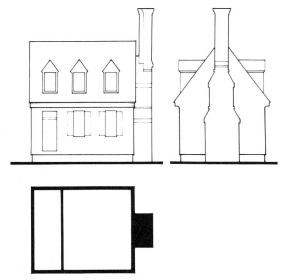

Fig. 4.4 Plan Type II had a passage opposite the chimney containing a stair to the second floor. This allowed more privacy than the one-room plan.

This was commonly called a hall-parlor plan in which the larger room had an exterior door and a stairway to the second floor. In the eighteenth century a large room entered directly from the outside was called a hall, in contrast to the present usage of the term. The smaller room was usually a master bedchamber.

In the second variation, the partition was moved toward the chimney, providing one very large unheated room and a small heated one. This plan was used for many shops in the urban areas. The large unheated room was a public sales area. The heated room was the merchant's counting room which sometimes had an exterior door and a stairway leading to a dwelling upstairs.

This general plan type (except for the hall-parlor version) did not give the symmetrical facade which Virginia colonists favored. Yet the building could be easily expanded, becoming the third plan type.

The third plan type (Fig. 4.5) was most common for small rural houses. It had two heated rooms on the first floor, separated by a passage, and two or three bedchambers on the second floor. There were many variations of this plan. One such variation occurred with

one exterior and one interior chimney; another rare type placed the chimneys on the rear wall to provide for future wings, which made the plan U shaped.

The colonial builder was limited to about 20 feet in simple timber spans. If he wanted to make a building deeper, he would have to build a bearing wall in the middle of the building to support the wood timbers or go to a very expensive framing system using heavy wood girders and summer beams. A simple timber span was called a pile. Therefore, the fourth plan type was called a side passage-double pile plan (Fig. 4.6).[5] The rooms had corner fireplaces to economize with only one chimney stack. The gambrel roof building gave more usable space on the second floor than the gable roof and was easily adapted to this plan type. The gambrel roof, then called a Dutch roof, was built in Tidewater Virginia only after the mid-eighteenth century (Fig. 4.7). It came to Virginia from England and had steeper, lower slopes than the earlier gambrel roofs found in the northern colonies which were introduced from the Continent.[6]

The fifth plan type was the center passage-

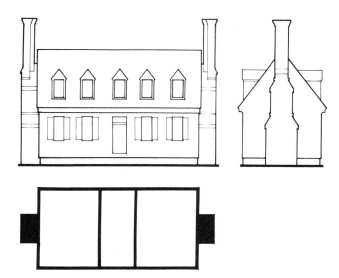

Fig. 4.5 Plan Type III incorporated two chimneys and a center passage, with two or three bedrooms on the second floor. Chimneys were sometimes placed on the interior.

Fig. 4.6   Plan Type IV is double-pile, or two rooms deep, with a side passage and a bearing wall dividing the front and back rooms.

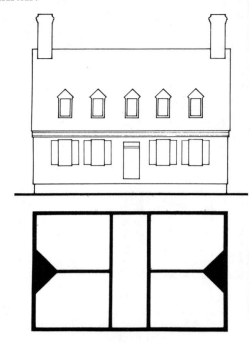

Fig. 4.8   Plan Type V incorporated two chimneys, bearing walls between front and back rooms, and center passage.

double pile plan sometimes one and a half stories high and sometimes two stories (Fig. 4.8). Its typical corner chimney placement produced asymmetrical rooms. After the Revolution, builders gradually centered the chimneys of the two main rooms on the end walls, keeping the corner fireplaces in the two rear rooms, and thus built four chimney stacks. In the nineteenth century, the center passage-double pile house had four separate chimneys, with one centered on the end wall of each room.

There were a few odd plan types but these five, with the variations mentioned, cover about 95 per cent of the frame houses built in Tidewater Virginia during the eighteenth century. As an example, Mount Vernon started out as one of these plan types.

MASONRY IN FRAME BUILDINGS

The typical frame house was usually built on a brick foundation wall, although many of the smaller ones were built with heavy wood sills laid directly on the ground. When this occurred the sills were called ground or mud sills. However, these buildings did not last long unless they were later jacked up and underpinned with brick foundations. Some were built on vertical wood posts set in the ground to carry the sill, but this was more common in the seventeenth century than the eighteenth.

Slightly more than half of the houses during

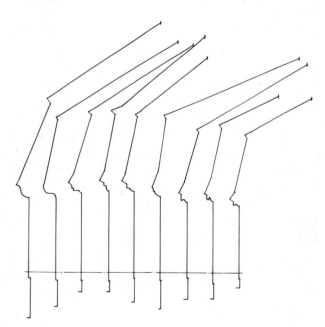

Fig. 4.7   Typical gambrel roof sections of Tidewater Virginia.

this period were built with cellars. The foundation walls were brick, usually laid in double English bond, although sometimes in single Flemish bond.[7] For buildings without cellars the foundations were seldom deeper than 18 inches below grade. They varied in thickness from 9 to 26 inches with the 9-inch foundations being used for smaller buildings without cellars. Houses up to 20 feet deep, with or without cellars, had 13-inch foundation walls; larger structures quite often had 18-inch walls. If the building was to be used for semi-public purposes, such as a tavern, the walls might be 26 inches thick.

Although most cellars had dirt floors, some were brick. Since stone was scarce in Tidewater it was rarely used for flooring. Cellar openings had wooden bars across them, usually horizontal, to keep out large animals and human encroachers. Glazed sash was very seldom used at this level. The openings were usually closed with interior batten shutters. Sloped cellar entrances with one or two doors were common.[8] Their brick steps were fitted with wood nosings which could bear heavy traffic such as rolling barrels that would have broken plain brick edges. Cellars were primarily intended for storage, although occasionally they were used for cooking. Detached kitchens, which were among the group of service outbuildings necessary for every dwelling, were predominant in Virginia.

Fig. 4.9 Exterior chimney. The Bracken House, Williamsburg, has typical massive exterior chimneys.

The bricks in Tidewater Virginia were obtained from two sources. Most were made from clay excavated for the cellar and burned on the site. In some heavily populated areas bricks could be purchased from the stock of a local brickmaker. There are many legends about bricks having been imported, but this does not appear to have been the case in any building I have examined.

Being handmade, the bricks were rough and uneven in texture. The colonial builder would have preferred his bricks smooth and perfect, but this was economically impossible. Because of their rough nature, the bricks were laid with mortar joints about 5/16 to 3/8 of an inch thick. The joints were then struck with a jointing tool against a straight edge to give an incised line about 1/8 of an inch wide and 1/8 of an inch deep. This gave a more regular appearance to the foundation walls and chimneys.

Mortar was generally composed of sand and lime made from burnt, crushed oyster shells, as limestone was nonexistent on the east coast of Virginia. The proper proportion was one part sand to one part lime. However, many bricklayers reduced the amount of lime, which caused subsequent brickwork failures.

In colonial times fireplaces were used both for heating and cooking; the chimneys were almost invariably brick (a few rural buildings had chimneys made of wood and clay). The sizes and shapes were determined by the size and location of the fireplaces, ovens, stills, coppers, and their flues. Exterior chimneys were generally laid in Flemish bond with sloping weatherings covered by bricks laid flat.[9] These sloping surfaces showed the configuration of the fireplaces, ovens, smoke chambers, and flues inside the chimney. The Flemish bond was changed to common bond in the chimney stack or the top vertical shaft. The chimney stack contained only flues, which usually resulted in its brickwork being only 4 inches thick. Flues were normally from 8 to 14 inches square, with the most common dimension being 12 inches square.

Most fireplaces or ovens used one flue; however, if they were wider than 4 feet, two flues were used. This contrasts with chimneys in New England which often had one very large flue. The flues in Virginia were generally parged with a very thin coat of brick mortar while being built. Intense heat caused this mortar to disintegrate, which often resulted in chimneys becoming hazardous. Obviously, exterior chimneys were the safest.

SIMPLE BRACED FRAME

A simple braced frame was used for most of the wood houses in eighteenth-century Tidewater (Fig. 4.10). It consisted of a sill laid on the foundation wall with joists. The joists were usually the same depth as the sill and tenoned into mortices in the sill. This sill was never fastened to the foundation. Corner posts were fastened to the sill by a mortice and tenon joint, and were held in place by heavy corner braces, morticed and tenoned both to the sill and the corner posts. Vertical studs were spaced between the corner posts; a plate was set on top of the studs and corner posts on the two long walls of the building. Second floor joists were set on the plates with a raising plate being nailed to the ends of these joists to receive the common rafters. A collar or wind beam joined the common rafters with a half-dovetail lap joint to keep the rafters from spreading. The rafters were fastened together at the top by an open mortice and tenon pegged joint.

Figure 4.10, an isometric drawing of a simple braced frame, has some unusual features seldom encountered in eighteenth-century Virginia construction. It shows a summer beam[10] in a building under 20 feet deep. This occurred rarely, here probably because timbers long enough for simple span joists were unobtainable. Another unusual feature is the awkward placement of the lower ends of the corner braces. In order to eliminate two adjacent mortices, these members were usually set at the far side of the studs instead of the near side.

Tidewater houses were constructed without sheathing. The window and door frames were fastened to the studs with lugs from their

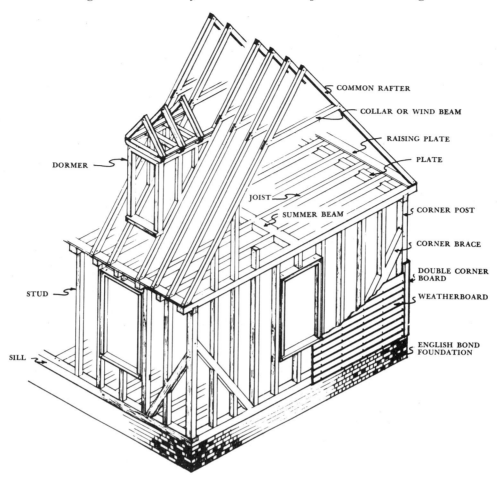

COMMON RAFTER
COLLAR OR WIND BEAM
RAISING PLATE
PLATE
CORNER POST
CORNER BRACE
DOUBLE CORNER BOARD
WEATHERBOARD
ENGLISH BOND FOUNDATION
DORMER
JOIST
SUMMER BEAM
STUD
SILL

Fig. 4.10 Simple braced frame, shown here with an unusual summer beam. This is a rare example, as most frame houses were either kept within the 20 feet limit or had interior bearing walls. These details are from the Brush-Everard House. Corner braces were joined to the corner posts at different levels to prevent weakening them at a single point. Use of a raising plate allowed the rafters to be spaced independently of the joists. One rafter has been omitted to allow for dormer.

heads and sills. In contrast, many buildings in the northern colonies used only widely spaced heavy posts with horizontal or vertical sheathing, to which windows and clapboards were fastened. Note the raising plate on top of the second floor joists. This allowed the rafters to follow a different spacing than the floor joists.

WIDE SPAN FRAMING

Simple braced frame construction was used on most of the frame houses in Tidewater Virginia, but some buildings required greater spans. Wetherburn's Tavern in Williamsburg is an excellent example of a wide span braced

frame (Figs. 4.11 and 4.12). It measures 26 feet by 45 feet and had non-loadbearing interior walls. It appears to be a double pile structure but in reality it is a single pile built with a system of heavy beams and girders. Although expensive, this system provided a particularly sound framework.

The framing was begun by laying a heavy wood sill of heart yellow pine, poplar, or oak upon the foundation wall (Fig. 4.13). Its size depended upon two factors: first, the span determined the thickness—a dimension chosen from the builder's previous experience and his reference to available handbooks. Secondly, size was determined by procurable tim-

61

Fig. 4.11 Wetherburn's Tavern, Williamsburg. Wide span braced frame construction
was used on this building; details are shown in the subsequent framing drawings.

ber. Usually these sills were cut four from a log, pit sawn on three sides and normally hewn on the fourth .side. A 24- to 30-inch diameter log could furnish four sills and four joists. The sills in Wetherburn's Tavern were 8

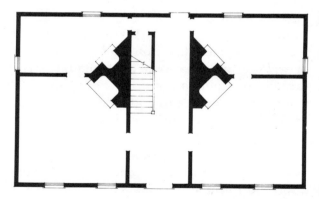

Fig. 4.12 First stage of Wetherburn's Tavern with center passage and interior chimneys. Later additions gave the facade shown in Fig. 4.11. Although two rooms deep, the structure had no interior longitudinal bearing walls; use of summer beams to achieve its depth is published here for the first time. (Author)

by 12 inches when cut, but reduced by shrinkage over the past 235 years to 7½ by 11 inches. The cross section of the sill has growth rings which reveal that it was a sapling shortly after Columbus discovered America. The sill length varies from 20 to 36 feet. Where spliced, the joint is either half lapped or is an open mortice and tenon joint with ¾-inch round wood pegs.

The end sill was set in place first, then the summer beams and girders were placed in position. After this the joists were fitted to the summer beams and front and rear sills. The joints between the sills, girders, summer beams, and floor joists were usually mortice and tenon wood pegged joints. There was a beveled shoulder above the tenon and sometimes a tusk below the tenon. All members were the same depth; therefore, over the foundation wall the tenons did not receive a vertical load because the members were also bearing on the brick foundation walls. When-

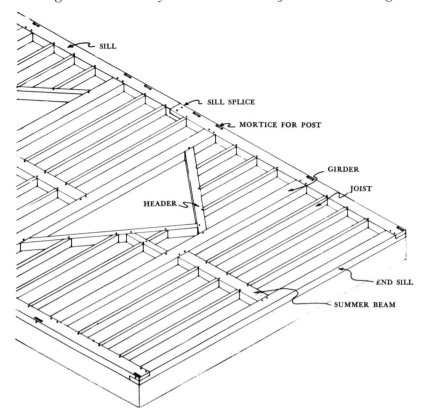

Fig. 4.13   Isometric of first floor framing. The sill was originally 8 by 12 inches, now shrunk to 7 ½ by 11 inches. Growth rings reveal that it was cut from a tree which had been a sapling when Columbus discovered America.

ever the wall was thicker than the width of the sill, the tusk below the tenon was unnecessary and omitted. Joints used in timber framing are shown in Figures 4.14 and 4.15.

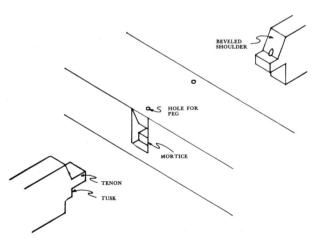

Fig. 4.14   Isometric drawing of shouldered tusk and tenon joint commonly used between sills, summer beams, girders and floor joists.

The girder spacing varied from 8 to 12 feet in this five bay building, 9 feet on centers. Girders and summers were the same depth as sills. Normally, joist spacing was 18 to 24 inches; however, on rare occasions it was as close as 14 inches. Joists were usually 3 or 4 inches wide and as deep as the sill. Here, they were 3 by 8 inches. The mortices for the vertical posts, studs, and braces were cut out with precision: this technique required very careful layout beforehand.

Next, the corner posts, braces, and intermediate posts were erected (Fig. 4.16). In common practice, the corner posts were usually 4 by 6 inches or 4 by 8 inches, with the wide face on the long side of the building. At Wetherburn's, the corner posts were 6 by 9 inches in section and were 10 feet, 4½ inches long plus tenons. In Virginia, corner posts were sometimes constructed of 20-inch square material with the interior corners adzed out to make

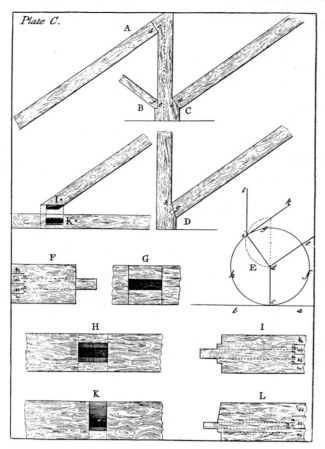

Fig. 4.15 Joints for timber framing. Plate C, *A Treatise on Carpentry*, Francis Price, London, 1733. Builders' handbooks such as this were commonly used by colonial carpenters and builders.

them L shaped to fit 6-inch interior walls. However, these were rare examples. L-shaped corner posts and T-shaped intermediate posts were used primarily after the Revolution, between 1780 and 1820.

The corner braces were usually the same size as the corner posts but sometimes they were 1 or 2 inches wider. The tops of the corner braces were framed into the upper third of the corner post, but the braces were unequal in height because two mortices at the same level would weaken the corner post. The braces on the end walls were usually set higher. The location of the feet, or the horizontal placement of corner braces, was determined by the stud locations. The intermediate posts on the front and rear walls were 6 by 9 inches and were located over the girders at about 9-foot spacing. One exception occurred

where one post on each end of the front wall was offset to clear a window. The corner posts, braces, and intermediate posts were tenoned into mortices at each end and were pegged.

Standard stud spacing was about 24 inches center-to-center, but varied from 18 to 28 inches. Spacing often varied from bay to bay and from story to story, depending upon fenestration. Studs were always spaced so that the vertical nailing pattern on the face of the weatherboards was a uniform width in each panel. The common size of studs was 3 by 4 inches; however, large structures had studs measuring 4 by 6 inches. Normally they were tenoned on both ends and fastened with wood pegs, although here the tenons in the mortices were nailed with a single nail. When interrupted by a corner brace, studs were nailed to the brace with only two nails. At the foot of a corner brace the stud usually rested on the corner brace itself, rather than being tenoned into the sill just beyond the brace; a mortice in this location would weaken the joint connecting the brace and sill. Thus, the stud spacing determined the slope of the brace, just as the spacing of the window determined the spacing of the studs. In Virginia, the corner brace always sloped down from the corner post to the sill.

Studs and other wood framing members were generally pit sawn on two, three, or four sides, and hewn on one or two sides, depending on their location in the log. Occasionally, water mill sawn members were used. Since Tidewater Virginia is very flat, it was without the waterfall necessary to operate sawmills which were so plentiful in the northern areas. Slaves worked the pit saws that furnished most of the building framing. They did much of the carpentry under the direction of a master carpenter who was responsible for all the layout and measuring.

On top of the studs and posts of the front and rear walls was placed a plate which was normally 4 by 6 or 4 by 8 inches. Here, however, it measured 6 by 9 inches. The mortices for the tenons of the studs and posts were worked and fitted on the ground. This was typical of most joints in the wood frame before

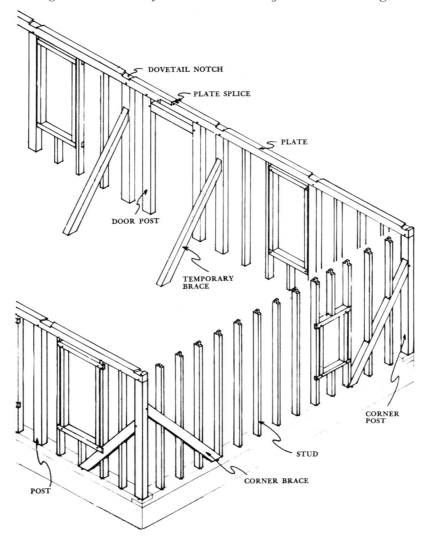

Fig. 4.16   Posts, braces and studs, with temporary bracing. Note how corner bracing interrupted the studs. Window spacing determined the stud spacing, which varied from bay to bay and story to story, according to exacting requirements.

they were hoisted into place; temporary interior braces steadied the rising framework. The plate was usually spliced in the area of the center bay near an intermediate post. Being high in the air and lacking continuous support, it was subject to stresses and the most common splice was a half lap-open mortice and tenon type which had a tenon on each piece (Fig. 4.17).

Girders were then hoisted and placed on the front and rear plates, which had been previously prepared with dovetail notches over the vertical posts (Fig. 4.18). These girders, 11

inches by 9½ inches by 28 feet, 6 inches long, weighed about 650 pounds each. They were by far the most difficult members to install, as the summer beams had to be inserted at the same time. It is easy to see why eighteenth-century builders avoided wide rooms whenever possible, since the framing requirements were so intricate and costly.

The dovetail lap, combined with the weight of the girders, tied the front and rear plates together without wooden pegs. Floor joists, here 3½ by 9½ inches, were inserted in the summer beams. Tusk and tenon joints, just as

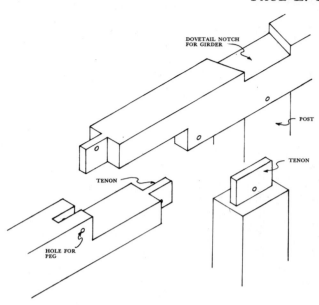

Fig. 4.17 Open mortice and tenon half lap splice, in isometric. Such joints were made for plates on the ground, before the subassembly was raised into place.

those on the first floor, were used at the summer beam but were notched over the plate without pegs. In some buildings the plates were also notched on the top to keep the joists from moving sideways. The projection of the girders and joists beyond the plates was very carefully calculated in order to fit the soffit and fascia of the cornice.

On top of the girders at the foot of the rafter, the principal rafters were 7¼ inches wide by 5½ inches deep, tapering upward to just under 5 inches at the ridge. These rafters were narrower than the girders and formed the sloping members of a modified queen post truss (Fig. 4.19). This truss was composed of two principal rafters—a wind or collar beam and the two queen posts. It was prefabricated on the ground, assembled on the second floor and raised in one piece. The rafters and posts had tenons that dropped into the girder mortices or the lower chord of the truss, and were then pegged. As each truss was erected, two horizontal purlins were tenoned into each sloping principal rafter. These purlins supported the very lightweight common rafters. Diagonal bracing was used in trusses for some of the larger buildings, but in the framing system described here, the purlins and the wood

shingle lath supplied the requisite bracing. The rafters at the ridge had an open mortice and tenon joint with one wood peg.

ROOF FRAMING AND ROOFING

In the eighteenth century, roof pitches were determined by geometric computations, but in a different manner from the method used today. With present-day use of a steel framing square, the roof pitch is set by a vertical rise of a certain number of inches to a base or horizontal dimension of 12 inches. The eighteenth-century carpenter had to use a simpler method of laying out his sloping roof members. This technique, almost forgotten today, made use of a simple proportion of the roof slope to the length of the attic floor joists. The length of the attic floor joists or beams equaled the depth of the building and the projections needed for the front and rear cornices. The colonial carpenter would divide the total length of the attic floor joist into a number of equal parts. If divided into four parts, he would then use a number of these parts, say three, and make the overall length of the rafter the resulting dimension. This would give a roof slope of 48 degrees and 30 minutes.

There were nine other proportional pitches commonly used in the eighteenth century: $5/6$, $4/5$, $7/9$, $2/3$, $5/8$, $3/5$, $7/12$, $4/7$, and $5/9$. The length of the rafter is the first dimension given, proportional to the number that the base width is divided. From my experience I have found that almost all eighteenth-century roofs were built according to the above system, with only one exception—the right-angle roof which had a slope of 45 degrees and a ridge angle of 90 degrees. Even this roof pitch could be calculated by the formula given above, the proportion being $17/24$.

The carpenter in the eighteenth century also installed the wood floor and roof framing for brick buildings. These features were similar to those in frame buildings, with the exception that a larger variety of roof trusses were used in brick buildings. The most common truss was the king post truss, with the queen post next in popularity. Handbooks of the

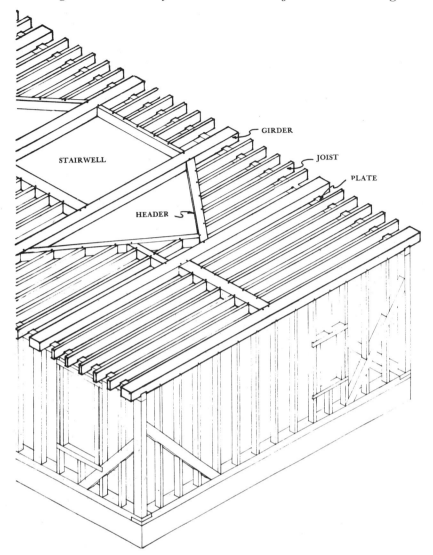

GIRDER

STAIRWELL

JOIST

PLATE

HEADER

Fig. 4.18 Second floor framing in isometric showing the girders, which weighed about 650 pounds apiece, the dovetail lap which tied the front and rear plates together without need of pegging. Projection of girders and joists beyond the plates was calculated to fit the desired soffit and fascia dimensions of the cornice.

period illustrated many traditional truss types, yet the colonial carpenter was always experimenting in an effort to develop a special type for each particular application.

Three methods were used to cover the roof. A temporary covering of oak clapboards was sometimes used to span the spaces between two rafters. These were lapped horizontally with ends lapped by the adjacent clapboard. This resulted in a straight joint from cornice to ridge at alternate rafters. After a year or

two, this temporary roofing was replaced by either being entirely removed and replaced with wood shingles, or by covering the clapboards by nailing shingles directly to them.

The second and most common method of roof coverings involved nailing shingle laths to the tops of the rafters. These were strips of hand-sawn wood about 1 inch thick and 3 inches wide, spaced approximately 6 inches on center. They were placed closer together at the first two feet above the eaves; quite often

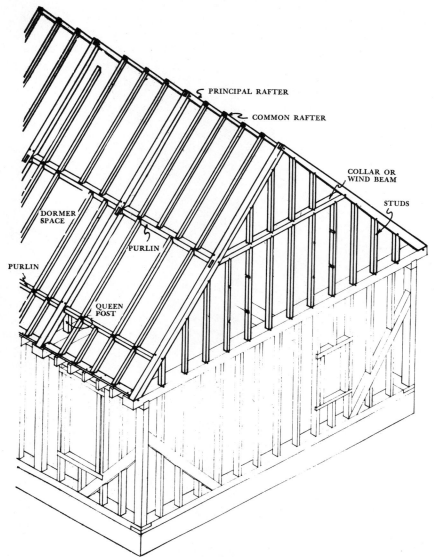

Fig. 4.19 Rafters and posts were framed into a series of modified queen post trusses. As each truss was raised and put into place, horizontal purlins were added to support the common rafters. Some large buildings had diagonal braces in the roofing system, but at Wetherburn's Tavern, the purlins and the wood shingle lath sufficed as bracing elements.

the first two laths touched. Wood shingles were then nailed on the laths. Shingle laths allowed the shingles to breathe due to the air space beneath them. This increased their longevity and reduced rot on the underside, which caused the first failure in wood shingle roofing.

Nailing wide sheathing boards to the rafters was the third method used in roof covering. There was very little space between them; the shingles were nailed directly to the boards.

However, this system was not preferred because it did not allow the shingles to breathe. Shingle lath was usually sawn from yellow pine but oak and other woods were also used. Roof sheathing was pine, poplar, and sometimes gum.

Shingles were rived (split) from logs about 18 inches long. This was the common shingle length in Virginia, although some have been found up to 24 inches in length. The shingles were tapered in thickness from about 5/8 inch or ⅞ inch at the butt end to a feather edge at the other end. The shingles were dressed with a draw knife which the craftsman used to achieve the final tapering in a uniform manner. Widths varied from 3½ inches to 8 inches. The butt end was quite often square. A better practice was to round the corners by chopping them with two or three blows of a shingling hatchet or by sawing off the corners. The round butt shingles cost more to prepare and they lasted longer because they did not become warped by the heat of the sun as fast as the square butt shingles.

Atlantic white cedar *(Chamaecyparis thyoides)* commonly called juniper, was the first choice for shingles, followed by cypress, yellow pine, oak, and chestnut. Oak shingles tended to warp or split. Cypress had the ability to withstand deterioration from water. Shingles were sometimes painted red to increase their durability and some were tarred.

Shingles were nailed to the roof beginning at the eaves (Fig. 4.20). The starting course was two shingles thick with the lower ones be-

ing only 8 or 10 inches long. The lower shingles of the starter course were full butt thickness and were always square. This was true even if round butt shingles were used for the rest of the roof. The shingles overhung the crown moulding of the cornice from 1½ to 2 inches. The use of the shorter bottom shingles in the starter course caused the roof slope to "kick" up slightly in order that the upper shingles of this course might lie flat against the lath or sheathing.

Above the starter course, shingles were normally laid with an exposure of about 6 inches, or one third of their length. This resulted in the roof being entirely covered with three thicknesses of shingles. If shorter or longer shingles were used, the spacing was varied accordingly.

When these three thicknesses of shingles met at the roof ridge, they were lapped over each other; the top course of shingles on one slope projected higher than its counterpart on the other slope. This was called "combing." The extended shingle course in the comb was

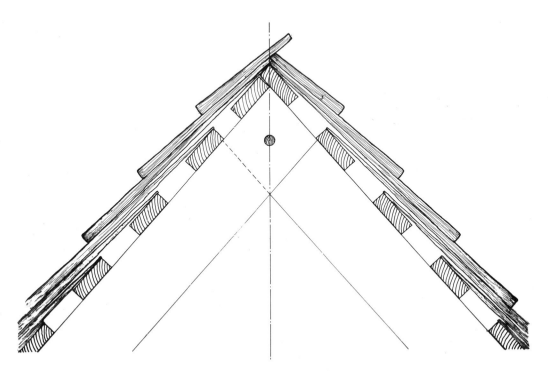

Fig. 4.20 Typical ridge details. Without using a ridgepole, rafters were mortised and tenoned to each other and pegged. Spaced laths for nailing shingles allowed the roof to "breathe" and dry out. The top row of shingles is "combed" to close the joint.

always angled away from the orientation of the most violent storms prevalent in the vicinity. In Tidewater Virginia the worst storms generally came from the northeast; therefore, the shingles were combed toward the south and west.

Shingles of hip roofs were "fantailed" at the hips by tapering the width of the shingles in order that no joint would coincide with another. Fantailed shingles were wider at the butt end. Shingles at roof valleys were curved; this is called a woven or swept valley (Fig. 4.21). A tapered board, wider at the top, was placed in the valley to help form this curve. The shingles in the valley were fanned, as they were narrower at the butt end. This was the reverse of fantailing.

Metal flashing was too expensive for ordinary use, but an alternate system of waterproofing was devised by installing a tapered board about 10 inches wide along the sloping edges of dormers and chimneys. These boards were tapered from ¾-inch thickness to a feather edge and were nailed to the lath or sheathing. The slant of the shingles attached to this board diverted rainwater away from vulnerable openings. These tapered boards were called "tilting fillets."

FINISHING THE BUILDING

After the carpenter completed the framing of any wood building, he then covered the exterior walls with weatherboards and installed the cornice before he covered the roof. Eave gutters were almost nonexistent in eighteenth century Tidewater. Gutters were generally ground-level brick gutters or drip strips. However, wood guttering occasionally appeared at the eaves of laundry buildings in order to collect preferred rainwater for washing clothes. The few wood gutters that did exist were either two boards nailed together in a V shape or they were hewn logs.

Doorways were simple; ornamental frontispieces were frequently added in the nineteenth century as doorways were altered and decorated in later architectural styles. About

Fig. 4.21 Roof valley shingling detail. This rare survival at "Bowman's Folly," Accomac County, shows the roofer's method of forming valleys before the advent of metal flashings.

one third of the colonial doorways had transoms which were usually one light high. Occasionally they were two lights high; these brightened dim interior passages. Panelled doors were installed on all but the smallest buildings. Pairs of doors were the choice for doorways over 4 feet wide. Dutch doors and glazed doors were practically nonexistent; porches, too, were rare but were added to almost every colonial building during the nineteenth century.

The windows of most frame houses had panelled shutters on the outside; interior shutters were used for brick houses. Some buildings had batten shutters and a lesser number had louvered shutters in which the slats were fixed. Louvered shutters with movable slats were introduced in Virginia in the nineteenth century. Most windows had double hung sash, three lights wide. The sash were fairly thin, ranging from approximately 1 to 1¼ inches thick. The muntins were wide, averaging 1¼ to 1⅝ inches. After the Revolution muntins gradually became narrower until they reached ultimate attenuation during the Federal period. The top sash were generally fixed in colonial buildings, although some did operate with lead weights. Some bottom sash were not counterbalanced and they had to be propped open with sticks or other devices.

Four-light-wide windows which were so common in the north were rare in Tidewater Virginia. The normal width was three panes but the window heights varied. If they were divided into unequal sash, the builder had to choose how to arrange the sash. Aesthetically, it was more pleasing to have the larger sash at the bottom of the window, but it was more economical to put the larger sash at the top due to less lead being required to counterbalance the smaller lower sash. Large fixed sash and bay windows were only used on shop buildings.

Once the exterior of the building was closed in and secured from the elements, the colonial carpenter turned his attention to completing the interior. Brick nogging was sometimes placed in the void spaces between the studs and posts of the exterior walls. The primary purpose of nogging was ratproofing instead of insulation, although it was installed in more severe climates as insulation. Brick nogging was usually composed of "samel" bricks[11] inserted between framing members. Sometimes the bricks were set in very sandy mortar, sometimes substituting clay for mortar, and sometimes they were laid without mortar.

All interior woodwork, such as flooring, door and window trim, mantels, stairs, cornices, and baseboards, was installed before the building was plastered. Interior flooring was edge grain, heart yellow pine with the boards measuring from 5 to 8 inches wide. The upper surface was hand planed but the lower face was left rough sawn. The upper surface of the floor joists was either rough hewn or sawn, and each floorboard had to be hewn on the lower surface to fit each joist. Joints between floorboards were either tongue-and-groove, splined, shiplap, or butt joints. A few very fine houses also had horizontal dowels in their floor joints. The floorboards were generally face nailed with rectangular T- or L-head nails; some, however, were fastened with rose head nails. A very small percentage were blind nailed with the nail going through the tongue of the tongue-and-groove joint. For economy, sometimes the builder would substitute a proportion of sapwood in the second floor flooring; this was poor construction even though it provided wider boards. Heart yellow pine floorboards were almost equal to hardwood and needed no wax, paint, or other preservative. All other interior woodwork was painted, unless the material was a hardwood such as walnut or oak. The colonists had discovered that unpainted softwood deteriorated in the humid climate of Tidewater Virginia.

Baseboards, door trim, and window trim existed in every house, and usually a chair rail or chair board was placed about 3 feet above the floor. In the nineteenth century, the typical chair rail was lowered to about 24 or 30 inches from the floor due to changing furniture styles. Cornices also existed in some, but by no means all, buildings; many had panelling below the chair rail. Most fireplaces had simple wood trim, with or without mantel shelves. Some chimney breasts were panelled from floor to ceiling, and a few of the more expensive houses boasted full panelling throughout the interior. The average house, however, had plain plaster walls and ceilings which were periodically whitewashed. After the end of the eighteenth century, tints were added to the whitewash; wallpaper was only used occasionally and fancy plaster work was limited to a few fine houses.

Stairs were considered utilitarian objects and only a few bore ornamentation. Stairways generally had handrails only on the side opposite the wall, although many wall-mounted handrails have now been added to colonial buildings for modern safety. The smallest structures, including shops and outbuildings as well as houses, had stairs that were little better than fixed ladders. In all buildings, the size of the passage was dictated by the living requirements allocated to other rooms. This, in turn, dictated the size and type of staircase.

After installing the wood trim, the carpenter prepared the walls and ceilings for the plasterer. He nailed split yellow pine lath to the framing. These lath were about 1/4 inch thick, 1 to 1 1/2 inches wide, and from 4 to 8 feet long. They were nailed to the framing with small, sharp-pointed rose head nails which left about 1/4 inch between each lath in order that the plasterer could form a key to hold it in place. The plasterer would then install a two-coat plaster with a total thickness of about 3/4 inch. The base coat was a sandy plaster with a small amount of oyster shell lime, which generally included some animal hair mixed in as a binder. The final coat was very thin, composed of almost pure lime, and was very white.

The interior woodwork was then ready for the painter who, in the better houses, had already primed the trim with a paint (called "Spanish brown") before the carpenter installed it. The interior woodwork was painted with one, and sometimes two, coats of lead and oil paint. Most of the paint ingredients were imported from England. Once painted, the house was ready for occupancy. The carpenter and his fellow craftsmen could take pride in the accomplished execution of their workmanship.

The "art of building" was esteemed in the eighteenth century,[12] both by men of taste and culture who appreciated the aesthetic refinements of architecture, and by the skilled craftsmen who exercised their expertise in the practical employment of construction. Carpentry was a master craft in the eighteenth century, requiring versatile abilities, disciplined training, and honest and proficient application. Mr. R. Campbell, who published *The London Tradesman* in 1747 for the instruction of youth in their choice of profession, aptly described the qualities and qualifications prerequisite to the trade:

The Carpenter is the next [among] Person[s] of Consequence in the Employ of the Architect. The Carpenter is employed in the Wooden-Work, from the Foundation to the Top. In Works where the Foundation is supposed soft, the Carpenter drives Piles down to support the Edifice. In Brick-Works he places Bearers, where the chief Weight of the Building lies: He lays the Joists, Girders, and Rafters in Flooring, and when the outward Case is built, he puts on the Roof and prepares it for the Slater. This is the proper Business of a House-Carpenter. He ought to have a solid Judgment in Matters of this Kind, to be able to act not only by the common mechanical Principles of his Art, but to strike out of the common Road when the Case requires it; as it frequently does in propping of old decayed Buildings; Strength is the chief of his Study, and to dispose his Work in such a Manner as that which is designed for the Support of a Building may not, by its Weight, overturn it. It requires a strong robust Body and hail Constitution. He must read *English,* write a tolerable Hand, and know how to Design his Work. He must understand as Much Geometry as relates to Mensuration of Solids and Superficies. This Business is by no means despicable in respect to its Profits: . . .

NOTES

1. *Weatherboards* were sawn, tapered boards about 8 inches wide and 10 to 20 feet in length. They were hand planed on the exposed surface with a bead worked along the lower edge. They were face nailed to the frame with large rose head nails.

   Clapboards were split or rived, instead of sawn boards. They were much shorter in length and narrower than weatherboards, having an exposure of 4 to 4½ inches. Clapboards were used extensively in the northern colonies but seldom in Virginia, except for temporary roofing, and only occasionally for the sidewalls of small buildings.

2. For the type of nails used, see "Nail Chronology as an Aid to Dating Old Buildings," by Lee H. Nelson, American Association for State and Local History (1315 Eighth Avenue, South, Nashville, Tennessee 37203), Technical Leaflet 48, reprint, *History News*, XXIV, no. 11 (November, 1968).

3. *Rake Boards* were boards nailed to the weatherboards at the juncture of the wall and the sloping roof on the gable end of a building. Rake boards usually terminated at the lower ends as part of a decorative *End Board* (a board that stops the end of the cornice) or they formed their own end board.

4. *Corner Boards* were vertical boards nailed on the external corners of frame buildings. They could be either single or double. If single, one board was nailed on the long side of the building. If double, a board was nailed to both sides of the corner. Their purpose was to stop the ends of the weatherboards.

5. Historical references to "double pile" appear in several publications. Sir Roger Pratt used the term in the seventeenth century, according to R. T. Gunther, *The Architecture of Sir Roger Pratt* (Oxford, 1928). Portions of this text are quoted in Nathaniel Lloyd, *A History of the English House* (London: 1931), p. 103.

   See also Hugh Braun, *The Restoration of Old Houses* (London: 1954) p. 26; John Harris, *Georgian Country Houses* (London: 1968) p. 1; R. W. Brunskill, *Illustrated Handbook of Vernacular Architecture* (New York: 1971) p. 104.

6. For information on northern gambrel roof slopes *see* A. L. Kocher, "Gambrel Slopes of Northern New Jersey," *Architecture*, LV, no. 2 (February, 1927) pp. 61–66.

7. *English bond* was the earliest and strongest brick bond. Bricks were laid in alternating courses of one course all stretchers and the next course all headers. When the bond was the same on both sides of the wall, it was known as double English bond. *Flemish bond* was more decorative, but not quite as strong. The bricks of each course were alternating headers and stretchers, with the headers centered over the stretchers below. Single Flemish bond would be Flemish bond on the exterior and English bond on the interior of the wall.

   American bond was unknown in the Tidewater area until well after the Revolution, although it was used in Pennsylvania and New England by the mid-eighteenth century. It consisted of one course of all headers, and from three to seven courses of stretchers before repeating the header course.

8. Cellar entrances are commonly called *bulkheads* today. The word bulkhead was not used in this manner in the eighteenth century.

9. *Weathering* — a sloping brick surface to throw off rainwater, incorrectly called a haunch.

10. A *summer beam* was a single span of timber that carried the ends of several floor joists. It could span from wall to wall or might rest on one wall and be carried by a girder at the other end. The term came from *somme*, the French word meaning load or burden.

11. *Samel bricks* were the bricks that were too far from the fire in a brick clamp or kiln to be burned enough to be serviceable as building brick. They would disintegrate into powder after many years. The rate of disintegration depended upon the amount of heat received in the kiln, and the subsequent exposure to moisture.

12. For the United States we have no overall technological study of wooden buildings. Nothing to compare with what John I. Rempel, *Building with Wood* (Toronto, 1967) did for Ontario. Some New England types were carefully examined long ago by Norman Morrison Isham and J. Frederick Kelly. Notable special studies include that pioneer effort of Harold R. Shurtleff, *The Log Cabin Myth* (Cambridge, 1939) a documentary landmark unfortunately never carried very far forward. We need more special studies with the technical competence of John Fitchen's *The New World Dutch Barn* (Syracuse, 1968). To underpin future thinking we should have basic information on the lumber trade of which a worthy example is John Hebron Moore, *Andrew Brown and Cypress Lumbering in the Old Southwest* (Baton Rouge, 1967) and the source material offered in recent issues of the Association for Preservation Technology's *Bulletin*. [Mr. Buchanan's essay raises questions about the assembly of house frames undreamed of by most Early American buffs. — Ed.]

CHAPTER
5

Brick and Stone: Handicraft to Machine

HARLEY J. MCKEE

Emeritus Professor at Syracuse University and a Fellow of the American Institute of Architects, MR. MCKEE is past President of the Association for Preservation Technology, author of *An Introduction to Early American Masonry* (1973) and shorter works.

This study is concerned primarily with ways in which brick and stone units were formed, and only incidentally with the building of walls. Particular attention is directed to the invention and application of machinery to assist in the processes of production.[1]

The nineteenth century inherited a long tradition of work processes in which craftsmen were efficiently organized according to their skills. The rate at which one man performed his kind of work was necessarily adjusted to the output of others on the same team. When the introduction of a machine into one part of the process either replaced a man or increased his productivity, this often required additional hand-workmen for the other operations, but the industrial process as a whole was able to continue in the traditional manner. Early machines, moved by hand or by a single horse, were not always able to produce more than the men they replaced. Investment in such machinery was, therefore, a questionable economy.[2] When water or steam power was applied to larger machines, the rate of productivity increased. Brickmaking machines were able to form denser, harder units than the handmade varieties, so that machine-made bricks gained acceptance for superior quality.

The great bulk of masonry materials had to be used near their source because of the cost of transportation. Stone building was virtually limited to the regions in which suitable material could be quarried. Fine material such as marble, however, for the trim and ornamental parts of a building, had a high value in proportion to its weight and bulk and was often shipped a considerable distance. The making of these special items became a large industry, concentrated in a few favored cities; it often included funerary monuments as well as building trim.

Brickmaking was more widely dispersed than stone quarrying and working because suitable clay was found almost everywhere. Few localities failed to support one or more brickyards, which were often seasonal or intermittent in their operation and which made common bricks. Fire bricks, ornamental bricks and pressed bricks for facing were usually made in the centers that manufactured fine pottery, which depended upon clay of better quality.[3] In both the stone and brick industries, one should distinguish between mass production of common material and processing of finer material for selected purposes.

The basic operations of the stone industry consisted, first, in extracting suitable material from the ground *(quarrying)* after disposing of inferior material lying above it. Large blocks taken out of the quarry were subdivided and rough-dressed to the approximate required size before being taken away from the site. Workmen at a quarry, regardless of their exact duties, were called *quarriers.* Methods of quarrying varied considerably according to geological conditions and the means available. Following centuries-old methods, workmen cut trenches around the blocks to be removed, using picks when the stone was soft enough (Fig. 5.2). To remove thick blocks, deep trenches were required, wide enough to admit the workmen. The ancient Egyptians had trenched hard stones like granite with pounding-balls attached to poles, but by the eighteenth century granite and softer stones alike were detached from the face of the quarry by cutting a series of holes with a drill or *jumper.* This tool was a long iron or steel bar with a chisel-like cutting edge. One man held the jumper in position and rotated it slightly between blows, while from one to three others struck it in succession with sledgehammers. Tapered iron wedges were then placed in the holes and driven tight, causing the rock to split.[4] Sometimes wooden wedges were employed, as in ancient times, their force being increased by wetting the wood after the wedges had been driven into place.

Naturally occurring horizontal seams, called *lifts,* were exploited whenever possible. If these seams were not present in the rock, it was necessary to drill horizontal holes in addition to the vertical ones. Blasting with black powder was sometimes practiced, especially when removing an inferior stone overburden and when procuring stone for crushing or for lime-burning. Blasting caused shock waves and damaged the rock, however, so it was rarely used close to material desired for building blocks.

After being detached from the face of the quarry, blocks were turned over by levers or jacks and hoisted with simple derricks, employing compound pulleys or windlasses. Stones were dragged along the ground on sledges called *stone boats,* or loaded onto heavy carts or simple carriages that ran on rails.

Rough dressing at the quarry was done with sledgehammers or heavy picks, making the stone easier to transport. The sequence is shown in Figure 5.3. Sometimes these *quarry-faced* blocks were used without further dressing, but when they were given a better finish the work was done in a *stone yard* or at the building site. The men who did this were called *stonecutters* or *finishers.* A further distinction was made between the workmen who finished plain surfaces and those who cut mouldings and carved ornamental pieces. Stone destined for trim, especially that receiving a polished surface, was usually finished in a *stone mill* located either at the stone yard or in a separate manufacturing plant.[5]

The simplest stone finish was obtained by *breaking* the material. This method was particularly adaptable to sedimentary stones that came from the quarry in relatively thin slabs, provided naturally with satisfactory top and bottom surfaces or *beds,* and to stones that had been split, such as gneiss. A slab was placed over the edge of a straight support and struck with a heavy hammer; sandstone and limestone generally broke off cleanly, leaving a nearly plane face. Broken faces can be seen in walls constructed by pre-European peoples in the southwestern states, who had no metal tools, as well as in the eighteenth-century buildings of the eastern states.

Stonecutters used several tools in sequence as they proceeded to finish a block; this was done in several stages, at any one of which the work could be terminated, depending on the texture that was desired and the builder's budget. Tools leave characteristic marks on the face of a stone, although those used last generally efface the marks made by the first ones. As a rule, builders required freestone

Fig. 5.1   Coquina Stone at St. Augustine, Florida. A stratum of shell rock suitable for building was reported on Anastasia Island as early as 1580. Much of colonial St. Augustine is built of coquina masonry. Shown here are *(a)* the quarries, *(b)* a *garita* on the Castillo San Marcos, *(c)* Interior of the Castillo, built 1672–1695, *(d)* The City Gates built 1808 and *(e)* The Torre de Matanzas on the Matanzas River completed 1742. Drawn by Harry Fenn for William Cullen Bryant, ed., *Picturesque America,* Vol I, New York, 1874.

Fig. 5.2 Brownstone was quarried at Portland, Connecticut, as early as 1665. At the right, three workmen with sledgehammers drive a row of wedges to detach a large stone from the quarry; others on top drill a row of holes into which wedges will be driven to split it into smaller pieces. A simple machine, apparently for drilling rock, is powered by one man turning a wheel. Horizontal traces on the face of the cliff indicate previous levels of quarrying activity. George P. Merrill, *Stones for Building and Decoration*, third edition (New York: Wiley, 1910).

for corners, door and window jambs, arches, shaped or moulded details, and ornamental carvings. The term *freestone* refers to any kind that can readily be cut in all directions and usually implies soft material.[6] When freestone was not available, rough masonry walls were trimmed with wood.

The method of finishing a stone was chosen to fit particular circumstances and stonecutters did not always follow orthodox practices. Two representative cases must suffice here to indicate the nature of these processes.[7] A relatively soft sandstone or limestone might have its greater irregularities removed with a *pick*. This implement had a pointed cutting end, with which the workman made furrows several inches long in the stone, usually parallel to each other but sometimes irregularly. Some picks had one cutting edge like an adze and they were used in the manner of an adze.[8] The

irregularities that remained after *scabbling* (*see* Fig. 5.3) with a pick might be further reduced with a tooth chisel, which the workman tapped with a wooden mallet or an iron hammer, depending on the hardness of the stone. Chisels were held at an acute angle to the surface. When a smoother surface was desired, a wide chisel with a straight cutting edge, called a *drove* or a *tool* depending on its width, was used to remove the small ridges left by the serrated edge of the tooth chisel. The stonecutter took care to test the flatness of the surface by sighting and by using a straight edge.

Hard limestone or sandstone, and especially granite, were usually finished by *hammering* with a stone *axe* (also called *pean-hammer*). First, however, major irregularities were removed with a *point* (sometimes called *punch*), a tool resembling a narrow chisel. Axes of various weights were used, the blade striking the

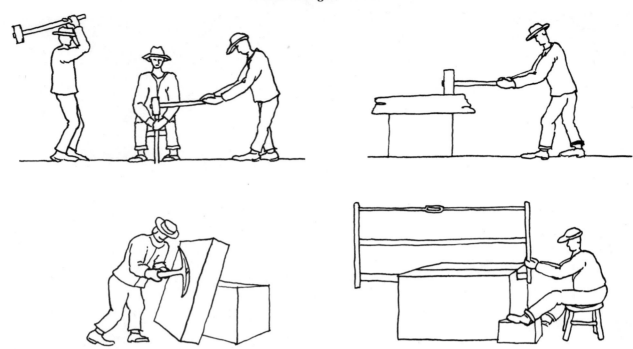

Fig. 5.3   Working stone by hand. For drilling rock, the man in the middle holds the drill and rotates it between blows struck alternately by two men with sledgehammers *(upper left)*. Man at upper right breaks a stone to obtain a moderately even face. *Scabbling (lower left)* is dressing a stone with a pick. Block *(lower right)* is sawn into slabs. *(Drawing by author)*

Fig. 5.4   Sandstone finished by hand, nineteenth century. Quoins *(right)* were *bush hammered*. The other blocks have a coarser finish, obtained with a *point*, a kind of pointed chisel. The plinth *(bottom)* and the margins of all stones were dressed with a *chisel*. Such finishes, and the textures obtained by hand tools, were virtually impossible to produce by machines. *(Photograph by author)*

stone in a hammering motion. Successive passes with the axe were made so that cuts were at right angles to the preceding ones during the process of hammering. *Rubbing* with an abrasive block produced a smoother surface, if desired, and successively finer abrasives were used in *polishing*.

SAWING STONE

Stone could be *sawn* by hand, using a smooth soft iron or copper blade mounted in a frame. After a small groove had been made with a point or chisel, wet sand was fed in under the saw blade to accomplish the cutting. Sawing was a slow process, especially when the stone was hard; its chief usefulness was in dividing blocks of fine material, such as marble, into thin slabs for wainscoting and trim. A sawn face was rubbed smooth with abrasive and often polished.

Sawing was the earliest process to which power-driven machinery was applied. There is some evidence to indicate that water power drove stone saws for the Romans before the end of the fourth century. It is not clear to me just when machine-sawing of stone was begun in the United States because the term *mill* did not necessarily imply power-driven machinery. In 1783, William Shepherd, who was about to erect a mill for sawing and polishing marble in the vicinity of Philadelphia, requested the patronage of the State Assembly.[9] In 1804, Eben W. Judd established a water-powered marble mill in Middlebury, Vermont.[10] The sawing machinery was said to have been devised by him.

Marble saws of the early nineteenth century utilized soft iron blades stretched in a horizontal frame. Several blades or *gangs* in a frame, which was given a reciprocating motion, enabled one machine to saw several slabs simultaneously.[11] Similar machines were used in Scotland and England, so it is probable that the ones in the United States were modeled after them; perhaps some were imported. James Tulloch's saw shown in Figure 5.5, is a good representative of its type.

Circular saws were used to cut marble slabs into smaller strips for trim (Fig. 5.6). Isaac D. Kirk of Philadelphia used a machine of this type, which he patented in 1832.[12] When there was but one saw blade in this machine it could be turned by a man, who was thus able to cut about 30 square feet of marble per day. That was said to be about four times as fast as a hand saw. The real advantages of a circular saw, however, were gained when several blades were turned by water or steam power. Machines of this type were sometimes called *ripping beds*.

After the discovery in Brazil, in 1842, of a variety of dark colored diamond, these hard stones began to be used for cutting purposes. Rotary saws with industrial diamonds set into the edge were exhibited at the Centennial Exposition of 1876.[13] One of them, called the Stone Monarch, was built by Branch, Crooks & Co. of St. Louis; it had a blade 5½ feet in diameter set with 84 black diamonds. A steam engine turned the blade at high speed, to cut through a sandstone block 6 feet long by 2 feet, 5 inches deep in 10 minutes.

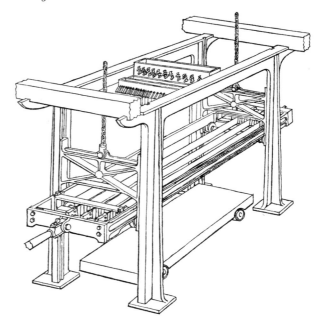

Fig. 5.5 James Tulloch's saw, patented in England in 1824, had a cast-iron frame 10 to 14 feet long, 4 to 5 feet wide and 8 to 12 feet high, securely fastened to the structure of a mill building. The saw frame, suspended from counter-weighted chains, ran on rollers; the length of stroke was about 18 inches. Near the end of each stroke this saw frame was lifted enough to facilitate the flow of wet sand abrasive beneath the edge of the saw blades. The block of stone being sawn was placed on a movable carriage below. *Journal of the Franklin Institute (JFI)*, XXI, Third Series, No. 4 (April, 1851) p. 246.

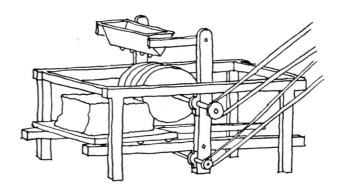

Fig. 5.6 Isaac D. Kirk's circular saw had a wooden frame and several soft iron or copper blades about 30 inches in diameter. The stone was placed on a carriage, which was advanced by a slowly rotating worm gear that engaged a toothed rack. A hopper at the top contained sand and water, which trickled down into the saw-kerf. This machine could be adapted for cutting mouldings; a cast-iron cylinder of suitable profile was substituted for the saw blades, and marble could be ground into shape with the aid of wet sand abrasive. Polishing was accomplished in a similar manner. Redrawn from *JFI*, XI, New Series, No. 5 (May, 1833) p. 325.

Machinery was used to smooth and polish sawn stone pieces at an early date, apparently being suggested by grindstones and discs used to polish gems. The term *lap*, often applied to the revolving horizontal *sanding plates* in early nineteenth-century mills, on which small pieces of marble were finished, suggests a lapidary origin. This device was a circular table of cast iron several feet in diameter, which revolved at a moderate rate of speed. Wet sand was spread on top of the plate and the stones to be finished were placed on it face downward. Wooden stops just above the revolving plate divided the work area into two or more sectors, each attended by one workman who moved the stones from time to time to ensure uniformity of grinding. Small pieces were held by the workmen when their edges were being ground.

To grind and polish large stones, a different kind of machine was required. Heavy cast-iron rubbing plates were placed on top of the stone and moved about in a compound motion calculated to assure equal wear to every part of the surface. A good example of this type is James Tulloch's grinding bed, shown in Figure 5.7. Other machines, such as the one used by Barton N. Fyler of Orange County, Vermont, patented in 1830, made use of a grinding wheel 3 feet in diameter. This wheel rotated on a vertical axis and worked over the top face of the stone being ground, which was mounted on a movable carriage. Fyler's wheel was made of wood and had strips of iron driven into it, even with the surface.[14] The combination of wood and iron held the abrasive sand very well. A machine patented by Adrian Olcott of Somerset County, New Jersey, in 1850, placed the stones above the rubbing bed and moved them about by chains suspended from a rotating wheel.[15]

In the 1850s, some mills that manufactured funerary monuments possessed machines to saw obelisks and other nonrectangular forms. In order to cut two nonparallel sides at the same time it was necessary to devise some way of pulling the saw blades along divergent paths, a motion that the simple frame could not accomplish. One such machine was pa-

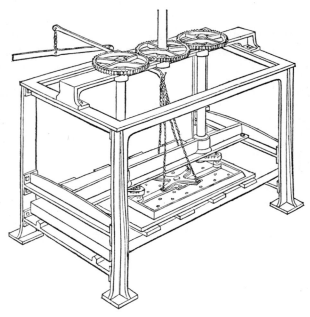

Fig. 5.7 Grinding beds were used in smoothing and polishing large slabs of marble and other stone. This machine, patented by James Tulloch, was employed in a marble mill in London. It is a well-developed example of a type common to marble mills in the United States. Holes in the grinding plate permitted abrasive material to flow through onto the stone. This plate was counterweighted; it could be adjusted for pressure and lifted to permit inspection of the work. *JFI*, XXI, Third Series, No. 4 (April, 1851) p. 253.

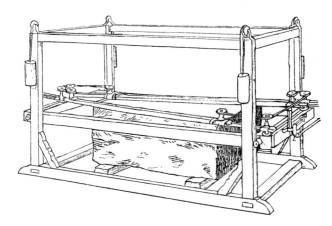

Fig. 5.8 Patent drawing of Schrag and Kammerhueber's stone saw, for obelisks and other monumental shapes whose sides were not parallel to each other. It could cut two faces at any desired angle in a single operation. This patent depended on flexible belts to apply power to the saw blades; other inventors depended on a variety of ingenious devices. (*Ann. Rep. Comm. of Patents, 1856. Patent No. 14296*)

tented by Philip Schrag and Joseph Kammer-hueber of Washington, D. C., in 1856. It resembled an ordinary reciprocating frame saw in some respects, but the saw blades were pulled by flexible belts that passed between vertical rollers. The rollers could be adjusted to guide the saws at any desired angle.

In the mining and quarrying industries, where work depended greatly on the use of drills, considerable thought was given to machines that might replace drilling by hand. It was difficult, however, to devise effective ways of rotating a drill between strokes and causing it to advance exactly the distance it had cut away on the preceding blow. Some machines of the 1840s and 1850s relied on the manual turning of a crank, which operated a drill similar to the traditional jumper. Others utilized the principle of cam-driven drills in which the impact was modified by a spring. Some employed weighted drills raised by a cam and allowed to fall by their own weight. Still others had jaw-like grips that held and released the drill at the proper instants. There were also various ingenious driving-wheel mechanisms. In 1857, Josephus Echols of Columbus, Georgia, patented a hydraulic-powered drill in which jets of water struck cups mounted on a shaft, driving it up and down, alternately. The apparatus was mounted on an adjustable frame to permit drilling in any direction.[16]

None of these devices proved to be successful and they were soon abandoned in favor of steam and pneumatic drills. This type appears to have come into use in the United States in the early 1870s, after having been introduced into Europe a decade earlier.[17] Charles Burleigh of Massachusetts had a drill, by 1873, in which a single cylinder was operated by steam or air at a pressure of about 80 pounds per square inch; it made 300 strokes per minute. Such drills, which quickly became standard, were called *gadders* or *gadding machines* when they were operated horizontally.

The first successful steam machine for cutting channels in quarry rock was made in 1863 by George J. Wardwell of Rutland, Vermont (Fig. 5.9). Channeling machines quickly gained favor in marble, limestone and sand-

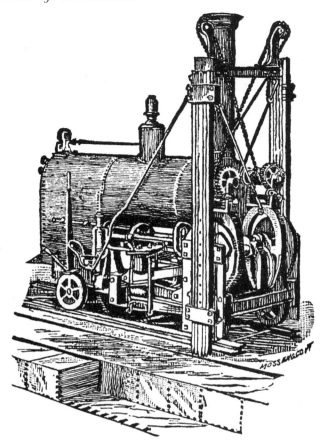

Fig. 5.9 Wardwell's steam channeling machine, made in 1863, was the first successful one for cutting channels in quarry rock. It resembled a small railway locomotive, moving back and forth on a track. Steel cutting bars at each side, lifted by the revolution of a driving wheel and allowed to fall by gravity, made 150 strokes per minute; the machine moved forward half an inch at each stroke to chip out a shallow groove about two inches wide. In marble or limestone, Wardwell's machine was able to cut 40 to 80 square feet of trench per day. George P. Merrill, *Stones for Building and Decoration*, third edition (New York: Wiley, 1910) p. 389.

stone quarries, being widely used by about 1880; they were not employed in granite quarries. Channeling by machine, although much faster than trenching by hand, was still expensive and so it was not done where drilling sufficed.[18]

Before 1860, many efforts were made to develop satisfactory machines for finishing the surface of ashlar blocks, but with little success. In these attempts, hand operations with a chisel or hammer were generally imitated, not because other principles were unknown but because the means were lacking to make and provide sufficient power to move the kind of enormous planing machines that began to be used in stone mills near the end of the nineteenth century. Inventors were stimulated by occasional reports of successful machines in Great Britain and other European countries.[19] In the United States between 1830 and 1860, experimental machines were equipped with batteries of chisels, picks or hammers. Some were actuated by cams and others were mounted on horizontal revolving cylinders; those with chisels or picks were arranged so as to strike the stone at an angle. They were difficult to maintain in operation: cutters broke or got out of alignment and even when they worked properly they had to be removed and frequently sharpened. They could not compete with skilled hand craftsmen.

MAKING BRICK

Traditional ways of manufacturing bricks were well developed before the age of written history. Basically, there were three distinct stages in making unburned bricks: preparing the earth or clay, shaping it into lumps or blocks, and drying them. To obtain a stronger and more durable material, bricks were baked or "burnt." Each of these four basic processes required its own distinctive skills and equipment; the industry was organized into corresponding divisions.

First, the clay had to be selected and prepared; if the material at hand did not contain enough sandy matter, sand was added.[20] Large stones and foreign objects were removed. *Tempering* was accomplished by the addition of water and thorough mixing. Sometimes this was done with a shovel and sometimes in a *ring pit,* a circular vat in which a heavy wheel was pulled around the central pole by a horse or mule. (Fig. 5.10). The next operation was *moulding.* A mould was simply a wooden box of the proper size, without a top and usually without a bottom. Before receiving the clay, the mould was dipped in water *(slop moulding)* or dusted with sand *(sand moulding)* to prevent the clay from sticking to

the sides. The *moulder* took up a lump weighing 8 or 9 pounds and threw it into the mould, forcing it toward the ends and into the corners. He then scraped off surplus clay at the top with a straight piece of wood. A boy, called an *off-bearer,* then carried the filled mould over to the drying floor, set it down, and picked up the empty mould, leaving the raw brick on the floor to dry for several hours. When the raw bricks were dry enough to be handled, they were turned up on edge and left until dry enough to be moved and placed in low stacks, called *hacks,* preferably under cover to protect them from rain. When sufficiently dry, after several days or weeks, they were taken to a *kiln* and stacked inside, to be *fired* or *burned.*

In either a temporary *kiln,* called a *clamp,* or a permanent kiln, usually holding from 20,000 to 50,000 bricks, a wood or coal fire maintained a temperature of about 1800 degrees Fahrenheit for several days. After cooling, the burned bricks were removed from the kiln, sorted, and stored or taken away to be used in construction. The technology of firing, a factor of major importance, will not be described here.[21]

Workmen in a brickyard performed specialized duties and they were organized into teams or *gangs.* A gang might comprise a *temperer,* who prepared the clay; a *wheeler,* who conveyed it to the moulding table and also, with the aid of the temperer, removed the partially dried raw bricks to the hacks; a *moulder;* and an *off-bearer.* The last, whose work was not so heavy as that of the others, was normally a boy. A satisfactory moulder could produce 5,000 or more bricks in a day; this was enough to keep the other members of the gang occupied.[22] Sometimes a second boy was employed to carry light loads and perform other duties. Sometimes the whole gang set the raw bricks in the kiln but in large yards this work was done by *setters.* Removal and sorting was done by the gang in small brickyards, but in large ones there were workmen who devoted their major efforts to this operation.

In an industry so well organized, machinery could readily be introduced to perform some

Fig. 5.10 At the upper right of this 1823 English print, clay is being tempered in a *ring pit*, which is turned by a horse. The other drawings show bricks being hauled and stacked to make temporary kilns, called *clamps*. These methods were commonly followed in the United States as well as in England. Drawn and etched by W. H. Pyne. *(Photocopy courtesy of the Smithsonian Institution)*

Fig. 5.11 *(a)* Hand molded bricks in wall ca. 1732 illustrate how variations of clay, mixing, molding and firing give characteristic irregularities. Other variations are attributable to the bricklayer. Joints were wide to provide a resilient cushion of mortar and to allow for variations in brick size. Detail at the Short House, Newbury, Mass. *(Photo by the author)*
(b) Machine made bricks in wall ca. 1850 illustrates how hard, uniform size, straight edge brick could be laid with a thin mortar joint. The bright red brick made at Philadelphia was widely admired and shipped to other sites. Detail of house at 313 S. American Street. *(Photo by John T. Chew, Jr.)*

parts of the process while others continued to be done by hand in the traditional way. Not until the end of the nineteenth century did the entire manufacture become mechanized, and even then small brickyards continued to employ manual labor for much of the process.

Tempering was the easiest and first operation to be done by machine; the ring pit, in fact, might be classified as one. The *pug-mill,* too, was a device that had been in common use in pottery plants throughout the eighteenth century. It began to be employed to temper clay for brickmaking early in the nineteenth century. Its name is derived from *pug,* a word that denoted tempered clay. A pug-mill consisted of a cylindrical metal barrel, usually placed upright, in which a revolving shaft carried bars or paddles between similar bars fixed to the circumference. Clay, sand and water were put in at the top and thoroughly kneaded between the rotating and fixed pieces. Pug-mills were often provided with a conical lower section in which a helical screw, attached to the same shaft, forced the tempered clay out through an orifice, compressing it somewhat in transit. At first, pug-mills were turned by a draft animal harnessed to one end of a long pole or *sweep,* the other end of it attached to the central shaft. Later, steam power was employed.

The principle of the pug-mill is simple but clay is a temperamental material. There are several kinds, each differing from others in absorbency, stickiness, plasticity and chemical composition. The clay found in a given locality or *clay bank* differs, therefore, from that found elsewhere. Experience and patience were required to supervise the action of even the simplest clay-working machine. One that worked well in Washington might not prove satisfactory in Philadelphia or New York, and vice versa.

A visitor to Philadelphia in 1804 wrote in his journal ". . . also viewed the Brick works and their machines for making them. . ."[23] Lacking more definite information, one assumes that the machines to which he referred were pug-mills, but it is possible that brick-moulding machines might have been in use there. Apollos Kinsley of Connecticut had patented a machine for moulding brick in 1793. It made use of a charger to force clay into moulds, which were moved successively into position on a revolving horizontal table.[24]

By 1819 there was a brick-moulding machine in operation near Washington, D.C. It is the earliest one for which there are known drawings (Fig. 5.12).[25] This machine, turned by a single horse, could mould 30,000 bricks

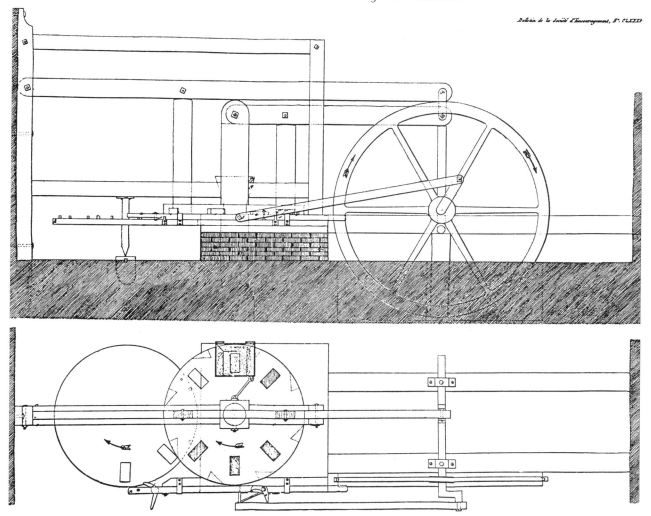

Fig. 5.12 Doolittle's brickmaking machine was being used at Washington, D. C., in 1819. Two circular tables, one containing eight molds, were rotated one-eighth of a turn at each revolution of the large driving wheel. One mold was filled with clay from a hopper, the material in a full mold was compressed by a plunger, and one formed brick was ejected from its mold onto the table at the left, from which it was retrieved by a workman and carried off. Such a machine, made largely of timber, must have been difficult to keep in exact alignment. *Bulletin de la Société d'Encouragement pour l'Industrie Nationale*, XVIII (1819). *(Reproduced by courtesy of The Smithsonian Institution and Prof. Carroll Pursell)*

in a 12-hour day, according to the statement of Mr. Doolittle, the owner. The bricks were said to be so dry when taken from the moulds that they could be fired immediately.[26] If the account is accurate, so-called dry clay[27] must have been used instead of tempered clay.

One authority states that in 1829 "The manufacture of bricks by machinery was successfully commenced in New York. The machines made twenty-five thousand bricks per diem of twelve hours, ready for the fire as

soon as they left the machine. They sold readily at five dollars to eight dollars per thousand."[28]

In 1835, Nathan Sawyer of Mount Vernon, Ohio, and Washington, D.C., patented a machine for moulding and pressing brick from dry clay. Pulverized clay was put into the mould with a shovel; by 1838 he had provided a hopper for that purpose.[29] By 1835 Sawyer's machine began to make bricks in a yard at Washington and it continued in use for at least

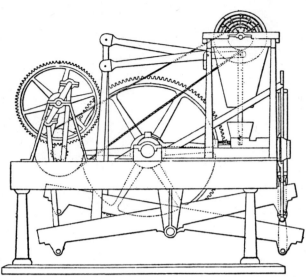

Fig. 5.13 Nathan Sawyer's brick machine was an improvement over his earlier patents. Dry clay was loaded into a hopper (not shown in the drawing). It then went through the pulverizer *(upper right)* where rapidly revolving wheels ground it up, and dropped into molds *(right center)*. Clay in the molds was then compressed by two pistons working from an eccentric one the shaft of the large driving wheel *(center)*, and forced through a tapered duct. As the piston retreated, a section of clay equal to the thickness of a brick was cut off by a vibrating knife and delivered to a shelf, from which it was removed by a workman. The pistons worked alternately, producing two bricks at each cycle. *(U.S. Patent Office, Specification No. 3,768, dated September 27, 1844)*

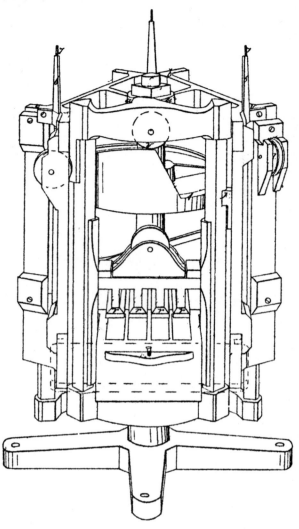

Fig. 5.14 Stephen Ustick's brick machine. The outer framework, carrying three sets of molds, revolves around a stationary vertical central shaft. Fixed to the shaft are two "cam wheels" that actuate two sets of pistons, upper and lower, to compress dry clay between them in the molds. The lower pistons, after the bricks are formed, lift them up for removal. This machine, made of cast iron, could be operated by horse power or by other means. *(U.S. Patent Office, Specification No. 1,045, dated December 28, 1838)*

16 years.[30] An improved model, patented in 1844 and shown in Figure 5.13, may reflect some basic features of his earlier machines.

Stephen Ustick, of Philadelphia, patented a machine to mould and press brick in 1838 (Fig. 5.14). He took out another patent in 1847[31] and still another in 1854.[32] All of these machines differed considerably from each other. It appears probable that some of them were actually employed in making bricks, for in 1857 he was selling "Ustick's Patent Brick Machine" at No. 72 South Third Street.[33]

The dry-brick machines described above all depended on compressing the clay by means of a plunger or piston. A different principle was applied by J. Parsons Owen of Cincinnati.[34] Thomas Culbertson, of the same city, made improvements to Owen's machine, which he patented in 1846 (Fig. 5.15). A model was exhibited at the Franklin Institute in

Philadelphia, where it received a favorable report by the Committee on Science and the Arts for its "merits of simplicity, ingenuity, and utility." Culbertson later moved to Philadelphia, where, with George Scott, he continued to improve the machine by applying heat to the pressing surfaces.[35]

John W. Crary, originally from Cincinnati, learned brickmaking in his father's brickyard

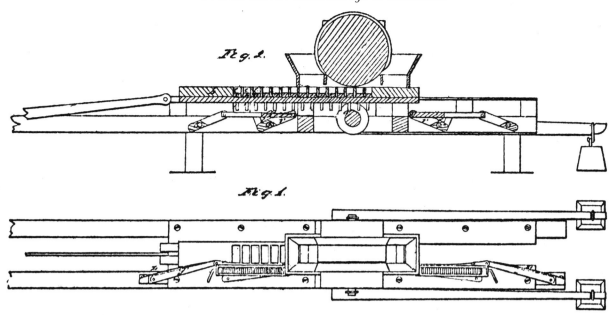

Fig. 5.15 Thomas Culbertson's brick machine is an improved version of one invented by J. Parsons Owen. Iron molds passed under a hopper, from which they were filled with dry clay, and were run between two rollers. They were then carried back between the rollers, thus receiving a second pressing before being ejected and taken off by an attendant. When clay was compressed by a roller, it often rebounded after the roller had passed, leaving a wavy or corrugated surface instead of a flat one. This fundamental difficulty was not readily overcome in machines of this type. *(U.S. Patent Office, Specification No. 4, 521, dated May 16, 1846)*

there during the early 1830s and then moved south, where he superintended brickmaking with a dry-press machine for the United States Customs House in New Orleans in 1850 and 1851. This machine produced 20,000 bricks per day at the yard on Biloxi Bay. Crary also made bricks from tempered clay at the same place. In 1858 he "put up a dry-press machine" to make bricks for Fort Jefferson on Dry Tortugas Island, Florida. That same year, he patented a machine of the roller type. Crary was a strong advocate of the dry-clay process and considered that proper firing was the key to successful brickmaking.[36]

The Narragansett Brick Company, in Rhode Island, began using brick-pressing machines in about 1850.[37] Walker and Cutting were manufacturing bricks in Chicago, Illinois, by 1859, with Adams' Patent Brick Machinery.[38]

A number of brick-moulding machines made use of moulds sliding horizontally along a table; this principle can be seen in patent drawings for William A. Jordan, of Thibodeaux, Louisiana. Machines of this general type were still being used in small brickyards at the end of the nineteenth century, some of them being run by horsepower. The same yards sometimes continued to temper clay in a horse-powered ring pit. A moulding machine of this type, using soft tempered clay (*soft mud*), running on steam power, could make 5,000 bricks per hour:

Four men are required to tend the machine. A "molder" who scrapes off the top of the mold as it is delivered from the machine and watches the consistency of the tempered clay, to see that it keeps uniform; a "mold lander" who takes the mold from the delivery table and places it on the truck; a "sander" who sands the molds before putting them in the machine, and a boy to watch the machine and stop it when necessary. Beside this there are four "truckmen" who wheel the bricks from the machine to the yard, where they are dumped on the drying floor by two "mold setters." In the afternoon these men are employed in hacking the bricks and wheeling the dry ones to the kiln.[39]

This description, written in 1900, reveals that methods of work common in the middle of the nineteenth century were still being followed outside the main centers of progress.

In another type of moulding and pressing machine, the moulds were placed around the periphery of a wheel or cylinder; as it revolved on a horizontal axis the moulds in turn came in contact with a large pressing roller. Ferdinand Zisemann of St. Louis, Missouri, patented a machine of this type in 1849 (Fig. 5.16). Others devised a pair of moulding cylinders rotating in opposite directions, arranged so

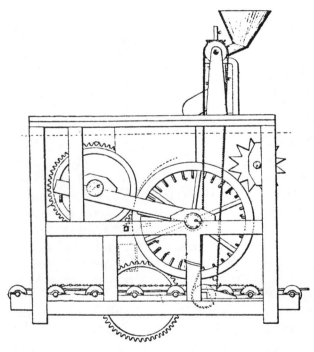

Fig. 5.16 Ferdinand Zisemann's cylindrical brick machine had a large molding wheel (right) and the pressing roller at the upper left. Clay from a hopper at the top (center) filled the molds as the wheel turned counterclockwise. As each filled mold moved downward, its movable bottom was pushed outward and the formed brick expelled onto a moving belt that carried it off. As each empty mold passed the sprocket (upper right) the bottom was pushed back into position, ready to receive another filling of clay. The apparatus at the top right is a sander, which dusted empty molds as they passed. (U.S. Patent Office, Specification No. 6,876, dated November 13, 1849)

that the surface between moulds on one cylinder pressed the clay into a mould on the other cylinder.[40] This type can be represented by Zachariah M. Paul's machine, patented in 1854.

An important family of brickmaking machines operated on the principle of extruding a continuous bar or prism of clay, which was then cut into individual bricks.[41] Stiff tempered clay, called *stiff mud*, was required by this method, which is represented by the Chambers, Brother & Company's machine (Fig. 5.17). During the 1850s, Cyrus Chambers, Jr., of Philadelphia, experimented with this process and the company, which manufactured steam engines, book and news folding machines, shafting, pulleys and other mechanical items, made some brick machines and began using them in a yard at Pea Shore in about 1862. The Chambers machine came into widespread use during the next several years.[42] Chambers, Brother & Company also used a *brick dryer,* in which 8 to 12 hours sufficed to make bricks ready for firing in the kiln. The dryer consisted of tunnels about 40 feet long, through which warm air was forced past carloads of bricks. The drying operation was continuous, moist bricks entering one end and dry ones leaving the other; thus it was mechanization of one more stage in the brickmaking process: a step toward the modern continuous kiln.

An extrusion-type or *die machine* for making brick and tile was also manufactured by H. Brewer & Company, Tecumseh, Michigan, which had begun making brick machines in 1849.[43]

The term *pressed bricks* is sometimes used indiscriminately to describe bricks of uniform appearance. In fact, there is no strict line of demarcation between machine-moulded and pressed bricks. Some moulding machines exerted considerable force on the clay in iron moulds and the resulting bricks can properly be called pressed. Some machines, even at an early date, moulded the clay at one stage of the operation and compressed it further at another stage. Pressed bricks had sharper corners and were more regular than those

Fig. 5.17 Operation was continuous in Chambers' brick machine. Clay was dumped into a chute from the second floor level, sand and water being added as required. It dropped through into the tempering device (a horizontal pug-mill), where it was mixed and kneaded. A helical screw blade at the left of the pug-mill next forced the clay through a die that formed it into a prism of dimensions determined by the length and width of a brick. This prism of clay was carried along on an endless belt to the cutting knife, which, at each revolution, cut off a length equal to the thickness of one brick. Depending on the dimensions of the clay prism, bricks could be *side cut* or *end-cut*. While cutting, the knife moved sideways at the same rate as the clay prism. Other cutting devices were introduced at a later date: wires on a wheel, or helical blades. An endless belt then carried the bricks farther to the left, through a sanding box, so they could be stacked without adhering to each other. The raw bricks were then wheeled to the hacks or drying sheds. Catalog, *Bricks and Brickmaking Machinery by Chambers, Brother & Co.,* 3rd ed. (Philadelphia: 1867). *(Courtesy of The Smithsonian Institution)*

moulded by hand; in general they were also more dense. Early in the nineteenth century a few bricks appear to have been pressed in hand-operated machines after they had been removed from hand moulds, before drying. This process was called *repressing.* There were United States patents for repressing machines by 1812 or earlier.[44] Hand-moulded bricks that had been repressed were sometimes called machine-made bricks in the early nineteenth century.

Many hand presses worked on the principle of the *toggle-joint.* One of them, patented by Thomas W. Smith of Alexandria, Virginia (then in the District of Columbia), in 1841, represents one application of this principle (Fig. 5.18). Other presses depended on a piston moved by a rotating shaft and cam. Some, such as the one patented by Collins B. Baker and E. Gifford of Troy, New York, were double-acting. Because the action was continuous, workmen attending such a press had to coordinate their motions with that of the machine. Ornamental bricks, few of which were made until after 1860, were pressed in hand machines against a metal die which shaped the plastic clay into the desired ornamental pattern.

TERRA COTTA

Terra cotta was being manufactured in the United States by the early 1850s. The Trinity Building in New York City, built in 1852 from the design of Richard Upjohn, contained ornamental terra cotta, as did the Hotel St. Denis, built in the following year, whose architect was James Renwick. There are houses in Bull Street, Charleston, South Carolina, with terra cotta trim that had been obtained in New York during the 1850s.[45] Architectural terra cotta was being manufactured in Trenton, New Jersey, in 1855,[46] in Louisville, Kentucky, by 1859,[47] and by 1860 in Cincinnati, Ohio.[48] Dealers advertised the material for architectural use: keystones, brackets, columns, caps for doorways and windows, friezes and cornices.[49]

Terra cotta was made from clays that had been selected and prepared with greater care

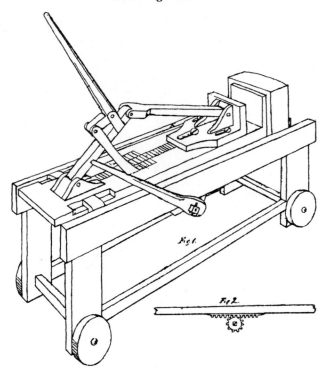

Fig. 5.18 Thomas W. Smith's Brick Press. A raw brick in a plastic state was placed in the open space at the right. A push on the lower lever moved the sliding top toward the right, carrying the brick into a mold. Then a push on the upper lever forced the follower (F) toward the mold, the pressure increasing to a maximum as the "elbow" joint straightened out. The top lever was then pulled back and a pull on the lower lever brought the brick out of the mold to its original position, from where it was removed. *(U.S. Patent Office, Specification No. 1,959 dated January 30, 1841)*

than clay for making bricks, and it was fired at a higher temperature—between 2000° F. and 2300° F. in down-draft kilns designed especially for pottery manufacture. Ornamental pieces of conventional repetitive design were shaped by pressing clay by hand into cast plaster moulds. Free ornament and relief sculptures were modeled directly by the artists. Except for grinding, mixing and tempering of clay there were few machines involved in the production of terra cotta. Small pieces were sometimes pressed in the same manner as ornamental bricks, from which terra cotta was arbitrarily distinguished by virtue of being larger than 8 inches square. The greatest popularity of terra cotta in American architecture occurred after 1890, when it came to be widely applied to tall office buildings and all types of public buildings.

As in other historical developments, the progression from handicraft to machine in the masonry industries does not lend itself to a simple nor clear generalization. I have presented a few small pieces of the story that I hope will still seem appropriate after further research affords a wider basis for judgment. It is evident that machines made only a small portion of the bricks and finished even fewer common building stones before 1860. The experimentation, however, revealed which types of machines would and would not work satisfactorily. It led to a better understanding of the nature of masonry materials and the ways in which they react to mechanical processes. Without this insight, it is difficult to see how the more rapid progress near the end of the nineteenth century could have been achieved.

(a) Small stone hammer.

(b) Stonecutter's hammer and point, which was a pointed chisel used for working hard stones, especially granite. This steel hammer weighs about four pounds.

(c) Stoneworking chisels, left to right: pitching chisel with blunt edge used to establish the edges when shaping a rough stone into a rectangular piece; point for working hard stones and for rough shaping of softer stones; tooth chisel for smoothing soft stones; wide chisel called a *marble plane.*

(e) Stone hammer for knocking off projections and breaking stones; often called a *splitting hammer.*

(d) Mason's leather mittens.

Fig. 5.19 Tools and Equipment at the Mercer Museum, Doylestown, Pennsylvania. *(Photos a to g courtesy of the Mercer Museum, Doylestown, Pennsylvania. John T. Chew, Jr., photographer, 1975)*

(f) Brick molds. The one at right is wood, the other is iron, possibly made for use with a machine.

(g) Wooden hods were used to carry bricks or mortar. Its weight was balanced on the workman's shoulder.

NOTES

1. This paper should be considered an interim report, not a definitive survey of the subject.

2. "A first rate moulder has been known to deliver from 10,000 to 11,000 bricks in the course of a long summer's day, but the average produce is not more than half this number. If, however, the average daily produce of one moulder be estimated only at 5,000 bricks, it is quite evident that the project of moulding them by expensive machinery, complicated, and therefore liable to want frequent repairs, cannot but be a most ruinous speculation." Lecture on pottery by A. Aikins, F.L.S., F.G.S., delivered at the Royal Institution. *Jour. Franklin Inst.*, XI, new series, No. 2 (February, 1833) pp. 134–135.

Moulding by machinery.—The moulding of bricks is consequently a very cheap process. In England, where labor is dearest, the cost of moulding is about ⅓ of a penny for 22 bricks; in Germany, 67 bricks may be moulded and partially dried for the same money, calculating the daily wages of the moulder as 1s. 3d., of the helper 10d. and the produce of their joint labor at 5000 bricks. The cheapness and simplicity of the hand-process render it difficult to devise machinery that shall supersede it; the produce of the moulder is nearly equal to that of a machine, and the prime cost and expense of keeping machinery in order are only likely to be remunerated where circumstances are very favorable, and a great outlay of capital is desirable. These combined circumstances seldom occur together. The interest of capital, the necessity of keeping a machine constantly at work, and the cost of the motive power which such machines require, demand a constant and very extensive market for the produce, if the undertaking is to succeed. So large a consumption is seldom possible, as the price of the bricks will not admit of carriage to a distance; it is, therefore, not surprising that machinery is so seldom employed by brick-makers.*

[* The rapidity with which cities are springing up in the United States, the expensiveness of manual labor, the cheapness of fuel, to drive steam engines, and the steady demand for bricks, cause much use to be made of brick machines in this country.—*Am. Ed.*]

Chemical Technology; or, Chemistry Applied to the Arts and to Manufactures:

by Dr. F. Knapp, Professor at the University of Giessen. Translated and Edited, with numerous Notes and Additions: by Dr. Edmund Ronalds, Lecturer on Chemistry at the Middlesex Hospital, and Dr. Thomas Richardson, of Newcastle-on-Tyne. First American Edition, with Notes and Additions, by Professor Walter R. Johnson, of Washington, D.C. Volume II. Philadelphia: Lea and Blanchard, 1849. pp. 318–319.

3. This paper does not attempt to describe the complex relationship between different branches of the ceramic industries.

4. Hand drilling was practiced in the vicinity of Philadelphia by 1757, and at Lynn, Massachusetts, by 1786.

The following description of hand channeling is taken from an article by Fred W. Bishop in *Greater Brecksville News* (Ohio), Sept. 7, 1945, by courtesy of James S. Jackson. It tells about work in the Peninsula Quarries, where it was occasionally necessary to cut a channel about two feet wide, six to eight feet deep, in the rock.

But it is vividly recalled that Pete Horrigan—older brother of Mike who lived in Peninsula more recently—dug at least one such channel [with a stone pick]. It took him months to complete the task and he usually stood in the trench below the level of the rock, was covered with stone dust from hat-crown to shoe-sole, and had a small sponge tied under his nose to prevent the inhaling of grit which was a menace to the lungs of all stone workers.

The advent of the channeling machine [in the late 1880s] did away with such tedious labor.

5. Advertisements displayed various names for these establishments: marble mill, marble works, marble and stone steam saw mill, steam marble works, stone company, etc.

6. Stone retains some ground water, called *quarry sap*, for several months after being quarried. Stone was usually finished before drying out, after which it became appreciably harder. England and France had old quarries of good freestone, some of them having been opened by the Romans. It was difficult to locate freestone in the United States before the science of geology was born; colonists in America often complained about the lack of freestone, not then realizing that it was really abundant.

7. See Harley J. McKee, *Introduction to Early American Masonry* (Washington, D.C.: National Trust for Historic Preservation, 1973), Chapter I, for a brief description of several finishing techniques.

Rubble for unpretentious houses and utilitarian buildings often made use of stones as they came from the field or quarry, except for splitting and breaking the pieces to size. Additional finish might have consisted of knocking off projections with a hammer or rough-dressing with a pick.

Ashlar, or cut stone work, required that the pieces be squared. Straight *margins* were cut with a chisel or a point, to define the desired surface, and the area between them was then cut with chisels or worked with hammers or axes. Next, the other surfaces of the piece, successively, were dressed in a similar manner, the *face* destined for exposure in the wall often being more highly finished than the *back, beds* or end *joints*.

8. This tool was sometimes used in America but it is more characteristic of western Asia and eastern Mediterranean countries. Adze-like stonecutting tools were common there in ancient times and their use persists to a limited extent to this day. The author has observed tooth-adzes being used in Iran but this practice is said to be rapidly dying out.

9. J. Thomas Scharf and Thompson Westcott, *History of Philadelphia* (Philadelphia: L. H. Everts & Co., 1884) Vol. 3, p. 2229.

10. Samuel Swift, *History of the Town of Middlebury* (Middlebury, Vt.: A. H. Copeland, 1859, p. 335, quoting Professor Frederic Hall, *Statistical Account of the Town of Middlebury*, 1821).

The marble in this village, which is now wrought on a large scale, and extensively diffused over the country, was discovered by Eben W. Judd, the present principal proprietor, as early as the year 1802. A building on a limited plan was erected, and machinery for sawing the marble (the idea of which had its origin in the inventive mind of the proprietor) was then first put in operation. In 1806, a new and commodious building, two stories high, and destined to comprise sixty saws, to be moved by water, was erected. In 1808, this enlarged establishment went into operation, and has continued to the present day.

The saws are made of soft iron, without teeth, and are similar in form to those, which are used in sawing marble by hand, in the large cities of Europe.

There was a steam saw mill in Cincinnati, Ohio in 1826; it was described in *Jour. Franklin Inst.*, Vol. III, No. 4 (April, 1827), p. 265.

Steam Mill for Sawing Stone. This establishment has just been made in the western part of the city, between Front and Columbia streets, and is owned by Mr. Alvin Washburn. The main building is 32 by 50 feet, 3 stories high, with one wing 20 by 40 feet, and is built of wood. It has a steam engine of 18 horse power. The first story is occupied with the machinery for sawing *free stone.* From the experiments already made, the proprietor feels confident of being enabled to saw 120,000 feet of stone per annum; and upon such terms as to make a signal reduction in the price. . .

11. A gang saw was used by Macdonald and Leslie in Aberdeen, Scotland, to cut slabs of granite. The mill had a 14 horsepower steam engine.

The saws are, as usual in such works, of soft iron plates, secured in a frame, and operate on the stone by means of quartz-sand and water, applied as in slicing marble. No emery is requisite in these operations, the particles of siliceous sand being sufficient to cut the quartz, the hardest material in the granite. Frequently, fourteen saws are used in a single frame, and occasionally they have had as many as eighteen employed at once on a single block of stone. The progress of the work, of course, is slow; it requiring a whole day to cut a groove two-thirds of an inch in depth, in the granite. The slabs, when cut, are polished by moving one over the other, by appropriate machinery; siliceous sand being first interposed, and then emery, of various degrees of fineness, until the requisite degree of lustre is obtained.

Jour. Franklin Inst., V, third series, No. 5 (May, 1843), p. 329, Professor Traill, "On the Introduction into Scotland of Granite, for Ornamental Purposes, by Messrs. Macdonald and Leslie, of Aberdeen."

12. Kirk assigned the patent rights to Richard S. Risley, of Philadelphia. *Jour. Franklin Inst.,* XI, new series, No. 5 (May, 1833) p. 327.

13. J. S. Ingram, *The Centennial Exposition Described and Illustrated* (Philadelphia: Hubbard Bros., 1876) pp. 187, 194–195, 353–354.

14. *Jour, Franklin Inst.,* V. new series, No. 4 (April, 1830) pp. 235–237.

15. *Jour. Franklin Inst.,* XX, third series, No. 3 (September, 1850) p. 175, Patent No. 7530.

16. *Ann. Rep. Comm. of Patents,* 1857. Patent No. 14,495.

17. The principle of the steam-hammer had been suggested as early as 1784 by James Watt but it was developed for practical use by James Nasmyth in

1844. Pneumatic drills were used in excavating the Mont Cenis Tunnel under the Alps in the 1860's. Charles Singer et al, *A History of Technology*, Volume V (New York and London: Oxford University Press, 1958) pp. 516–517, 527.

A steam engine was adapted to drilling by Joseph J. Couch, of Philadelphia, who attached a drill rod to the piston. His invention, which was patented in 1852, still made use of a clamping and releasing device similar to earlier mechanical drills. It was, however, an important step toward the principle of pneumatic action. *Jour. Franklin Inst.*, XXV, third series, No. 1 (January, 1853) p. 22.

18. George P. Merrill, *Stones for Building and Decoration*, 3rd ed. (New York: Wiley, 1910) pp. 389, 404.

When a channeling machine was operated at too great a speed it "stunned" the rock. The term *stunning* was used by quarrymen to describe impact fractures that could extend a foot or more into the rock and thus waste otherwise good material. Oliver Bowles, *The Stone Industries* (New York: McGraw-Hill, 1934) p. 83.

19. One report described how James Hunter, in 1835, was using five planing machines, all run from one six-horsepower steam engine, each of which could finish several blocks simultaneously in a comparatively short period of time. Hunter worked at Leys Mill Quarries, Arbroath, Scotland. He obtained an English patent for his machine in March, 1835. He also had a boring machine that was used to drill trenail holes in rock, for work on the new harbor at Arbroath. *Jour. Franklin Inst.*, XVII, new series, No. 2 (February, 1836) pp. 139–141). For several years after this publication, the number of patents in the United States for stoneworking machines increased perceptibly.

Some stories must have appeared inaccurate to any intelligent reader. One told about a man named Muller, a native of Strassburg, who had invented a machine powered by one dog on a treadmill, that could shape as many stones in a day as 40 stonecutters working by hand. *The Civil Engineer and Architect's Journal* (London), IV (1841) June, p. 196, c. 2.

20. Clay alone is too plastic; it shrinks and cracks when drying and warps in the kiln. A mixture of about 30 per cent sand and 70 per cent clay was considered best for ordinary brickmaking purposes. Brickmakers liked to find sources that were naturally of the right composition, to save time and labor.

21. Firing or "burning" of bricks took many technical factors into account. In brief, they concerned the type of kiln, fuel, time, temperature and atmosphere.

Temporary kilns consisted of the dried raw bricks themselves, built into piles with tunnels extending through them, in which fires were built. The outside of the pile was plastered with mud to help retain the heat. Bricks nearest the fire were most intensely heated, while those at the top were but slightly burned.

Permanent kilns were rectangular or "beehive" shaped, built with thick walls of bricks laid in mortar, closed at the top. There was a furnace chamber either below or at one side, from which heat entered the kiln through openings. Dried raw bricks were loosely stacked inside. The draft of hot gases rose upward in ordinary kilns. Down-draft kilns came to be widely used in the late nineteenth century; the heat in them was more uniform than in older kilns.

When the fuel was wood, its kind and condition affected the temperature in the kiln, as well as the quantity required and the rapidity of burning. Coal eventually replaced wood as a fuel. Sometimes coal dust was sprinkled among the bricks as they were stacked in the kiln. In early times, brush and straw often served as the only fuel. All of these materials had their own characteristic ways of burning that demanded careful attention to the regulation of air intake and smoke exhaust.

During the period we are considering, there was no accurate way to measure the temperature inside the kiln; its determination rested on the experience of the man in charge. In general, heat was built up gradually, but this rate as well as the ultimate temperature had to be varied according to the composition of the clay. If 5 to 7 per cent of ferric oxide was present, the bricks would burn hard at about 2000°F. and would be red. The presence of lime in the clay allowed little leeway between underburning and overburning. Control over the air admitted to the kiln not only affected the rate of combustion of the fuel, but it helped determine the reactions (oxidizing or reducing) at the surface of the bricks and consequently their color.

22. According to this assumption, each member of the gang had to handle some 20 tons of clay during a 12-hour work day.

23. *The Pa. Mag. of Hist. and Biog.*, LXIV, No. 1 (January, 1940); Bayrd Still, "To the West on Business in 1804, an account, with excerpts from his journal, of James Foord's trip to Kentucky in 1804," p. 5.

24. *Smithsonian Journal of History*, Spring, 1968; Carroll Pursell, "Parallelograms of Perfect Form: Some Early American Brick-making Machines."

In Europe, a machine invented by Hostemberg in 1807 was used to make bricks in St. Petersburg, according to Dr. F. Knapp, *op. cit.*

25. Carroll Pursell, *op. cit.* An engraved plate was published in *Bulletin de la Société d'Encouragement pour l'Industrie Nationale*, XVIII (1819).

It should be noted that information regarding inventions patented in the United States before 1836 is scanty, because of the disastrous fire in the Patent Office that occurred then.

26. Robert Hunt, ed., *Ure's Dictionary of Arts, Manufactures, and Mines*, 6th ed. (London: Longmans, Green, and Co., 1867) Vol. 1, p. 479. These statements appear to have been repeated from the 4th edition (1831).

27. The term "dry" was applied to clay containing about 7 per cent moisture; that was as dry as the material could be found in a clay bank. It was considered economical to utilize clay in its natural condition whenever feasible, for brickmaking. It requires considerable pressure to make dry clay cohere, when moulding it.

28. John Leander Bishop, *A History of American Manufacture from 1608 to 1860*, 3rd ed. (Philadelphia: Edward Young & Co., 1868; reprint by Johnson Reprint Corporation) Vol. 2, p. 340.

29. U. S. Patent Office, Specification No. 881, dated August 13, 1838.

30. Carroll Pursell, *op. cit.*
Jour. Franklin Inst., XVI, new series, No. 5 (November, 1835) p. 316. "One of the machines is at work in this city, (Washington,) and upon weighing the bricks made by it, we have found them, when baked, to weigh upon an average nearly eight ounces more than good stock brick of the same size, and from the same yard, made in the usual way." It made from 16 to 24 bricks per minute.

31. *Jour. Franklin Inst.*, XVI, third series, No. 4 (October, 1848) p. 227.

32. *Jour. Franklin Inst.*, XXVII, third series, No. 6 (June, 1854) pp. 383–384.

33. *Business Directory of Philadelphia*, 1857: "Ustick's Patent Brick Machine. Making bricks from untempered clay."

34. Patent No. 4123, dated July 26, 1845. See *Jour. Franklin Inst.*, XI, third series, No. 6 (June, 1845) pp. 395–396.

35. *Jour. Franklin Inst.*, XX, third series, No. 2 (August, 1850) pp. 121–122. Patent No. 7,453.

36. J. W. Crary, Sr., *Sixty Years a Brickmaker* (Indianapolis, Ind.: T. A. Randall & Co., 1890), *passim.* Courtesy of Charles E. Peterson.

37. Nathaniel F. Potter was president of the company; partners included William Tallman and James Bucklin. *The Victorian Society in America Newsletter,* Vol. 5, No. 7 (Winter, 1973).

38. *Newburgh Gazette* (New York) Feb. 8, 1859, p. 4, c. 3. Courtesy of Charles E. Peterson.

39. Heinrich Ries, *Clays of New York:* Bulletin of the New York State Museum, No. 35, Vol. 7, June 1900 (Albany, N.Y.: 1900) p. 662 and plate facing p. 661.

40. Henry Robert Salmon obtained an English patent, dated May 8, 1830, for a machine with a pair of mould cylinders. See *Jour. Franklin Inst.,* VI, new series, No. 4 (October, 1830) pp. 243–244.

41. Such machines were in use by the late 1830s in Great Britain. The Marquis of Tweeddale (1787–1876) invented and used one in about 1839, that could make 30 bricks per minute. *The Civil Engineer and Architect's Journal, Scientific and Railway Gazette* (London). Vol. II (1839) p. 371, c. 2.

    H. Chamberlain's machine appears to have been in use by this time, as well.

42. According to the Chambers' catalog (1867), which was brought to the attention of the author by Robert M. Vogel, their machines were used by Simon Barnard, of Hestonville, Pa., in 1863; John M. Buist, of Philadelphia, in 1863; the Improved Brick Company, of Detroit, in 1864; the Washington Steam Brick Works, Washington D.C., in 1864; Boyce, Tennant & Co. Chicago Vicinity, in 1866, and others. The first houses erected with Chambers' bricks were built in Camden, N.J. in 1862 or 1863, by William S. Scull.

43. Carroll Pursell, *op. cit.*

    This company was also making mold machines in the 1890s, perhaps long before then. *Catalog, H. Brewer & Co.,* c. 1892. Courtesy of the Onondaga Historical Association.

44. Julius Willerd, of Baltimore, held a patent for a brick press, which expired in 1826. *Jour. Franklin Inst.,* II, No. 1 (July, 1826) p. 63.

    See also *The Civil Engineer and Architect's Journal,* Vol. III (May, 1840) p. 160: "Bakewell's Patent Brick Machine. . . .This press has been in use for many years, more particularly in the midland counties. . .it is not much known in London."

45. These houses were pointed out to the author in 1963 by the late Samuel Gaillard Stoney.

46. The Trenton Terra-Cotta Works, Trenton, N. J., were established in 1855, by a Mr. Lynch. E. M. Woodward and John F. Hageman, *History of Burlington and Mercer Counties, New Jersey* (Philadelphia: Everts & Peck, 1883) p. 696.

47. P. Bannon's Falls City Terra-Cotta Works, Louisville. Files of Charles E. Peterson.

48. Geo. W. Hawes, *Ohio State Gazetteer and Business Directory for 1860–61* (Indianapolis, Ind.: 1860): advertisements by Willard & Co. Steam Terra Cotta Works and Chas. D. Foote & Co. Queen City Terra Cotta Works. This reference was supplied by Charles E. Peterson.

49. *The New York City Directory for 1854 and 1855,* p. 20 (adv.). Roche Brothers, agents for the Hudson River Pottery. This advertisement mentions the terra cotta work on the Trinity Building, implying that it had been supplied by their firm.

    A. J. Downing, *The Architecture of Country Houses* (New York: 1851) p. 329; James Lee and Co., New York and Boston; chimney tops.

    New York Daily Tribune, May 1, 1856: Long Island Pottery, New York; chimney tops.

    Harley J. McKee, *op. cit.,* pp. 54–55: Ambrose Tellier, 1194 Broadway, New York City—dealer in terra cotta, listed in 1859 business directory; Louis Scharf and William Gilinger Whitemarsh Township, Montgomery County, Pa.—manufacturers of terra cotta; Moorehead's Terra Cotta Works, Montgomery County Pa., established in 1866; Alfred Hall, Perth Amboy, N. J., began making plain and ornamental terra cotta in about 1876.

CHAPTER
6

Pioneer British Contributions to Structural Iron and Concrete: 1770-1855

R. J. M. SUTHERLAND

MR. SUTHERLAND, a practicing engineer in London, is a Fellow of the Institution of Structural Engineers and Convenor of its committee on history.

This chapter is about the development of structural iron and concrete as used in buildings in one country over a limited but critical period. It does not cover bridges, except to the extent that bridge technology has influenced building construction. The period—1770 to 1855—was chosen because these dates virtually bracket the change in design from intuition to calculation, and in construction from craft to industry. The centre of change was Britain, although by no means all the innovations of the period can be claimed by the British. However, at this time Britain did provide an unrivalled setting in which new structural ideas could thrive and also gave birth to some remarkably bold and inventive engineers.

Note: The majority of this paper is based on published, often contemporary, information or personal observation. However, thanks are due to Rear Admiral S. L. McArdle and to the staff of the Department of the Environment for unpublished data on buildings in Portsmouth Dockyard. I am also indebted to the Department and to Professor A. W. Skempton for permission to reproduce many of the figures in this chapter.

These men were a new breed quite unlike their continental counterparts and surprisingly little influenced by them.

The national split was much wider than can be explained by language or by the English Channel or even by the French Revolution, although Britain's era of major structural innovation was still young when the Bastille fell and the country remained in a varying state of isolation for 25 years until Waterloo. Differences in temperament and social background were much more significant than geography or particular events. In France throughout the eighteenth century, engineering was an intellectual subject in the hands of well-educated men dependent on a powerful court: the dominant École des Ponts et Chaussées dates from 1747. In Britain engineering was the work of artisans; science was an intellectual subject, but not engineering.

The position of Britain in relation to the Continent can best be shown in two related diagrams. The first, Figure 6.1, shows the development of structural theory which can be seen to have taken place almost wholly on the Continent, largely in France. There were few notable contributions from Britain between the work of Robert Hooke in the seventeenth century and that of Eaton Hodgkinson toward the middle of the nineteenth. A possible exception was Thomas Young, whom most

Pioneer British Contributions to Structural Iron and Concrete: 1770-1855

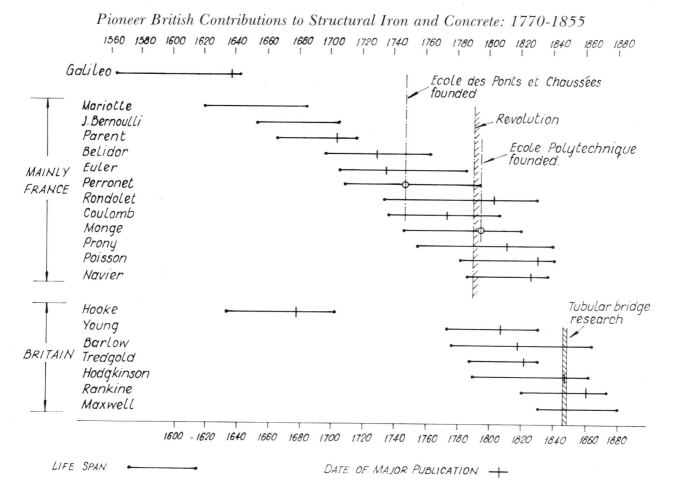

Fig. 6.1   The development of structural theory, which was dominated by France until nearly the middle of the nineteenth century.

engineers could not understand. The *ad hoc* material testers like Barlow and Tredgold did much immediately useful work but had only limited scientific understanding and no more than a scrappy idea of the broader developments elsewhere.

The second diagram, Figure 6.2, shows the lifespans of some of the greatest British engineers (born roughly 1700 – 1800) in relation to the educational opportunities of their time. In Britain, engineering at university level came at least 100 years later than in France, and the men named here were virtually wholly self-educated.[1]

British engineers were undoubtedly limited by their relative ignorance of mathematics, but they were also freed by this. Without the rigour of an early scientific training behind them, their starting assumption with any

problem could be "It can be done and I will do it," rather than the more intellectual "Can it be done and, if so, why hasn't it been done before?" These men were tough, ingenious and creative, blessed with a recent and growing abundance of iron, and to a lesser extent cement, but still largely unhampered by precedent. They grew in stature as the swing from agriculture to manufacturing increased the demand for their skills. Their achievements were great and, perhaps understandably, so were some of their mistakes.

The research for the Menai and Conway tubular bridges is shown in both diagrams (Figs. 6 – 1 and 6 – 2) because it was here that the ingredient of advanced theory was added for the first time — by Eaton Hodgkinson — to the flair and ingenuity of traditional British engineering. The result was a spectacular suc-

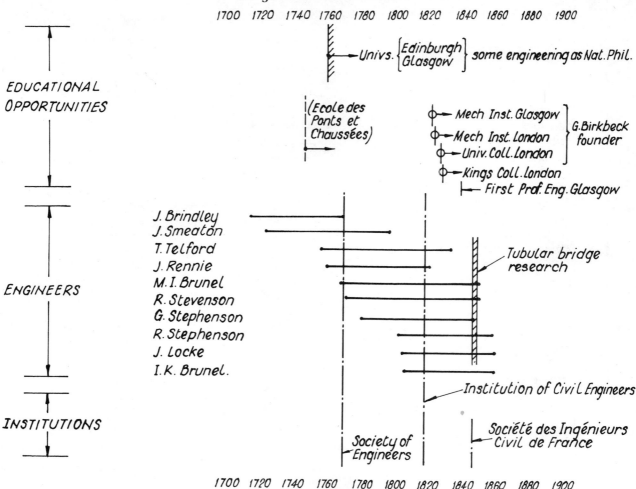

Fig. 6.2 Life spans of some of the greatest British Engineers in relation to education opportunities of their time.

cess which, although outside the scope of this chapter, cannot be ignored because of its effect on all branches of structural engineering.

It would be wrong to imply that all engineering development in the eighteenth and early nineteenth centuries took place in either France or Britain. There is need for a comparative analysis of contributions from all countries at the time, including, for instance, American timber engineering, iron in Russia, and truss forms and suspension bridges again in the USA. However, it is likely that such a study would prove that, at least up to the French Revolution, most theoretical thinking on the European Continent was closely related to, if not centred on, France and that for many years after about 1780 the development of structural iron was dominated by Britain.

Any study of the British contribution to structural form cannot be divorced from a review, however brief, of the availability of materials.

Before 1770, virtually all walls were timber framed or built of brick and stone in lime mortar with masonry vaults or timber for floors and roof structures; lead, tile, slate or thatch completed the weather seal. The new materials which changed all this were principally cast iron, wrought iron and cement.

It is significant for engineering that the lead in actually developing and producing these new materials at the time was taken in England, although many of the original ideas came from elsewhere earlier. It is also notable that during this period timber was not developed as a structural material in Britain as much as in Germany or in the USA. Britain had limited stocks of modest-sized trees but plentiful supplies of iron, which the timber-rich countries were then desperately short of, or had not started to extract.

Iron was known in Britain as early as about 450 B.C.[2] and much earlier elsewhere, notably in China and India. Early British iron was all wrought; that is, heated in charcoal furnaces until it was pasty and could be worked and purified by hammering but never intentionally melted.

Cast iron, known in China in the sixth century A.D.[3] if not before, is thought to have been introduced into Britain in the fifteenth century from Germany, where the development of an improved furnace, the *Stuckofen,* made it practicable to heat the material enough for it to flow. Even then, castings could only be fairly small and of variable quality until 1709[4] when Abraham Darby succeeded in using coke instead of charcoal for smelting; the quality and size of castings then increased enough for big structural members to be worth considering.

Although wrought iron was available long before the period considered here, it became effectively a new material as a result of developments by the Cranages, Henry Cort, S. B. Rogers, Joseph Hall and others between the 1760s and 1830s.[5] During this period it advanced from a material used only for small forgings for tie rods, cramps and other fixings to one capable of being rolled at a competitive price and in large quantities into rails, tees, angles, plates and, toward the end of the period, joists.

Apart from lime, deposits of stronger natural cementitious materials have been found in Europe for centuries: pozzolanas in the south since Roman times; trass, its northern equivalent, in the seventeenth century if not before.

Toward the end of the eighteenth century and well into the nineteenth, Britain was bom-barded by extravagant claims for artificial cements, products constantly improving but so varied as almost to defy classification. These included, amongst others, hydraulic limes, Roman cements (as Parker's) and Portland cements. Aspdin's patent of 1824 was only one landmark in the steady development throughout the nineteenth century from the promotional approach of the patent medicine era to the establishment of the reliable product known as Portland cement throughout the world today.[6] The significance of each advance in cement technology may have been small, but with a growing supply of cements of increasing strength, new methods of using masonry and concrete continued to present themselves.

Although in the 1740s Benjamin Huntsman, working near Sheffield, developed a furnace hot enough to melt and cast a steel of uniform quality, the necessarily small scale of the operation precluded its use in structures.[7] It was not until the introduction of the Bessemer Converter in 1855–56 that structural steel became a real possibility: even after that, it took several decades for steel to displace wrought iron.

DEVELOPMENTS IN CONSTRUCTION

Each new structural development was not necessarily confined either to cast iron, wrought iron or concrete. All of these materials were tried in combination with each other and with the more traditional timber, brick and stone in proportions or ever increasing variety at least up to the 1850s. Figure 6.3 shows some of the more notable structural forms or groupings used in Britain in this period. It is by no means all-embracing; developments fanned out in many directions, each growing or waning and affecting the other in an extremely complex way until eventually one material, wrought iron, and one technique, rivetting, came out on top. Any attempt to show even the salient achievements in each direction is likely to degenerate into a boring and incomprehensible list. To make some pat-

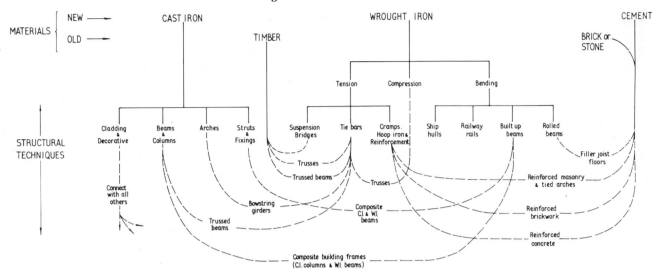

Fig. 6.3   Combination of structural materials and techniques tried in period 1770–1855.

tern out of the British achievements, they are divided here into cast iron, wrought iron, reinforced masonry and reinforced concrete.

Cast Iron

Rogue structural uses of cast iron date from the mid-seventeenth century. The beam over the Coalbrookdale furnace cast by Basil Brooke, with the date 1638 on it,[8] iron pipes at Versailles in the 1660s;[9] the cast iron columns of 1752 in the monastery kitchen at Alcobaca in Portugal; the cast iron beams of 1768–72 in St. Petersburg:[10] all can doubtless be matched by others. However, it was probably the celebrated 100-foot Iron Bridge near Coalbrookdale built by Abraham Darby III between 1777 and 1781[11] which really established cast iron as a structural material in Britain. Apart from being the pioneer cast iron bridge, it is interesting for its details which have more in common with timber-framed construction than the masonry arch form it was replacing. Iron Bridge was soon followed by many other cast iron arch bridges in which the arch became flatter, the structural logic increasingly clear and the form—still derivative—closer and closer to that of masonry. Telford's beautiful Mythe Bridge at Tewkesbury (170-foot span, 1823–26)[12] was perhaps the peak of this development before form became lost in embellishment.

Although Iron Bridge may have been the first example of cast iron to catch the public imagination, regular use of the material in buildings antedated it. St. Anne's Church, Liverpool, with cast iron gallery columns, 1770–72,[13] was soon followed by St. James's Liverpool, about 1774,[14] Tetbury 1777–81[15] and a whole stream of examples of church and domestic architecture where cast iron was chosen for economy, slenderness and the easy reproduction of complex (especially Gothic) detailing. The iron was simply a substitute for timber in beams and for masonry or timber in columns. The slenderness achieved was sometimes startling, as at Tetbury (Fig. 6.4) where the roof columns are encased in timber, the real diameter being less than half the apparent one.

The earliest uses of cast iron in buildings were for visual effect as much as for economy. It was in the mills of the 1790s and early 1800s that its more utilitarian qualities were proved and it was developed on an engineering basis. The destruction of the great Albion Mills by fire in 1791, only 5 years after completion, was just one of many mill fires of the period, but the scale of the disaster probably did more to encourage the development of fire-resistant construction than any other event. William Strutt made the first step forward with his Derby Mill of 1792,[16] which retained the tradi-

Fig. 6.4   Tetbury Church, Gloucestershire, 1777–81; cast iron columns less than half their apparent diameter encased in decorative timber. *(Sutherland photo)*

to say that the wedge-shaped thickening at the bottom of these beams was due to an appreciation of the relative weakness of cast iron in tension. But with no comparable increase in section at the top, where the beams are continuous over intermediate supports, one must admit that the logic of the design is at best incomplete. The bottom thickening was probably intended mainly to provide a springing for the brick arches and the same may be true of the Boulton and Watt inverted T Section of 1799 to 1801 at Salford. Whatever his initial understanding, Bage did develop a theory of beam design which was correct enough to relate well to load tests and he used a consciously improved design of section for a mill at Leeds in 1802–03.[17]

After Bage, many contributions were made to the design of cast iron beams, all based on the use of larger tension than compression flanges, culminating about 1830 in Eaton Hodgkinson's "ideal" section.[18]

One improvement on the Bage and Strutt mills was again based on cast iron columns and main beams, but with the masonry jack arches replaced by close-spaced iron joists slotted into the main beams (Fig 6.6).[19] The stabilising outer walls were still of heavy masonry, as in other examples of this construction, e.g., M. Brunel's pioneer sawmill at Chatham (1813–14).[20] Even in the early nineteenth-century prefabricated barracks using this technique, there always seemed to be stone or brickwork left.

Next, the beams themselves became progressively bigger and their use was extended into all forms of building, often to provide wide open spaces in public rooms. This was the case with the complex pierced beam of 41-foot span in Fig 6.7, designed in 1824 by J. Rastrick for Smirke's use at the British Museum. These may not have been the longest cast iron beams actually used in building but must be near the extreme dimensions. Many engineers considered that 30–40 feet was the practical limit for cast iron while a few were apparently willing to consider spans of up to 60 feet.[21] The brittle nature of cast iron for beams was always recognised and proof test-

tional heavy timber beams but substituted brick arches with iron ties for timber floor planking and cast iron columns for timber ones. Further, the timber beams were protected from fire with plaster below and with brick tiles and sand above. The Derby Mill has been demolished but the detailing has survived at Milford (1793) and at Belper (1795), as shown in Figure 6.5 along with several later developments.

The next advance toward the full iron frame was made by Strutt's friend Charles Bage who in 1796–97 built a mill at Shrewsbury following Strutt's principles; this, for the first time, had iron beams (spanning 9–10 feet) in place of timber ones. It would be nice

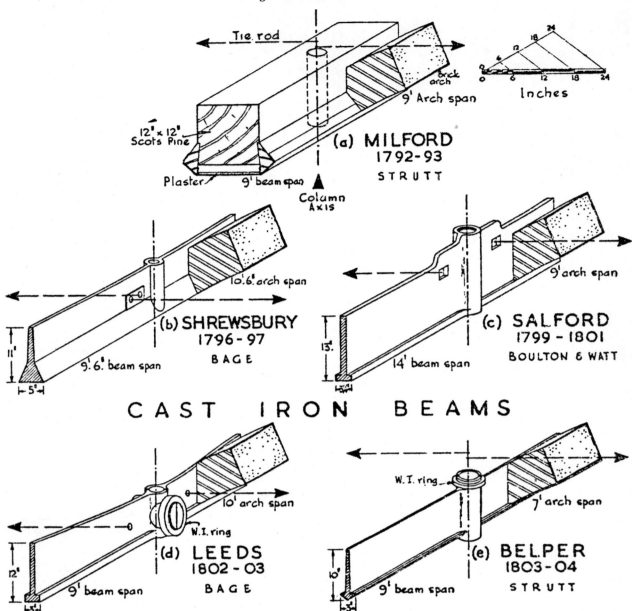

Fig. 6.5 Development of iron framing in British mills 1792–1803 (From "William Strut's Cotton Mills 1793–1812," by Johnson & Skempton. *Courtesy of Prof. A. W. Skempton*)

ing was regularly demanded; in many cases, every beam for a project was tested.

For columns, no one seemed to doubt the adequacy of cast iron. Sometimes it was chosen for strength, its economy depending, as in the mills, on slender sections. Elsewhere, for architectural expression, the members were greatly over-sized, such as John Nash's great Doric columns at Buckingham Palace (1825) and Carlton House Terrace (1827).[22] The Corn Exchange (Fig 6.8) was a happy compromise: the cast iron is fully modelled as masonry, but with the section more slender than would have been reasonable in stone.[23]

So far, cast iron in buildings had been seen mainly as a material for components: columns alone, or beams and columns in predominantly masonry construction, where the overall

Fig. 6.6 Cast iron beam and column construction at John Watney's distillery, Wandsworth, London. *(Sutherland photo)*

stability and much else depended on brick or stone. What about the complete cast iron building? The extent to which Count Demidov's all-iron house in St. Petersburg, which Casanova visited in 1765,[24] was structurally iron may never be known; there were moves in this direction in Britain much later. J. D. Bunning's celebrated London Coal Exchange[25] of 1847–49 is often referred to as the climax of cast iron in architecture, but

here even the 80-foot cast iron dome and the three tiers of cast iron galleries on cast iron columns are really just part of a mainly masonry building.

The roof over the Fish Market at Hungerford must have been one of the very earliest free-standing building structures in Britain wholly of cast iron. This was designed by Charles Fowler with ironwork from Messrs. Bramah's works and erected in 1835. It was

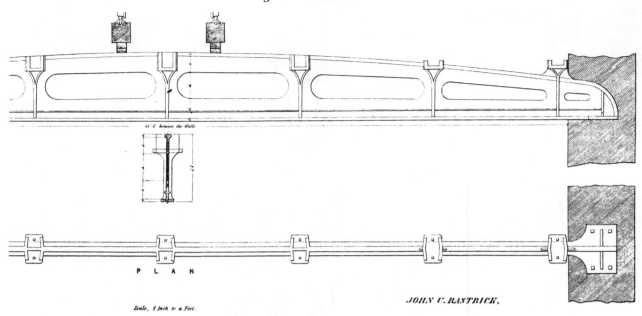

Fig. 6.7   Cast iron beam of 41-foot span by J. Rastrick for British Museum; note provision for test load. (*From* Royal Commission on Application of Iron to Railway Structures, *1849*)

effectively a series of portal frames and, although surrounded by masonry buildings, these only sheltered it without providing any support. It was demolished in 1862.[26]

Another complete cast iron structure, still surviving and completely clear of other buildings, is the old fire station at Portsmouth Naval Base which dates from 1844 (Fig. 6.9). This building (165 feet long by 35 feet wide internally, with columns on a 12-foot-square grid) looks as if it has a frame designed to depend for stability on the stiffness of its joists.[27]

The large conservatories, like those designed by Turner in the 1840s, have all the appearance of being wholly cast iron buildings, but in virtually every case they combined wrought and cast iron and thus belong in the next section.

Wrought Iron

In Europe wrought iron as a structural material came both earlier than cast iron and more gradually.[28] It was chosen for its incomparable tensile strength but, because of high cost, was mainly used in combination with cheaper bulk materials: at first for straps and cramps: then for ties to arches in the fifteenth century if not before; later for chains such as those around the dome of St. Paul's Cathedral.

In Britain and America, before the 1830s, the nearest wrought iron came to being able to span gaps on its own was in suspension bridges. In France, on the other hand, in the decade before the Revolution some really remarkable developments were going on with roofs and floors supported wholly by wrought iron. The earliest example is thought to be the wrought iron trussed roof with a central lantern over a room some 50 feet square in the Louvre. This was designed by Soufflot in 1779 and completed in 1781 by Brebillon after Soufflot's death.[29]

In the 1780s the French (Ango, St. Fart, etc.) developed hollow pot construction, with the pots set in plaster or weak concrete as infill between wrought iron roof trusses or between shallow wrought iron beams lost in the thickness of floors(Fig. 6.10). One of the boldest of these early roofs was designed by Victor Louis for the Palais Royal Theatre, 1786.[30] Curved iron trusses spanned over 70 feet and supported a thin vault of pots on the top

Fig. 6.8 Corn Exchange, Sudbury, 1841. Slim cast iron columns detailed as for stone and cast iron window tracery. *(Sutherland photo)*

chord (Fig. 6.11). These are just examples of what became a firm tradition described at length by Eck in the 1830s and in the later editions of Rondelet,[31] not in historical terms but as current practice.

There are three features of this wrought iron and pot construction which are worth noting, both in relation to later developments and to concurrent practice in other countries. These are: 1) the ironwork, with its strapped or scarfed joints and frequent changes of section, was of the hand wrought type, following on from a much older tradition of decorative ironwork; 2) the structural form, although trussed, was essentially that of the tied arch but with uncertainties of logic in the placing of the members—even in the later examples; 3) the iron was considered as resisting all the compression as well as the tension with no composite action (other than lateral support) from the pot in-fill; this seems particularly clear from the connection of the top and bottom chords of the floor beams as shown in Figure 6.10.

Although these French developments were noticed in Britain and reputedly influenced Soane's work on the Bank of England and Strutt's mills in Derbyshire, they seem to have had no long-term effect on British structural forms. The two countries were advancing on separate paths and there was a clear difference of outlook. To the French "iron" meant "wrought iron," while to the British it meant "cast iron." In Britain cast iron did not lose its dominance until the late 1840s, although much earlier than this wrought iron started to edge its way more and more into favour either as an auxiliary material giving tensile strength to cast iron, masonry or timber, or as a direct substitute in the tensile members of trusses or frameworks.

Wrought iron combined with timber was used in Britain mainly as a substitute, as in roof trusses which in the early nineteenth century quite commonly had iron ties. As an "auxiliary," wrought iron had a very slight vogue in the same period in the form of flitch beams[32] and a rather greater, although still limited, popularity in trussed beams (timber beams strutted up from deflected bars fixed to their undersides).[33]

A vast multitude of combined forms of wrought and cast iron were tried out in Britain starting in the 1790s, again in some cases with the wrought iron considered as a substitute and in others as a built-in prop or even an improver of the actual performance of the cast iron. In the Wear Bridge of 1796 the castings were extensively strapped together with continuous wrought iron flats in a way which was presumably thought to combine the best

Fig. 6.9   Fire Station, Portsmouth Dockyard, 1844. Earliest all-iron building structure—
or not a building at all? *(Department of the Environment photo)*

of both materials and effectively improve the performance of the cast iron.[34]

Later, P. W. Barlow went further and tried to increase the tensile strength of cast iron beams by casting wrought strips in the bottom flange; this was reinforced cast iron but as a technique it never took hold.[35] A similar attempt to make wrought and cast iron act as one, this time for end-to-end connections, was by socketing.[36]

Trussed or assisted cast iron beams were quite widely used in Britain between about 1830 and the late 1840s with varied success. As with trussed timber beams, the deflected wrought iron rods propped the castings but in this case they were generally fixed on either side, rather than below the beams, and were positively tensioned as nowadays in pre-stressed concrete.

A major reason for using this type of construction was not just to increase bending strength but to provide effective joints where long beams could not readily be cast in one section. The story of the widespread structural misconception which lead some designers to fix the bars to raised sections at the ends of the beams, as shown in Fig. 6.12, is too long to give here;[37] it is enough to say that one such beam failed in a mill in the mid-1840s and another one, with much publicity, in Robert Stephenson's Dee Bridge in 1847. In spite of some errors, many trussed cast iron beams

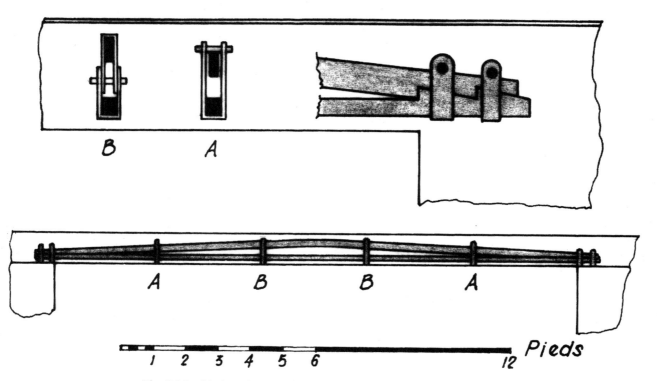

Fig. 6.10   M. Ango's wrought iron floor beam of the 1780s, France.

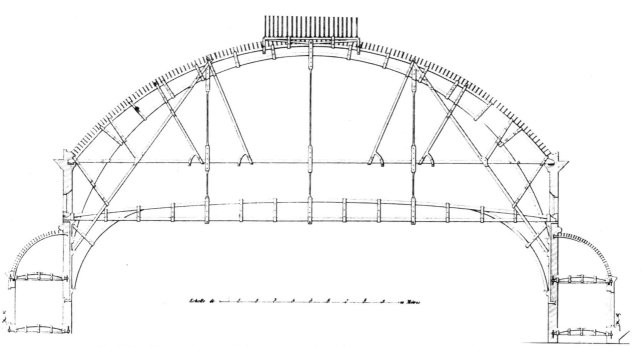

Fig. 6.11   Victor Louis's 70-foot span, wrought iron roof structure of 1786, France.

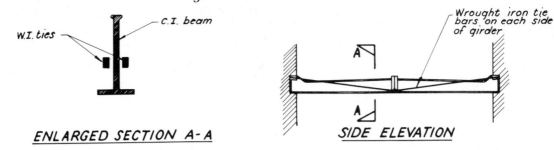

ENLARGED SECTION A-A   SIDE ELEVATION

(A) TRUSSED GIRDER (FOR MILL) OF TWO CASTINGS SPANNING ABOUT 30ft.

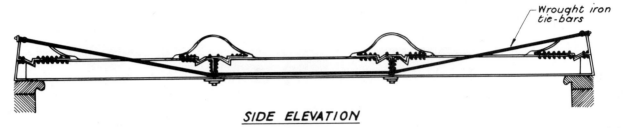

SIDE ELEVATION

(B) TRUSSED GIRDER OF THREE CASTINGS FOR RAILWAY BRIDGE OF NEARLY 100ft. SPAN
NOTE. CROSS-SECTION SIMILAR TO THAT IN EXAMPLE (A)

Fig. 6.12 Cast iron beams trussed with wrought iron rods in illogical form, 1840s; both collapsed.

made good theoretical sense, although in almost all cases the practical effectiveness of the bars must be open to doubt.

One of the finest examples of "good" trussed beams still exists[38] in what is now No. 6 Boathouse in the Naval Dockyard at Portsmouth; this dates from 1845.[39] It is a magnificent three story building some 120 feet wide with masonry outer walls and two lines of internal cast iron columns. It is the climax of the iron mill type of building pioneered by Strutt and Bage in the 1790s and 1800s, but now with a clear span at least three times as great. This was achieved by using 40-foot beams, each cast in one piece but strengthened with wrought iron trussing bars logically placed and faultlessly detailed (Fig. 6.13).[40] Not only was this building the climax of the original mill type structure but it must have been virtually the end of the line, the next real advance being the introduction of full structural framing as in the Boat Store at Sheerness, described below.

Trusses have had a long history in timber

construction but there is no clear date for the first all-iron truss. The French tied arches of the late eighteenth century and the truss patterning[41] of cast iron arches in Britain around 1800 do not really qualify because they do not rely on an understanding of truss action. On the other hand, William Murdock's roof trusses of 1810[42] were wholly of iron and do show such an appreciation; the triangulation is complete and logical, with cast iron in compression and wrought in tension (see upper part of Fig 6.14). This form developed in complexity (as in the lower part of Fig 6.14) but still with faultless logic. However, by no means all the roof trusses of the period showed the same clarity of thought[43] although those at Euston Station designed by Charles Fox for Robert Stephenson in the mid 1830s were outstanding.[44] Here, perhaps for the first time, perfect truss form was used not in composite construction but wholly in wrought iron, with rolled tee sections in compression and rods for the ties (Fig 6.15). In spite of this precedent and the structures following on

Fig. 6.13   No. 6 Boathouse at Portsmouth Dockyard, 1845; well placed truss rods to cast iron beams. *(Department of the Environment photo)*

from it, composite cast and wrought trusses also continued to be popular well into the second half of the century.

Trusses with parallel top and bottom chords mostly date from the 1840s and owe much to

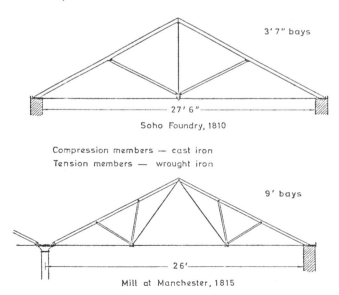

Fig. 6.14   Early iron roof trusses by Murdock. *(Sketch by Professor A. W. Skempton of original in Bolton & Watt Collection)*

the work of American pioneers like Whipple, Howe and Pratt. It is hard to understand why such trusses were not used earlier, considering that in the sixteenth century Palladio illustrated a truss bridge showing in elevation an apparent understanding of the function of each member, although not necessarily of the magnitude of the forces.[45] It is equally surprising that the designer of the Euston roof trusses should have been happy with "all cast" trusses for the 24-foot spans of the Great Exhibition Building as late as 1850s.[46] Whatever Fox's reasons (possibly speed), it is clear from details of the 48-and 72-foot composite trusses[47] for the same building that at least by 1850 the analysis of the forces in each member of a parallel chord truss was understood by him and doubtless many others.[48]

In the examples of composite cast and wrought iron frameworks given so far, there was a simple question of choice as to which material to use for which member. By the 1840s there was also a growing tendency for cast and wrought iron to be mixed more intimately, so that in any framework one member

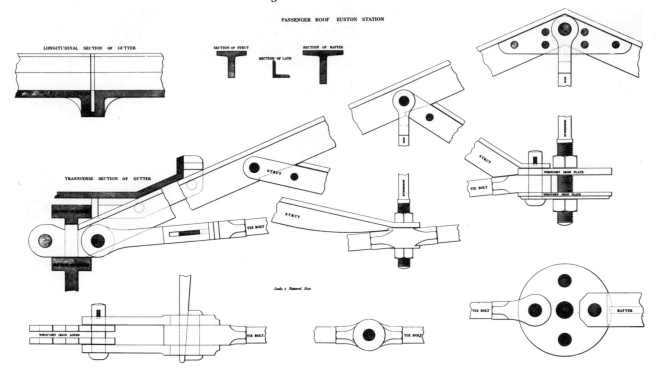

Fig. 6.15   All wrought iron roof truss designed by Charles Fox for Euston Station in the mid 1830s. (*From Simms,* Public Works of Great Britain)

might contain both and that member would "flow" into the next, possibly masked by decoration so as to make it hard to see what type of iron was used at any point. In these cases the wrought iron was not just added to improve the cast iron or as an auxiliary to it: the two materials acted in an almost equal partnership. Examples of this construction are the curvilinear glasshouses of Richard Turner[49] such as the Palm House at Kew Gardens, and the superb iron roofs of 1845–47 over shipbuilding slips in the naval dockyards at Portsmouth (Fig. 6.16).[50]

The story of all wrought iron construction in Britain is the story of rolling and rivetting; neither was worthwhile without the other. The craft approach to wrought iron, which dominated structural iron in France for nearly 80 years after about 1770, had virtually no place at all in Britain.

British iron construction was industrialised, starting with cast iron, right from the beginning of the period. Although structural wrought iron first proved itself where cast iron was found inadequate, it only took over

when production and jointing techniques had advanced enough for it to be able to compete with cast iron both in price and in the quantity available. The two essential rolled sections were plates and angles, but others followed.

The rolling of angles and tees in Britain dates from some time in the first two decades of the nineteenth century[51] and of rails from the 1820s. Rolled iron plate had been used for boilers in the late eighteenth century and there is some evidence of rivetted iron plate in boat hulls around 1812; certainly the hull of the Aaron Manby of 1822 was of rolled iron.[52] In the next 25 years the iron ship industry developed quickly in Britain: in this period the bulb angle and bulb tee section were introduced—the latter often considered the father of the I beam (Fig. 6.17).[53]

There is a tantalising lack of certainty about the date when wrought iron beams first appeared in British buildings. As mentioned above, roof trusses wholly of wrought iron, including rolled tee sections, were used at Euston Station, completed in 1837. Two bridge decks are recorded as being built in rivetted

Fig. 6.16   Composite cast and wrought iron roof over shipbuilding slips at Portsmouth Dockyard 1845–47. *(Department of the Environment photo.)*

wrought iron in 1841,[54] but it is likely that it was toward the end of the 1840s before wrought iron beams in buildings came into anything approaching general use. Whatever the date of the first rivetted wrought iron floor beam in Britain, it is almost certain that it was preceded by one or more examples from Russia. For instance, Figure 6.18 shows the beam system used in 1838 for rebuilding part of the Winter Palace at St. Petersburg after the fire in the previous year.[55] The span here is about 40 feet clear and, although the beams had cast

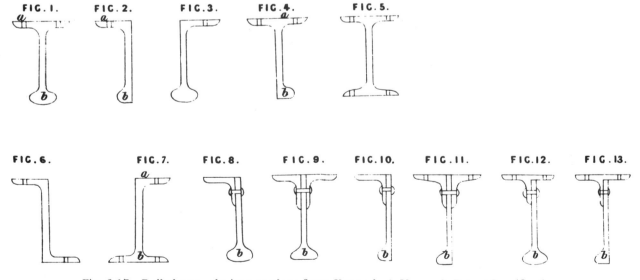

Fig. 6.17   Rolled wrought iron sections from Kennedy & Vernon's Patent Specification of 1844, including the I beam form.

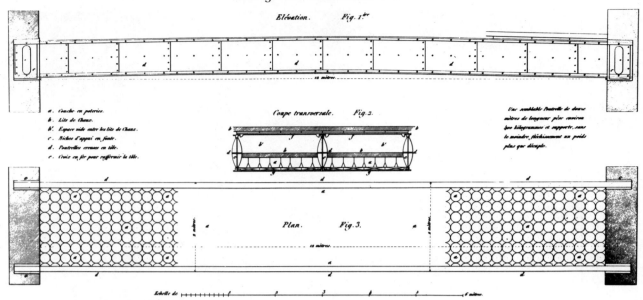

Fig. 6.18   Rivetted wrought iron of 40-foot span at Winter Palace in St. Petersburg,
1838.

iron stiffeners and end shoes, the whole bending strength depended upon the combined action of rivetted iron sheet and angle.

There were other claims of disputable relevance about this time, but what does seem clear is that the technique of making spanning structures out of rivetted wrought iron plate had been sorted out in Britain for ships' decks at least by early in the 1830s. The full method of analysing what we would now call plate girders was first established in the mid 1840s by Stephenson, Fairbairn and Eaton Hodgkinson in their crash research programme for the Britannia and Conway Tubular Bridges. This research was clearly the peak of British structural engineering and, apart from being in a broad sense the watershed between the intuitive and the analytical approach to design, it provided the world with a method of building which hardly changed for several decades.[56]

The rolled I beam was first used commercially in France in 1848–49 as a result of development by Zorès, which it seems was largely independent of the earlier work by Kennedy and others on ships' decks and rails.[57] The question one may ask is why, having got so near to the I beam, the British did not make it a production item earlier. The most likely answer is that it was not thought necessary; industry had become geared to rivetting angles onto plates, which gave a beam of almost identical properties to a solid I beam. Why change? For the French, the I beam represented a much greater advance on what they had been doing before.

Reinforced Masonry

The technique of using iron ties and cramps in masonry—at least 500 years old in Europe and dating back much further in the East—reached its climax in France in the late eighteenth century, with construction like Soufflot's portico of 1770–72 to Sainte Geneviève (later the Panthéon) in Paris.[58] As illustrated in Figure 6.19, the wrought iron is so complex that it looks like a concrete reinforcement drawing of today.

Soufflot, and Rondelet after him, both understood well enough the need to tie a flat arch but they did not seem to have seen this construction as virtually equivalent to a reinforced beam. Marc Isambard Brunel certainly did see this and he is generally thought to have been the first engineer to do so. In 1835 in London, he carried out some loading tests on a 25-foot brick beam 17 courses deep, built with Roman cement and reinforced with hoop

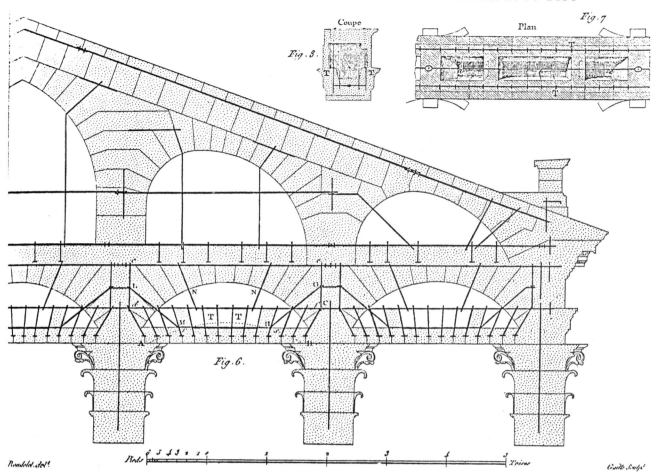

Fig. 6.19 Soufflot's reinforced masonry for the portico of Ste. Geneviève (Panthéon)
Paris 1770–72.

iron laid in the lower five bed joints. It carried a central load of 12 tons before collapse. This and subsequent tests are fully described by General Pasley.[59] In 1837 General Pasley himself made further tests on reinforced brick beams, this time 18 in. wide, 12 in. (4 courses) deep and spanning 10 feet. He built two beams in cement mortar, one with and one without reinforcement, and a third reinforced beam using lime mortar. The reinforced beams each contained five lengths of hoop iron: two in the top joint, two in the bottom joint and one in the middle, thus indicating some uncertainty—a doubt not shared by Brunel earlier—as to where best to put the reinforcement in a simple beam. The reinforced beam with cement mortar carried about ten times as much load as either of the other two, thus establishing the need for good

bond and shear strength if reinforcement is to be effective.[60]

A further test made in 1838 at Brunel's suggestion by Messrs. Francis, the cement manufacturers, was witnessed by a distinguished band of spectators. A slightly shorter span beam, 19 courses deep and similarly reinforced to Brunel's beam, it failed at nearly 23 tons applied load, almost twice the load which broke Brunel's first test beam.[61]

Although proved by these and other tests, as well as being used in the shafts of Brunel's Thames Tunnel and demonstrated at the 1851 Exhibition, reinforced brickwork did not catch on. For spanning gaps, cast iron or later rivetted wrought iron were far too strong as rivals. However, the use of hoop iron as an uncalculated tie did become quite common, especially in the walls of terrace houses.

Reinforced Concrete

Mass concrete had been used for foundations in France in the eighteenth century under reasonably controlled conditions; in the 1820s Smirke used it successfully for footings to buildings in Britain,[62] but until the early nineteenth century there had been little incentive to use it as a substitute for brick and stone. The move toward better fire resistance in the 1840s and 1850s, helped by the improving quality of cement, first gave concrete a chance to come up out of the ground—not so much for walls, which one might expect, but for floors.

It was found to be an excellent material for filling in between structural beams or ribs whether lightened by pots, as in the French *"poteries et fer"* construction, or used in mass. Whatever name was used, it is not always clear whether this in-fill was concrete or plaster. The technique was certainly a long way from reinforced concrete, at least until the late 1840s. Then Lambot's boat (1848) and soon after his floor system, Monier's reinforced flower pots (1849) and F. Coignet's filler joist floor all pointed to the French leading the field in the commercial development of reinforced concrete as we now know it.[63] Patents for Lambot's floor system (wires or bars in concrete) as well as for Coignet's filler joist floors were applied for in England in 1855. However, up to that time there is no evidence that any of these Frenchmen really understood the principle that the concrete withstood all the compression due to bending with the iron placed so as to resist the tension in the most efficient way. It was at this point that Britain could have snatched the initiative.

In October, 1854, William Wilkinson of Newcastle-on-Tyne applied for a patent for his form of fireproof construction.[64] This system (Fig. 6.20) allowed for either shallow arched construction or flat concrete slabs reinforced with wire rope knotted at the ends for anchorage and looped in a way which clearly shows that he understood (and was probably the first to do so) just where to place the iron for the best bending resistance, both at mid-span and over the central supports. There is nothing to suggest that Wilkinson was aware of what was going on in France—anyway, he appears to have been ahead of them in his thinking—and there is equally no evidence that his work influenced the later development of reinforced concrete in France. Wilkinson was on his own. It is not even clear how much he built, but one small structure of his embodying these principles survived until 1954.[65]

CONCLUSION

The late 1840s marked the climax of a period of accelerating progress in British structural design when the scientific approach, comparatively newly learned at the Continental level, finally coalesced with native intuition and proved itself an asset. The new structural ideas of the time, although mostly initiated for bridges, had all found their place in building by the mid 1850s.

The buildings incorporating the new ideas of the 1840s were often remarkable in themselves, although generally less spectacular than the bridges of the same period. Virtually all the notable achievements in actual building depended on iron; innovations in the use of timber, masonry and concrete had little physical impact.

There has been no space to catalogue the successes nor to describe them[66] except in relation to the growth of particular techniques. However, there is one building barely mentioned so far but of such particular significance compared with practice today that it deserves to be made an exception, even though it comes just outside our period of consideration. That is Godfrey Greene's boatstore at Sheerness,[67] built in 1858–60 and shown in Figure 6.21. It was possibly the first multistorey building in the world with a complete "moment frame" in iron where all the vertical loads were carried by the columns and the whole lateral stability depended on the stiffness of the joints with no help from masonry or diagonal bracing.

Whether a "first" or not,[68] this building

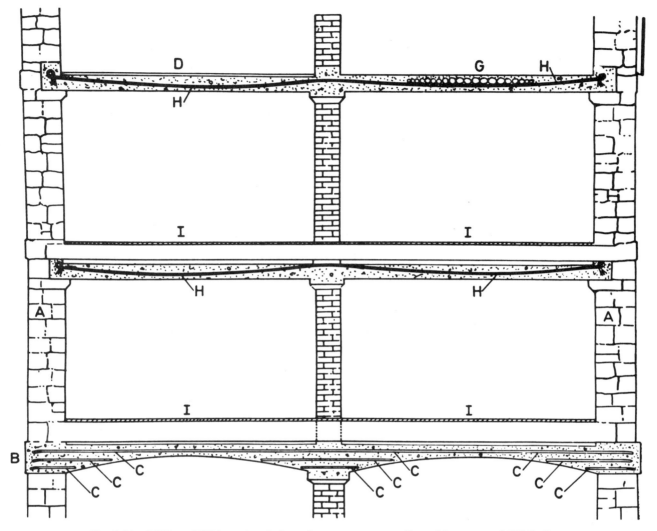

Fig. 6.20   William Wilkinson's reinforced concrete system (from his patent of 1854). One building in this system was demolished in 1954 but others may still exist.

should have had a major effect on future construction. In practice it seems to have been almost unnoticed—like Wilkinson's reinforced concrete—until quite recently. It is very doubtful if the high-rise pioneers of Chicago in the 1870s to 1890s were even aware of it.

From the late 1850s onward, Britain enjoyed several decades of a shade too complacent scientific consolidation and commercial exploitation, with little need for new ideas for structures. Wrought iron, still frequently combined with cast iron, could do anything. There was no call for reinforced concrete or for "engineered" timber.

Innovation went abroad, where needs were

greater. Roebling, for instance, found an economical solution at Niagara to the problem of the long span railway bridge, using only wire and timber in a country still relatively short of iron—much to the admiration of Robert Stephenson. Later in the century, for similar reasons, the French turned reinforced concrete into a viable commercial technique and then had the pleasure of being able to export it to the British, who could so easily have developed it from Wilkinson's ideas if they had seen the advantage earlier.

In the last quarter century, with increasing economic constraints, there has been a marked return to innovation in British build-

Fig. 6.21   Godfrey Greene's boatstore at Sheerness of 1858–60. Full structural framing with columns at 30 foot centres. *(Photo by Eric de Mare)*

ing methods. Perhaps the belief that invention depends on an incentive, if not always on necessity, is true. If so who knows what the future holds?

NOTES

1. It is also interesting to note that in France the professional institutions came many years after higher education in engineering, whereas in Britain the society or institution of day-to-day practitioners was found necessary well before the advent of university engineering.
2. H. R. Schubert, *History of the British Iron and Steel Industry,* 1957 p. 5.
3. Gloag & Bridgwater, *A History of Cast Iron in Architecture,* London, 1948, p. 3.
4. Schubert, *op. cit.* p. 331.
5. W. K. V. Gale, "Wrought Iron: a Valediction" *Trans. Newcomen Society,* Vol XXXVI, 1963.
6. For an account of the stages in the development of cements *see* P. E. Halstead, "The Early History of Portland Cement," *Trans. Newcomen Society,* Vol. XXXIV, 1961; and A. W. Skempton "Portland Cements 1843–87," *Trans. Newcomen Society,* Vol. XXXV, 1962.
7. Article by H. R. Schubert in *A History of Technology,* Volume IV, Oxford, 1958, p. 107.
8. Still in existence as part of the Ironbridge Gorge Museum.
9. Gloag and Bridgwater, *op. cit.,* p. 79.
10. Turpin Bannister, "The first iron framed Buildings" *Architectural Review,* April 1950; and references given in this article.
11. Still in existence, but now only as a footbridge. Barrie Trinder, "The Iron Bridge," published by the Ironbridge Gorge Museum Trust, 1973.
12. Still in full use, although strengthened.
13. Quentin Hughes *Seaport: Architecture and Townscape of Liverpool,* 1964.
14. Hughes, *op. cit.*
15. Marcus Whiffen, "In the Modern Gothic Manner," *Architectural Review,* July, 1944.
16. H. R. Johnson and A. W. Skempton "William Strutt's Cotton Mills 1793–
1812", *Trans. Newcomen Society,* Volume XXX, 1956.
17. The development of bending theory in relation to cast iron beams up to this time is described by Professor A. W. Skempton, "The Origin of Iron Beams," *Actes du VIIIᵉ Congrès International d' Histoire des Sciences,* Florence, 1956.
18. Eaton Hodgkinson "Theoretical and Experimental Research to ascertain the strength and best form of Iron Beams", *Memoirs of the Lit. and Phil. Soc. of Manchester,* Vol. V, (2nd Series) 1831.
19. Grain store at John Watney's distillery, Wandsworth, London, recently demolished. Exact date not known, possibly about 1830.
20. John Newman, "West Kent and the Weald", *The Buildings of England,* Penguin, 1969, p. 198.
21. Evidence given to Royal Commission on the Application of Iron to Rail-

ways Structures (London 1849).

22. Gloag and Bridgwater, *op. cit.,* pp. 120, 155.

23. The Corn Exchange, Sudbury (M. E. Kendall, Architect) 1841. This building was remodelled as a library a few years ago but without altering the basic structure.

24. Turpin Bannister, "The First Iron-Framed Buildings," *Architectural Review,* April 1950; and references given in this article.

25. H. R. Hitchcock, "London Coal Exchange," *Architectural Review,* May 1947.

26. Jeremy Taylor, "Charles Fowler; Master of Markets," *Architectural Review,* March 1964.

27. At present the building has diagonal tie rods between some of the columns in the direction of its shorter side but these rods, with fixings cut into the mouldings of the cast iron columns, look like an afterthought which leads one to speculate whether originally this may not have been a "building" at all. The cast iron structure was used to support a large water tank *(see* Pevsner and Lloyd, "Hampshire and the Isle of Wight," *The Buildings of England,* 1967, p. 414) and it is quite possible that it was planned as an open framework, the diagonal tie-rods being added for greater lateral strength when the cladding was put in. More research is needed here.

28. For an introduction to the early use of structural iron in India, China and classical Greece *see* S. B. Hamilton, "The Structural Use of Iron in Antiquity" *Newcomen Society,* Vol. XXXI, 1957.

29. Charles Gustave Eck, *Traité de Construction en Poteries et Fer,* Paris, 1836, p. 51. Also Turpin Bannister, *op. cit.*

30. *Ibid.* p. 50 and plate 31.

31. Jean Rondelet, "L'Art de Batir." *See also* Turpin Bannister, *op. cit.,* who gives the date of the Palais Royal Theatre as 1785–90.

32. Smeaton, who died in 1792, is reputed to have used wrought iron and timber flitch beams in the London Docks. See Bidder in discussion at Institution of Civil Engineers, April 1847 (PROC. ICE VI P 222)

33. This form of construction probably reached its logical climax in the "Paxton" gutters and secondary floor beams of the Great Exhibition Building of 1851 *(see* Downes and Cooper, *The Building erected in Hyde Park for the Great Exhibition 1851,* John Weale, London, 1852).

34. J. G. James, "Some Early Cast Iron Bridges," *The Engineer,* 29 January, 1965.

35. Described as "Gardner's Combined Girder" and apparently used by Bar-

low on the Reading and Reigate Railway; see letter of 18.4. 1848 in Appendix 1 of the Royal Commission on the Application of Iron to Railway Structures, 1849.

36. Socketing consisted of casting the ends of wrought iron bars into local thickenings on cast members. The technique was used in 1823–24 on the first iron railway bridge on George Stephenson's Stockton and Darlington Railway, but it is not certain whether this became at all widely accepted as a structural connection, especially in building. However this method of jointing was extensively used later for making iron bedsteads, which, if one looks at the details of Stephenson's bridge, may not seem a very large step onward. A section of the bridge is preserved in the Railway Museum at York and its construction is described by Walters and Sealey, "The First Iron Railway Bridge," *Architectural Review,* March 1963.

37. R. J. M. Sutherland, "The Introduction of Structural Wrought Iron," *Trans. Newcomen Society,* Vol. XXXVI, (1963–64), pp. 72–75.

38. Partly damaged by bombing in the 1939–45 war.

39. Pevsner and Lloyd give 1843 in "Hampshire and the Isle of Wight," *Buildings of England,* Penguin, 1967, p. 410.

40. The designer was confident enough to have the allowable load—40 tons distributed—cast on the side of each beam. However one cannot help wondering whether these large castings would not have carried their loads as well over the last 130 years without the truss bars.

41. There is some indication of this in the Iron Bridge at Coalbrookdale but it is much more pronounced in later bridges, such as Stourport (1806); *see* Figure 6 in J. G. James, "Some Early Cast Iron Bridges," *The Engineer,* 29 January 1969.

42. Drawings copied by Professor A. W. Skempton from original sketches in the Bolton and Watt Collection in the Birmingham Reference Library.

43. Taking some widely separated examples, there are the abattoir roof at Bourges (Cher) and several roofs in Russia described by C. L. G. Eck which have flimsy tie rods for both tension and compression *("Traite de l'Application du Fer de la Fonte et de la Tôle" Paris, 1841,* as apparently had the Philadelphia Gas Works of 1837 (Charles E. Peterson, "Iron in Early American Roofs," *The Smithsonian Journal of History,* Vol. 3, No. 3, Fall 1968, Fig. 12. For C. Barrie's new Houses of Parliament in London, cast iron was taken for compression members and wrought for tension

but the latter often depend on the dubious tensile strength of the cast iron lugs in the connecting pieces. (John Weale, *An Account of the Construction of the Iron Roof of the New Houses of Parliament,* London, 1944).

44. F. W. Simms, *The Public Works of Great Britain,* 1838, p. 3 and plate VII. These pioneer wrought iron trusses were all but copied again and again in many later stations up and down the country but at Euston the structurally elegant but modest train shed was set off by Hardwicke's vast Doric portico which clearly demonstrated in architectural terms the extent of the new belief in railways, as did the Great Hall built there a few years later. It is interesting that in most major railway stations after Euston the emphasis changed, the train sheds becoming more and more monumental and the ancilliary buildings, although large, more human in scale.

45. The plan form with the amount of wood in the bottom chord apparently reducing towards the centre is less convincing (See Leoni edition, London, 1721).

46. Apart from the many general descriptions of the Great Exhibition Building, the drawings in Downes and Cooper, *op cit.,* are invaluable.

47. The redundant compression diagonals were of nonstructural timber and simply to make the trusses look similar to the 24-foot cast iron ones. *See* Downes & Cooper, *op. cit.,* p. 9.

48. S. P. Timoshenko, *History of the Strength of Materials,* discusses the mathematical analysis of truss forces (p. 185 *et. seq.*) and supports the 1850 date.

49. Richard Turner, a Dublin iron founder, combined decorative cast iron with ribs of wrought tee section all braced together by iron bars within and tightened onto tubes of cast iron. The Palm House at Belfast (started 1839) was followed by the Dublin Conservatory of 1843, Regents Park Winter Garden (opened 1846) and, most famous of all, the great Palm House at Kew Gardens of 1845–48. Traditionally Decimus Burton had been given most of the credit for the last two of these, but John Hix has now gone a long way toward sorting this out; see his article, "Richard Turner: Glass Master" *Architectural Review,* November 1972; and his more recent book, *The Glass House."*

50. The construction of these, by Messrs. Baker of Lambeth, consists of iron principal roof members at 30 foot centres made up of wrought ribs spanning 85 feet with cast iron infill sections all joined with rivets and bolts. Longitudinal stability is given by "arched" cast iron beams while

roof purlins are extremely light composite trusses. *See* Frederic W. Cumberland, "Iron Roofs erected over Building Slips Nos. 3 and 4 in Her Majesty's Dockyard, Portsmouth," *Journal of Proceedings of the Royal Engineers*, 1846.

51. W. K. V. Gale, "The Rolling of Iron," *Trans. Newcomen Society*, Vol. XXXVII.

52. Sir. W. Fairbairn, *A treatise on iron shipbuilding, its History and Progress*, London, 1865.

53. Kennedy and Vernon's patent (No. 10143) of 1844 claims that they introduced not only bulb tees and bulb angles rolled in one piece but also "iron rolled in one piece, having flanges on each edge projecting on one or both sides." Further, Fig. 5 in the patent specification shows what looks like a modern I beam (reproduced in part in Fig. 6.17).

54. One at Ware, the form of which was a matter of angry dispute between Robert Stephenson and Fairbairn (*see* R. J. M. Sutherland, "The Introduction of Structural Wrought Iron," *Trans. Newcomen Society*, XXXVI), and the other at about the same time in Scotland.

55. Eck, *op. cit.* It would be interesting to know if this structure is still in existence.

56. The research is fully described in Edwin Clark *The Britannia & Conway Tubular Bridges*, London 1850.

57. Robert A. Jewett, "Structural Antecedents of the I beam, 1800–1850," *Technology & Culture*, Vol. 8, No. 3, July 1967.

58. Rondolet, *op. cit.*

59. Major General Sir C. W. Pasley, *Observations on limes, calcareous cements, mortars, stuccas and concrete*, John Weale (2nd Edition, 1847), p. 162 *et seq.*

60. Pasley, *op. cit.*

61. Pasley, *op. cit.*

62. J. Mordaunt Crook, "Sir Robert Smirke: a Pioneer of Concrete Construction," *Trans. Newcomen Society*, XXXVIII, 1965.

63. S. B. Hamilton, "A note on the History of Reinforced Concrete in Buildings," *National Building Studies Special Report No. 24* (H. M. S. O., London, 1956).

64. Patent No. 2293, 1854.

65. Joyce M. Brown, "W. B. Wilkinson (1819–1902) and his place in the History of Reinforced Concrete," *Trans. Newcomen Society*, XXXIX, 1967.

66. The railway stations which followed on from Euston are particularly well worth study; for instance, Paddington by I. K. Brunel and M. D. Wyatt of 1850, Kings Cross by L. Cubitt of 1851–52 (originally with arch ribs mainly of timber) and Lime Street, Liverpool, 1849–51 with curved roof trusses spanning 153 feet and designed by Richard Turner, the glass house builder. Also the buildings in the Naval dockyards where there is still a need for a complete study of the iron structures of the early and mid nineteenth century and of the extent to which men like John Rennie contributed to the actual structural forms. By the mid nineteenth century there were also many remarkable iron structures elsewhere such as Les Halles in Paris (1854 onward) and the ironwork of Bogardus in New York.

67. A. W. Skempton, "The Boat Store, Sheerness (1858–60) and its place in Structural History," *Trans. Newcomen Society*, XXXII, 1959–60.

68. The 1845 Fire Station at Portsmouth is a possible contender for this title, but even if originally a "building" it is hardly multistory; also it belongs more to the intuitive period whereas the Boat Store at Sheerness was consciously designed for stability through the detailing of the joints. The Great Exhibition Building had joints with some intended stiffness, but these were hardly designed in the present engineering sense and in its final form the frame was liberally tied together with diagonal bracing.

Building in the Age of Steam

ROBERT M. VOGEL

MR. VOGEL is Curator of the Division of Mechanical and Civil Engineering, the Smithsonian's National Museum of History and Technology, a cofounder of the Society for Industrial Archeology and editor of its *Newsletter*.

The literature of America's building industry forms an inverted pyramid with vast quantities of books on the exalted work of the architect at the top, fewer on the work of the various craftsmen (especially the carpenter) in the middle and, at the bottom, the pitiful handful on the work of the actual builder (contractor). Before the twentieth century, building appears to have been more a folk or oral tradition than a systematically transmitted art. Techniques were learned on the job, not from books.

The written history, even in the Age of Steam, 1800–1860, completely ignores small

Note: I must acknowledge, with deepest thanks, the vast contributions to the preparation of this paper made by Charles T. G. Looney, Professor Emeritus of Civil Engineering, University of Maryland, and presently a Smithsonian Research Associate. [This paper is a somewhat shortened version of Mr. Vogel's symposium presentation.—Ed.]

buildings, although it is clear that they were built with equipment no more exotic than hods, shovels, wheelbarrows and the horse-drawn scraper. On large works, for which the record is somewhat better, the Age of Steam proves to be an age of creeping progress in substituting power machinery for muscles of men and animals. Even this progress was limited, generally, to moving vast quantities of earth and water, driving immense numbers of pilings, and lifting extremely heavy building elements. Not until well into the nineteenth century, nearly beyond the period we are dealing with here, was much machinery of any kind applied to the construction of buildings.

It was large public works—bridges, tunnels, canals and railroads—rather than buildings that called for machine power to replace or supplement muscle power. America's use of muscle power during the period was little different than that of all the societies which preceeded it. Immigration, both spontaneous and directly recruited for the purpose, provided the muscle power. The circumstances were little different than those of the Middle Ages in Europe, when simple machines had been used to multiply muscle power of crews that erected the cathedrals. These simple machines had compensated for the vast hordes of slaves used in Egyptian and Roman construction,

Fig. 7.1 Medieval construction hoist, man powered. Trustees of the British Museum. From David Jackson, *Adventures in Engineering*, 1960. *(All photos in this chapter courtesy National Museum of History and Technology unless otherwise credited)*

and had been little improved over several centuries (Fig. 7.1).

Probably the greatest concentrated aggregation of men, animals and machinery ever gathered for a single constructive purpose except for the pyramids was that assembled in 1586 by the Italian architect Domenico Fontana to move the famed Vatican Obelisk (Fig. 7.2). With 900 men, 74 horses and a fantastic number of capstans and other gear, he succeeded in laying down, moving a distance of 275 yards, and reerecting the 350-ton monolith in the plaza in front of St. Peters. This was proof that with enough muscle and tackle there was practically no limit to the mass that could be raised and moved—given, of course, that cost and time were no object.

The extension of human and animal muscle through the use of machinery reached what must have been its epitome in the construction of the Neuilly Bridge over the Seine near Paris in 1768 (Fig. 7.3). Designed by Jean Rudolphe Perronet, director of the prestigious Ecole des

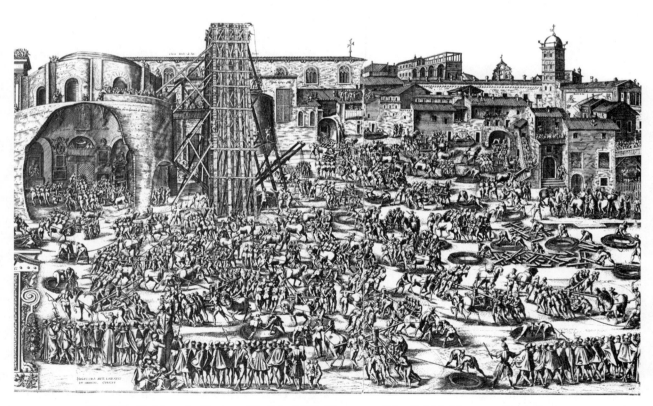

Fig. 7.2 Raising the Vatican Obelisk, 1586. Trustees of the British Museum. From Bern Dibner, *Raising the Obelisks*, 1952.

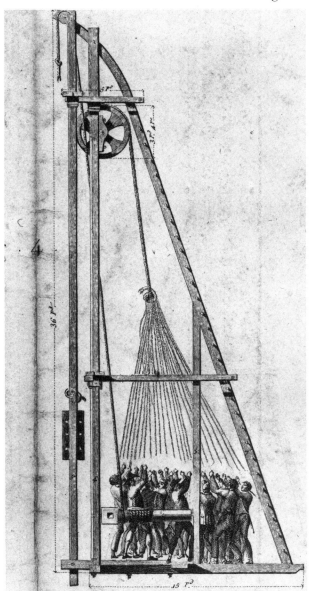

Fig. 7.3   Man-powered pile driver at Neuilly Bridge, 1768, Perronet.

ferdams, the single task that required a constant delivery of power and, unlike the hoisting, excavating, and pile driving, could be carried out for long periods of time at a single location. The water was removed by large scoop-wheels in the cofferdams, driven through gearing by current wheels in the river. The river flowed, and the wheels turned, day and night. No need to prime them when the water level in the cofferdams fell; no need for any human intervention except occasionally to adjust the height of the current wheels as the river rose or fell. A scheme wonderfully simple and simply wonderful![2]

By the end of the eighteenth century, the steam engine as both a pumping and driving device was well developed. In 1829 Zachariah Allen, Rhode Island textile manufacturer, scientist, inventor and civil engineer, wrote ". . . so necessary to the wants of life at the present day are the products of machinery, the Steam Engine has become almost as important as the plough."[3] And indeed, in many ways steam was a revolutionary force in the building industry during what we are viewing as the Age of Steam.

The railroad, of course, had the profound effect of making possible the cheap transportation of building materials and components from source to site. Steam power enabled both the mass extraction and mass production of building materials. From the sawmill to the brickworks to the iron furnace to the iron foundry to the sash, door and millwork plant to the hardware factory, once these establishments collectively were capable of operation by steam—independent of the limitations of water power and hand production—there was a nearly geometric rise in building construction.

The secondary effects of the Age of Steam on the building industry are interesting although well beyond the scope of this study. For example, the introduction of the rolled iron structural section at just about midcentury is a complex story: the rolled section was the direct descendant of the rolled iron railroad rail, the development of which was inspired by the steam locomotive, whose increasing weight

Ponts & Chaussées, the bridge established not only a record for refinement of design in stone masonry but a record for elegance of the construction equipment. Perronet left us a large folio volume,[1] perhaps the most complete record ever made of the method and plant used in erecting a structure up to that time. Of special interest to us is Perronet's exploitation of the one source of nonmuscle power available at the site—the flowing Seine—to perform the dewatering of the cof-

and sophistication were the result, to a large extent, of the availability of large, steam-powered machine tools and steam forging hammers. At the same time, the capacity of mills to produce iron rails and, later, iron structural sections was the direct result of increased blast furnace capacity resulting from steam-powered blowing and handling machinery and, most of all, from the use of steam engines to drive the rolling mills.[4]

But despite the rapid rise of steam power to 1860, the Age of Steam saw widely varying effects on different industries. Railroading, for example, quite obviously was fully dependent on steam, while ocean transportation, where an alternate source of energy existed, did not become wholly dependent on steam until well into the twentieth century. Manufacturing, by about 1820 almost fully dependent upon water power, did not depend equally on steam and water until about 1832.

Once steam power began to take hold, its rise was dramatic, while at the same time the use of human muscle power continued to rise very slowly: actually less, in proportion, than the rise in the population. Animal muscle power, however, also climbed sharply through the first half of the nineteenth century, but this is somewhat misleading as most of that use was on farms. I mention all this not to reflect what was happening in the building industry, but rather what was *not* happening.

Building materials and components, and their transportation, were, to be sure, heavily influenced by the introduction and rise of steam power and the general technological ferment of the nineteenth century. But when we come to the *actual construction* of buildings and structures, we discover that for the first half of the century time stood practically still so far as mechanization is concerned. We discover, to our astonishment, that conservatism among builders was grossly inconsistent with the technological upheaval of the times. Hand methods that had been classical for centuries were by far the rule, and where machinery occasionally was to be seen on a building site, in nearly all cases it was man or animal powered. There were a few notable excep-

tions. In erecting the famed High Bridge that carried the Croton Aqueduct across the Harlem River at New York (1839–42), most of the timber arch centering and the stone itself were hoisted by a steam powered crane designed by the contractor. In 1842 the capstone of the Bunker Hill Monument was hoisted into place by a steam hoist, which presumably had been in earlier use during construction. In 1850 much notoriety was given to the construction of Dr. Jayne's great granite building in Philadelphia and to the steam hoisting apparatus employed by the contractor.[5]

These may certainly be regarded as exceptional instances, and as it turns out, construction probably was the last major technological-industrial area in the world to employ steam power widely and consistently.

It may have been the exploding immigrant population, creating a nearly inexhaustible labor supply, coupled with the general shortage of skilled labor in America, that accounted for the industry's conservatism. While we have seen the modest use—at least on important works—of simple labor-extending machines through the eighteenth century, there were more structures built in the first half of the nineteenth century by pure man power than otherwise. Gosta Sandstrom, Swedish engineering historian, maintains that the majority of buildings standing in the Western world (and I suppose the Eastern as well) have at one time or another been carried on the backs of men and women.[6] Even the simplest machinery often was introduced on building sites against the opposition of the workers who, as traditional, feared for their jobs. Yet, if we had vast pools of labor from immigration, there always were even greater numbers left back home, so there was a somewhat greater impetus in this country than in Europe to employ labor-saving devices. Much of the powered construction equipment that ultimately did evolve was of American origin.

Now that I've devoted so much time to describing the singular lack of response by the American building industry to the Age of Steam, I'll mention a few early exceptions.

It was Oliver Evans, that American genius who, as a means of exploiting his newly invented high-pressure steam engine, appears to have been the first to apply power to a construction operation in America, assuming that dredging is an act of construction. Evans contracted with the Philadelphia Board of Health in 1804 to construct a steam dredge "for the purpose of cleansing the docks and wharves, removing obstructions in the navigation of the river, and other purposes . . ."[7] The result was the legendary *Orukter Amphibolos* or amphibious digger. It was simply a 12 by 30-foot scow with what today would be called a ladder dredge slung over one side, and a primitive paddle wheel for propulsion, both dredge and paddle wheel driven by an Evans engine and boiler.[8]

In the early years of American railroad development there was a scheme devised to avoid the time and expense of building up railroad embankments by laying the entire line on pilings whose length varied to provide a level line-of-way regardless of the terrain. A good many miles of railroad were laid down in this way, principally on the Erie and on the Syracuse & Utica, by means of a remarkable steam-powered combined double-pile driver and cutoff saw, invented by Smithe Cram in June 1838 and shown in Figure 7.4. Four of these machines were in use on the S & U somewhat earlier than this and as many as six on the Erie. Cram's machine imitated traditional pile drivers in employing the power to perform the same work that men and horses always had, but was distinctly successful, putting down the two lines of 12-inch pilings at once, cutting them off at the required elevation and moving ahead on the completed piles. Each machine was operated by eight men—far fewer than would have been required for a manual driver—and completed a mile of road per month.

Beginning in 1834 an aqueduct was built over the Potomac River at Georgetown, to extend the Chesapeake & Ohio Canal to Alexandria, Virginia. The depth of the pier foundations induced the contractors, and the Army's Corps of Topographical Engineers which succeeded them, to employ steam

Fig. 7.4   Cram's double-pile driver on the Syracuse & Utica R.R., 1838. John Weale, *Ensamples* [sic] *of Railway Making*, 1843.

power. At first the power was used solely for dewatering the cofferdams within which the piers were constructed, but later for exacavating as well. An account of the work left by Major William Turnbull,[9] the Army's resident engineer, gives tantalizingly little information about the steam plant, except to mention that there were two 20-horsepower engines and boilers mounted on a scow next to the cofferdam, presumably moved as the work progressed from pier to pier. There is no word on precisely why steam power was chosen, except a note that it was difficult to procure dependable labor, and steam would have been the only means of ensuring absolutely constant operation of the pumps.[10]

Before 1860, in construction or in *any* industry where the use of powered machinery was optional, its adoption depended, naturally, entirely upon the relative economic benefits of the various alternatives. In factories and mills individual machines were fixed in place; power transmission equipment could easily connect the prime mover with the individual machines, and it thus was feasible to drive the entire plant with a single large efficient prime mover: a steam engine or a water turbine. Construction presents opposite conditions. The work site is nonpermanent; power is required at constantly changing spots all over the site; power consumption varies continually—in hoisting, for example, being nil for most of the time and rising to a maximum only for short periods—and everything is exposed to the weather and lots of mud.

Since the needs for power on a building site are so fragmented, with excavation proceeding at a dozen places over a multiacre plot: a mortar mixer here; another there; a derrick somewhere else—there is literally no possibility of transmitting power to all these constantly changing locations from a large, central prime mover. If building construction machines are to be powered at all, they must be *self-powered* independent units. The implication of all this simply is that until the technology existed to make prime movers small, and thus portable, there could *be* no power on the building site. It was Trevithick in England and Evans in the

United States who, with their nearly simultaneous work in the first decade of the nineteenth century, raised boiler pressures and engine speeds. This work thereby so reduced the size of engine, boiler and necessary auxiliaries that the steam plant could, while still delivering useful quantities of power, become a free and independent entity: free to move over the Earth's surface. This was a vital act of liberation, for it made possible not only the locomotive and the steamship, but also, ultimately, a large family of powered construction machines: excavators, dredges, materials, hoists, mixers, earth movers, and a great variety of on-site tools.[11]

With the proven feasibility of powered construction machinery—at least for certain tasks—by about 1840, the inordinate lag in its widespread introduction would seem curious. But only until we look at the economics involved. Because of building's distinctive peculiarity of impermanence, the builder was confronted with a highly amorphous set of economic factors when he set out to assemble his plant for a project.

Through the entire nineteenth century, there was a major decision that he would be forced to make again and again: is it cheaper to perform a certain operation with men, or with animals, or with powered machines? The chief advantage of men was their wonderful flexibility: a man was quite clearly the most adaptable, multipurpose construction tool the builder had at his disposal—one that could lift, carry, climb, move along nearly any sort of irregular path and, with relatively simple tools, dig, assemble, lay stone, etc. If man is the most important single element in building construction even today, how much more was that so in the first half of the nineteenth century? There was, moreover, no direct capital expenditure for men: they could be hired and fired to suit the work and they didn't depreciate. Machines, however, could perform tasks far beyond the strength and endurance of men and animals. On major works in the wilderness, machines didn't require the construction of the elaborate communities of bunk houses, mess halls and the other amenities required for the contractor's human equipment and, as the early automobile ads informed us, when they weren't working they weren't eating.

I have become convinced that it was this lack of a clear case favoring men against animals against machines that produced the most conspicuous characteristic of the contractor's plant in the Age of Steam: the apparent total lack of its systematic progression from muscle to steam power.

If we regard the *Orukter Amphibolos* and such isolated uses of powered construction machinery as Cram's pile driver and the Potomac Aqueduct equipment as noble experiments that inspired few followers, where then does the trail really begin, without break? About where it should. The most time- and man-consuming work in all of construction was earth movement, and in the Age of Steam, which we could well call the Age of the Canal and Railroad, this constituted by far the greatest amount of work on such jobs. Excavation, then, was where the inducement to mechanize was greatest.

In 1836 William S. Otis of Canton, Massachusetts (not to be confused with Elisha G. Otis of elevator fame) patented what may be viewed as the first consistently used, powered construction machine. It has persisted to the present day in a form so nearly unchanged as to border on the miraculous. Unlike the early powered pile drivers, Otis did not simply substitute a steam engine for men in a preexisting, man-powered machine. Functionally it was without precedent. The inventor termed

Fig. 7.5 "American Steam Excavator of Otis, or, The Yankee Geologist," 1837.

it "the American Steam Excavator of Otis, or, the Yankee Geologist" (Fig. 7.5). In 1837 he had an experimental machine built and shortly went into commercial production. The machines were equipped with 15-horsepower, single-cylinder, high-pressure engines and cost $7500. They could excavate 1500 cubic yards of earth in a 12-hour day, where a fair day's work for a man was 12½ yards of *loosened* earth. The excavator, shown on site in Figure 7.6, did the work of 120 men. Comparative costs were on the order of 3¢ per yard for the shovel and 8¢ per yard for hand labor.

Otis' excavator succeeded, for it must have been obvious to even the most reactionary contractor that where large quantities of excavation were involved, the considerable capital investment in such a machine easily justified

itself. George Washington Whistler took several to Russia for building the Czar's Moscow–St. Petersburg railroad. The machine's complex movements—strongly anthropomorphic—and the accompanying sound of the engine straining under load, were fascinating to observers. The *Concord* (Mass.) *Courant* reported in 1846 that:

The great "lion" in these diggins, now a days, is the Excavator, in use upon the Northern Railroad . . . It is a curious and most ingenious combination of all the mechanical powers.—It goes by steam of course. At the end of a large beam is a sort of a huge iron shovel—which by means of pullies and screws, &c, is brought to bear upon the soil, then the steam is put on, and

Fig. 7.6   Otis steam shovel built by John Souther (Boston), at work in Needham, Massachusetts, loading gravel for Back Bay filling, ca. 1860. *(Boston Athenaeum photo)*

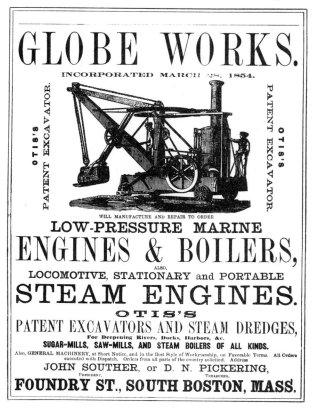

Fig. 7.7   Advertisement for Otis shovel by Souther, ca. 1864.

it "goes ahead," till it is full, and then it stops, as it should. Then it is cunningly whirled round till it's brought directly over a dirt cart, into which it empties itself in short metre. The car being filled is quickly moved away, and another takes its place, and so on indefinitely. The way it digs into the "bowels of the harmless earth," is a caution.

A bizarre byway in the development of the steam shovel was a transitional machine invented in 1852 by J.C. Osgood of Troy, New York (Fig. 7.8). In his answer to the Yankee Geologist, Osgood nearly perfectly imitated the basic functions of Otis' machine, but all powered by a pair of horses! It is doubtful that the machine was any sort of commercial success, for by the time they had overcome the internal friction of the excavator's own machinery, there could have been little of the horses' energy left to perform any useful excavation. It is reminiscent of Roland Emmet's

wonderful cartoon of a new, improved ditching machine, guaranteed by the plate to do the work of three men and a boy. Operating it, of course, are three men and a boy.

By the 1850s, a certain amount of fully mechanized, powered construction equipment was commercially available, but its use still was the exception rather than the rule. One contractor's manual of the period by civil engineer George Cole noted "The motive power used in the construction of public works is chiefly manual, and horse power. Steam power is sometimes substituted for the latter, and to a limited extent for the former. Manual labor is chiefly directed to the preparation of materials, and to the excavation of earth and rock, and their removal for short distances, and horse power for long distances."[12]

This is borne out by the verbal and graphic records of several projects in or near Washington, D.C. In 1848, a decade after steam had been used on the Potomac Aqueduct, all stone for the Washington Monument was hoisted by hand derricks. Four major Government works, between 1857 and 1861, under the superintendence of Major Montgomery C. Meigs of the Army Engineers, also serve as examples. Lest it be thought from what follows that Meigs was a classical conservative, he was not. An innovator, able constructor and creative civil engineer, Meigs would have employed machinery wherever he felt it would be an advantage. The projects were the Washington (water supply) Aqueduct, and extensions to the Post Office, the Treasury, and the Capitol.

For erection of the Capitol's immense cast-iron dome, Meigs devised a derrick boom of heroic proportions, centrally placed in the rotunda area (Fig. 7.11). It was supported on a vertically extensible platform raised in increments as the work progressed. The fixed horizontal boom reached beyond the base of the dome to pick up components. The derrick commanded the entire circular dome area: rotationally by pivoting of the mast and boom, and radially by changing the length of the two sets of hoisting tackle fixed at the boom's inner

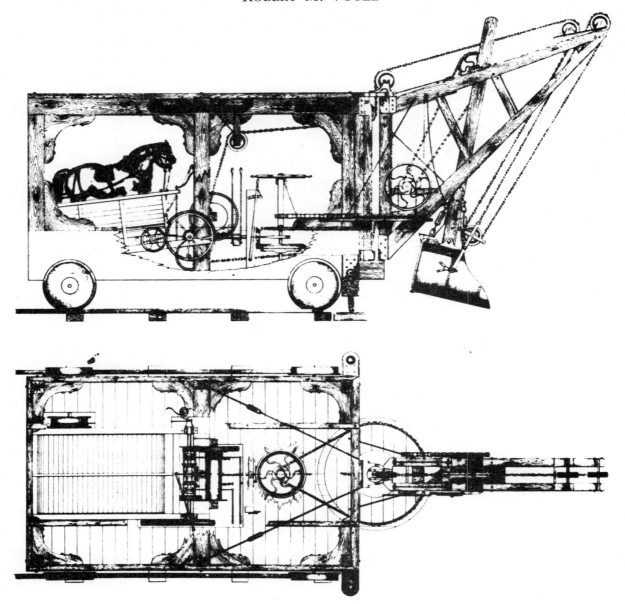

Fig. 7.8   Osgood's excavator, horse powered, 1852.

and outer ends. It was an eminently simple, effective and dependable mechanism, essentially a precursor of today's climbing radial tower cranes. For the Capitol's rectangular wing extensions, Meigs employed a group of traveling cranes and other lifting devices (Fig. 7.12).

Apparently Meigs totally rejected powered machinery, even on this immense project. I say apparently because there is conflicting evidence on the matter: reasonably hard *verbal* evidence that steam power was used in the

hoisting operations *vs.* even harder *graphic* evidence that it was not, at least not seriously. In 1855 Meigs reported to Congress[13] that steam powered cranes would raise the castings and other material from the ground to the base of the dome, "within reach of the great derrick" (no mention of steam for the great derrick itself) and that "the engine has been procured" for this purpose. A year later he reported[14] "the machinery was all complete and in place. One of the iron columns was hoisted to the top of the eastern portico, for

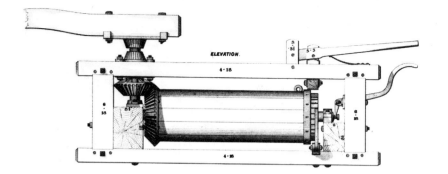

HORSE POWER.

Scale 1 F! to 1 Inch.

PLAN.

N.B. Patterns of Bevel Gearing similar to those represented in this plan of horse power will usually be found in the founderies throughout the civil-ized world, therefore there is no necessity for showing same in detail.

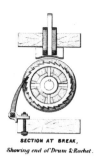

SECTION AT BREAK,
Showing end of Drum & Rachet.

Perspective Sketch of
HAND POWER.

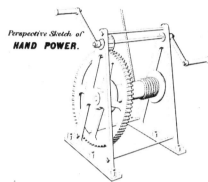

Fig. 7.9   Horse whim details from *The Contractors Book*, George Cole, 1855.

Fig. 7.10   Quarry crane details. Cole.

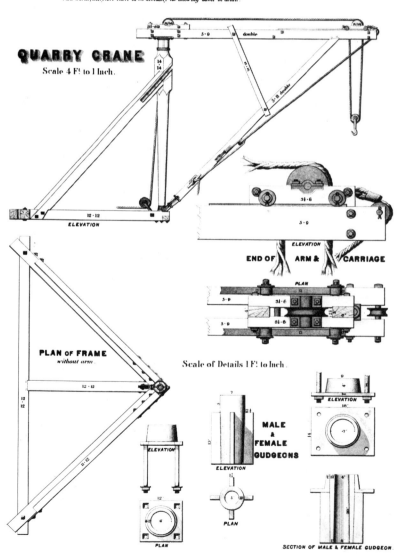

QUARRY CRANE

Scale 4 F! to 1 Inch.

double

ELEVATION

ELEVATION

END OF   ARM &   CARRIAGE

PLAN

Scale of Details 1 F! to Inch.

PLAN OF FRAME
without arm.

ELEVATION

ELEVATION

MALE
&
FEMALE
GUDGEONS

ELEVATION

PLAN

PLAN

PLAN

SECTION OF MALE & FEMALE GUDGEON

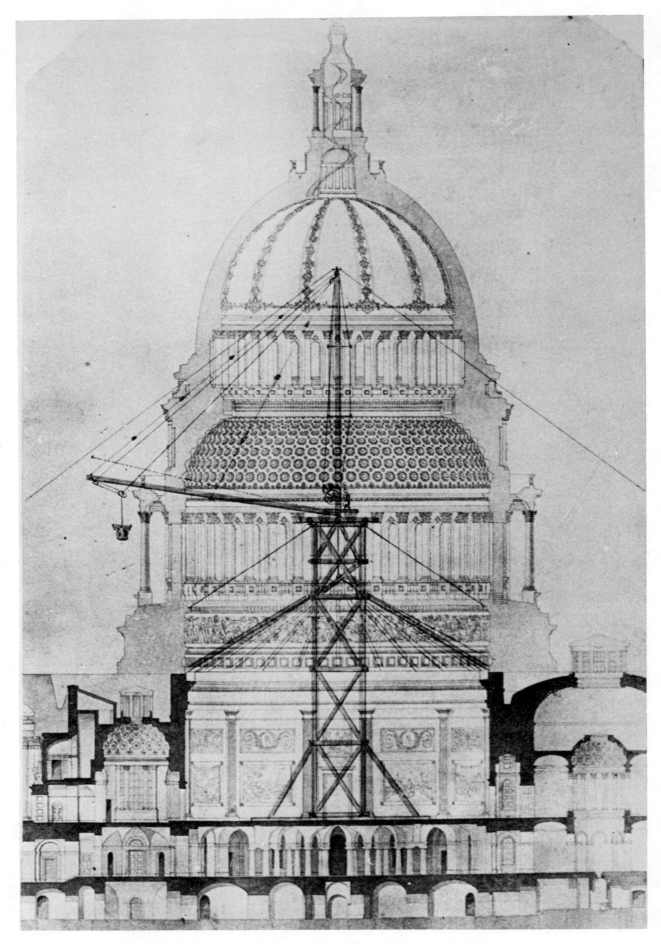

Fig. 7.11 Radial boom crane for erecting the U.S. Capitol dome. (*Architect of the Capitol*)

Fig. 7.12 Traveling crane for erecting the Capitol wings. (*Architect of the Capitol*)

absence in any of the Capitol photos (and the drawings) of so much as a wisp of steam or smoke, let alone any steam machinery, coal piles or other signs of steam power, except by its absence from the scene? There are photographs of the crabs (winches) used (Fig. 7.13), and of the cranes themselves, and long views of the work, but no sign of a steam plant. I can conclude only that Meigs had a brief go at steam hoisting at the outset of the project, found it wanting and dropped it.

I believe that Meigs found that muscle power was more efficient for two reasons: the speed of erection of the dome probably did not depend upon the speed with which its parts could be raised into place. Erection speed instead depended on parts delivery, connecting standing and new components, inspection and the like. Second, the two men who worked a hoist's winding crab could be employed elsewhere on the job between lifts. A steam plant would have required at least two men, an engineer and a fireman who would, like hoisting engineers today, have been specialists, idle much of the time. To say nothing of the space occupied by engine, boiler and coal piles; the nuisances of water supply and ash disposal; fire danger and the myriad other ills attendant upon the use of steam power. It is easy to imagine Meigs deciding, after a brief trial at the Capitol, that steam machinery being what it was—bulky, messy and not always dependable—he was better off without it.

the purpose of testing the working of the derricks and steam-hoisting apparatus, and showing the effect of the colonnade at that height." There is no question, then, that steam hoists were used, at least for secondary hoisting.

And the graphic evidence that it was not? Meigs photographically recorded all his projects, a superb record that gladdens the heart of the historian of building technology. There are overall views and details, plus views of the construction plant, testing equipment and other fine details. Nothing was ignored by his camera man. How then, can we explain the

Fig. 7.13 Hand crab of a Capitol crane.

Meigs conducted the work on the other buildings and the aqueduct, if we can believe the photographs, in similar fashion. Every aspect of the erection was of his own devising, the equipment all designed by him or under his direct supervision. He used extension boom cranes on the Treasury (Fig. 7.14) and the Post Office extension. The heaviest components to be moved and raised were the monolithic stone columns that were an invariable feature of all major Government buildings at the time and it appears that only for these did Meigs substitute animal muscles for men's.

For the Cabin John Bridge, principal structure on the Washington Aqueduct, Meigs used steam power where it was most efficient, to saw the large quantity of lumber required. The intermittent work of raising the stone for the span was, as usual, done with hand-powered travelers. We see, then, that Meigs's decisions as to the type of equipment used were based on practical considerations. Men's and animals' muscles were the most economic power sources for dryland construction where there was a consistently dependable labor supply. But under different conditions—hostile weather, a short construction season, the

Fig. 7.14 U.S. Treasury Building. Column raising. (*Division of Prints and Photographs, Library of Congress*)

Fig. 7.15   Victoria Bridge. Steam dredge used in excavating puddle chambers.

need for quick return on investment, and limited and unstable labor—a totally different approach was called for.

The most formidable assemblage of steam power for a North American construction project before 1860 was employed on the Grand Trunk Railway's Victoria Bridge across the St. Lawrence at Montreal, 1854–59. The bridge was a mile and a quarter long, with 25 tubular-girder spans. Steam was used to power the reciprocating and centrifugal cofferdam pumps, stone hoists, excavators and dredges (Fig. 7.15) and pile drivers. A bevy of steam tugs and locomotives attended the work. Steam shovels which dredged between the sheet-piling wall of the cofferdams were powered by separate engine units, the power transmitted by chain. The project was a constructional *tour de force,* mechanization being responsible for the remarkably short building time of five years.

POSTSCRIPT

I am forced to conclude that the Age of Steam had less impact on building construc-

tion than on other industries of the nineteenth century. The real problem lay in subdividing the prime movers' power to distribute it around the site. It would have been impractical to criss-cross a building site with a web of articulated shafts and ropes, joined with myriad universal joints, resembling a vast array of outsize dentist's drills. Until the middle of the century, this would have been the only means of applying the power of a single prime mover to a number of tools. Just after our arbitrary 1860 cutoff date for this book, however, the introduction of power transmission by compressed air further made possible this "subdivision" of power, and so revolutionized mechanized construction techniques.

In 1866 Charles Burleigh of Fitchburg, Massachusetts, brought a practical pneumatic rock drill (with an effective steam powered compressor) to the Hoosac Tunnel in northwest Massachusetts. The Hoosac had been under construction for over 10 years, but by drilling blast holes with pneumatic drills rather than by hand the remaining 70 per cent of the job was completed in less than another 10. The implication was important: the capability of transmitting centrally produced power to slave machinery and tools scattered indiscriminately over a construction site with utter flexibility of distance, location and movement.

The Victoria Bridge and the Hoosac Tunnel were watershed projects in the mechanization of construction, both completed just after the middle of the nineteenth century. Yet the seemingly random application of man, animal and steam power to building continued well into the twentieth century. The Transcontinental Railroad (mid to late 1860s) was built largely with hand labor. When work was resumed on the Washington Monument in 1880, an Otis steam elevator machine had been obtained for hoisting stone and, later, passengers. In the 1880s, handwork still was

the rule for the Northern Pacific Railway. But by the 1890s, labor costs and quick return on capital investment were pressing factors; considerable mechanization had occurred in urban building construction. The steam shovel had become both standardized and a standard item of the large contractor's plant, its form and function little changed from the Otis Excavator. But in 1909, beams for the Coolidge Mill in Manchester, New Hampshire, were raised by hand-powered "Mary-Anns."

To end on a personal note, it was with the keenest interest that I witnessed my first building operation in the year 1939: the excavation of cellar holes for a large housing project outside Philadelphia, with horse-drawn drag scrapers.

NOTES

1. J. R. Perronet, *Description des Projects . . . des Ponts de Neuilli* [sic], Paris, 1788.
2. But that was the extent of powered machinery. On the job were upwards of 900 men and 170 horses, operating a stunning array of finely designed pile drivers, stone hoists, dredges and other machinery, the description of which could in itself justify an entire symposium.
3. Zachariah Allen, *Science of Mechanics*, p. iii, from Carroll W. Pursell, Jr., *Early Stationary Steam Engines in America,* Washington, Smithsonian Institution Press, 1969, p. 89.
4. For an account of the all-important I-beam, see Robert A. Jewett, "Structural Antecedents of the I-beam, 1800–1900," *Technology and Culture,* Vol. 8 (July 1967) pp. 346–362; and "Solving the Puzzle of the First American Structural Rail-Beam," *Ibid.,* Vol. 10 (July 1969) pp. 371–391.
5. Erection of Dr. David Jayne's New Granite Building, a spectacular example of "prestige architecture" in 1849–50, enjoyed good coverage in the Philadelphia *Public Ledger.* Standing for over a century in the 200 block of Chestnut Street, it was designed by architect William Johnston (ca. 1811–1849) and completed after his death by the well-known Thomas U. Walter. Construction began about Nov. 1, 1849 under the direction of S.

K. Hoxie, the "celebrated granite man." The masonry of Quincy granite reached a height of 95 feet above the pavement. One slab—the largest ever brought to Philadelphia—was 27 feet, 4 inches long and weighed about 10 tons. Collins the rigger, "created quite a sensation" before crowds on the street.

We have no description of Hoxie's "steam-hoisting apparatus" but we know that by July 1, it was being used again to erect the huge Classic Revival iron gasometer on West Market Street. When underground water was encountered at the Jayne Building, the engine pumped it out. See also "AnteBellum Skyscraper," *Journal of the Society of Architectural Historians,* Vol. IX (Oct. 1950) pp. 27–28; and "The Jayne Building Again," *Ibid.,* Vol. V (March 1951) p. 25.—Ed.
6. Gosta E. Sandstrom, *Man the Builder.* New York, McGraw Hill, 1970.
7. Greville Bathe and Dorothy Bathe, *Oliver Evans,* Philadelphia, 1935, pp. 111, 112.
8. Although it is known that the Orukter did reach the Delaware via the Schuylkill River, there is no record of it actually being used for dredging. In England, however, a similar machine with a chain of buckets was used to excavate a vertical entrance shaft to Marc I. Brunel's Thames Tunnel, started in 1825. This was

powered by an inverted "V" engine.
9. William Turnbull, *Report on the Construction of the Piers of the Aqueduct.* Washington, 1873.
10. An excavating machine similar to Brunel's was tried on this project, with indifferent success; the engines were also used to drive a winding drum by which buckets of spoil were raised with hand ropes.
11. It was not until the internal combustion engine replaced the steam engine in the twentieth century that another such quantum leap was made in mechanizing the building industry.
12. George Cole, *The Contractors Book of Working Drawings of Tools and Machines used in Construction Canals, Rail Roads and Other Works, with Bills of Timber and Iron. Also Tables and Data for Calculating the Cost of Earth and Other Kinds of Work.* (Buffalo, N.Y.?) 1855. J. S. Vernam, Mechanical Engineer and Draughtsman, also appears on the title page.
13. Report submitted Nov. 16, 1885, appearing originally in U. S. Congress, 58th Congress, 2nd Session. H. R. Report No. 646; later in *Documentary History of the Construction and Development of the United States Capitol Building and Grounds,* Washington, 1904.
14. Report dated Nov. 13, 1856.

CHAPTER
8

Roofing for Early America

DIANA S. WAITE

Ms. WAITE is in charge of the National Register Program in the New York Office of Parks and Recreation, Division for Historic Preservation, Albany, and the author of several recent short works on building history.

No part of a building is more important than its roof. The roof not only shelters its inhabitants and their possessions against sun and rain and wind and dust. It makes the space underneath livable in summer heat and winter frost. It protects the walls and holds them together. Few structures can long survive a break in the overhead cover.

Availability, appearance, cost, durability and fire safety are among the factors that the owner, his architect and his builder have had to consider in selecting roofing. This survey will discuss briefly the principal types of roof coverings used in this country before the Civil War, especially in the American Northeast.[1]

Over the years a large number and variety of materials were tried, then accepted or rejected as materials for roofing. Among these were bark, sod, thatch, tile, board, shingles, slate, copper, iron, lead, tin, zinc, galvanized

iron, and the many so-called composition roofing materials. Each fulfilled the various qualities hoped for in a roofing material to a greater or lesser extent, but they did not gain equal acceptance and usage. Thatch and zinc, for example, were never as widely utilized as shingles and tinplate. The question that immediately arises is, of course, why some proved popular and some did not.

THATCH

Thatch, widely used in northwest Europe, was one of the first roof coverings to be used on this side of the Atlantic. An ordinance of 1648 mentioned that the houses in New Amsterdam were "for the most part built of Wood and thatched with Reed."[2] Records relating to early houses in New England suggest that thatch was frequently used there as well. Among the advantages of a thatch roof were its low cost, reasonable durability, effective insulating properties, and (to us today—at least) pleasing appearance.[3]

In most of the colonies, however, thatch was seldom used after the mid-seventeenth century. This was due in part to the availability and cheapness of wood shingles. It was probably also due to the fact that by this time the early colonists had realized that thatch did not hold up as well under the extremes of the American climate as it had under more moderate conditions in their homelands. But the demise of thatch in areas having any concentration of structures was due in large part to the danger from fire. Legislation limiting its use for this reason was enacted at an early date.[4] In rural areas and places where shingles were not available, thatch was used to some extent at least through the mid-nineteenth century.[5]

CERAMIC TILE

Concern about the combustibility of thatch evidently served to stimulate the adoption of tile roofing. For nearly two centuries—from 1656 through 1849—the fire and building codes in New Amsterdam and New York City specifically recommended the use of tile because of its fireproof qualities. Tile was also highly regarded because of its durability, ease of maintenance, and lack of thermal conductivity. For many early European settlers, tile was a familiar material, in both the functional and aesthetic sense.

During the seventeenth century some tiles were imported to New Amsterdam from Holland. Further up the Hudson the need for domestic manufacture of tiles had been recognized as early as the 1630s.[6] By the 1650s tile making was a going concern in Rensselaerwyck and tiles were being sent down to New Amsterdam.[7] The pantiles excavated at Fort Orange from the site of a house built at some time between 1649 and 1651 (Fig. 8.1) were fashioned with lugs to attach them to the scantlings, and were further secured with mortar.[8] In New Orleans and elsewhere, both pantiles and plain (flat) tiles were used well into the eighteenth century. In the years both before and after the Revolution, several tile yards were in operation in the New York area.[9] Even as late as 1802 a New Brunswick, New Jersey, manufacturer advertised that he had enlarged his works and could furnish any quantity of the very first quality tile at short notice.[10] Specimens of ceramic roofing in the form of "shingles" and pantiles at the Roanoke, North Carolina, settlement (1585) and at the early Virginia and Maryland capitals of Jamestown

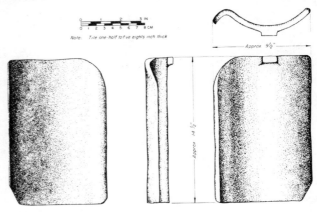

Fig. 8.1 Reconstruction of pantile excavated at Fort Orange, Albany, New York, dating from a house constructed at some time between 1649 and 1651. (*Drawn by G. Gillette, New York State Office of Parks and Recreation, Division for Historic Preservation*)

and St. Mary's have been excavated from ruins.[11] Later tiles were used in the German settlements of Berks County, Pennsylvania, at Zoar, Ohio, and in Missouri.

But due to a number of factors, during the next few decades tile rapidly fell out of favor in much of the Northeast. In rural areas and in fact everywhere but within cities officially laid out as fire districts, wood shingles were widely used. Unless there was a tile yard nearby, tile provided little competition as a roofing material for most buildings, since shingles were much lighter and more readily available at a fraction of the cost. Furthermore, in the city fire districts, the fire resistant quality of tile was no longer enough to guarantee its continued usage, for other fireproof materials had become available. One of these was slate. The others were sheet metals: copper, iron, tinplate, zinc, and galvanized iron. Also, by this time the appearance of tile roofing was becoming unfashionable. In 1830, for example, one American author referred to tile as being "clumsy" looking.[12] Since by their nature tiles had to be used on roofs with a fairly steep pitch, they were necessarily visible. So with their appearance meeting with disfavor and other, more pleasing, readily available, and reasonably priced fireproof materials obtainable, there seemed to be no reason to continue using tile—and this is apparently what happened.[13]

WOOD SHINGLES

Wood shingles had always been one of the most popular roofing materials in America. They were used on all types of buildings, even the very monumental. The first New York State Capitol, for instance, designed by the noted architect Philip Hooker and completed in 1809, was covered with pine shingles, the common species used in the Albany area. This fact was noted with considerable dismay by Horatio Gates Spafford, who wrote that the excellent slate "with which . . . the state abounds" should have been used.[14] Even houses as grand as the home of Philip Schuyler, completed in 1762, and Lorenzo, completed in 1808, were originally covered with shingles.[15]

The shingles used on these structures were made in the traditional manner—by hand. In this type of manufacture the grain of the wood was followed for the full length of the shingle. It was this characteristic that was responsible for the many qualities that made shingles so desirable. But it was not long until attempts were made to improve upon the efficiency of hand manufacture: as early as 1802 a United States patent was granted for a shingle machine. By 1850 many shingle machine patents had been awarded and shingle sawing became an important branch of the American lumber business.

While square-butt shingles were the most common, round-ended shingles were also frequently used before the nineteeth century.[16] They were used at Washington's Headquarters, an eighteeenth century building in Newburgh, New York (Fig. 8.2),[17] and on other

Fig. 8.2 Rounded butt shingles were utilized at Washington's Headquarters in Newburgh, New York. This nineteenth century lithograph by Currier and Ives is based on a view appearing in the *New York Mirror* of December, 1834, done after a painting by Robert W. Weir. The building was not being well maintained at that time, but the present roof is believed to date from the eighteenth century.

Germanic buildings in that vicinity. The curved shape may have been developed to help alleviate warping of the shingles and it may have been used in an attempt to suggest a tile roof. Later, during the mid-nineteenth century when ornamental roof surfaces became popular, the butts of shingles were again shaped.[18]

It is interesting to note that shingles and board roofing were used on the flats of roofs during the eighteenth century and later. In New York City, for example, shingles and boarding had been outlawed for roofing on new construction within the fire districts since 1791, but the flat of a roof was excepted. The flat might be covered, the ordinances noted, with board or shingles, provided that it did not "exceed two equal fifth parts of the space of such roof" and also that a "substantial balcony or balustrade" be erected around it.[19] This provision was carried along in the ordinances up through 1829.

One of the disadvantages of wood shingles was their short life expectancy, a problem that was recognized from an early date. As early as the 1690s, pine tar had been used along the eastern seaboard to help preserve shingles.[20] In other instances they were painted with oil paint in colors of Spanish brown, red, or occasionally green or blue, possibly to give the impression of slate.[21]

During the mid-nineteenth century, many new patent roofing compounds designed for built-up roofing systems were recommended for use on shingle roofs. The makers of Gline's Patent Roofing Paint, for instance, advertised that their paint would prove "very ornamental, not only giving an old decayed, mossy and warped shingle roof, the handsome, durable and uniform appearance of a slate roof," but would also save the expense of reshingling, since one coat was reported to be equal to a new layer of shingles.[22]

SLATE

Slate roofing was used in the Eastern cities before the Revolution. The seventeenth century building ordinances of New York and Boston specifying fireproof roofing encouraged the use of slate as well as tile. After the Revolution, usage became more common. It was estimated, for example, that by 1830 half the roofs in New York City were covered with slate.[23]

Wooden Boston seems to have been among the first towns to have slate roofs. Slate Island and Hangman's Island in Boston Bay (offshore of Hingham) were the local sources and instances of quarrying have been cited back to 1654.[24] The Thomas Hancock house, a famous landmark on Beacon Hill built in the 1730s, was a conspicuous example.[25] But the fact that the Carpenter House on Second Street, Philadelphia, built ca. 1698 was generally known as "The Slate Roof House" probably indicates that the material was uncommon there at that time.[26] One of the great country houses of the South, Whitehall, the residence of Governor Horatio Sharpe near Annapolis, built 1764–73, was roofed with slate, a large part of which still remains under a later tin roof.[27]

Slate, as noted, was valued for its fire resistant qualities. Early American advertisers also remarked upon its "handsome appearance,"[28] while others stated that it was the "most lasting, the cheapest and the best covering for buildings in use."[29] Others claimed that the rain water collected from slate roofs was so pure that it could be used for drinking or cooking without filtering. After tin roofing became popular, promoters of slate stressed the fact that slate roofs needed less maintenance than metal ones, which required frequent painting.

But, as with any material, there were drawbacks. Despite the earlier claims that slate was cheap, builders' guides indicated that slate consistently ranked among the more expensive materials. The cost per square of roof to be covered was compounded by the fact that slate was heavy itself and therefore required a more substantial, and consequently more expensive, framing system for support.[30] Preliminary investigation indicates that the vast percentage of the slate used through the first quarter of the nineteenth century came from Welsh quarries. Several quarries opened up in America during the end of the eighteenth century and the beginning of the nineteenth century,[31] and there are examples of notable American buildings being covered with native slate, such as the Virginia Capitol and the Old State House in Boston. But few of these early American quarries became successful enterprises during these early years.

The problem was one of both quality and economy. While some quarry men boasted that they could turn out slates equal to those of Welsh origin, the users complained that they did not withstand the American climate as well. In addition, the cost of American slate was high, often equaling or exceeding the cost of the Welsh product, whose quality was preferred.

A primary factor in the high cost of American slate was related to its weight. Not only

Fig. 8.3 "View of the Art of taking Slates out of Quarries, and of Slitting and Cutting them" from *The Universal Magazine of Knowledge and Pleasure* (London) May 1766, No. CCLXV, Vol. XXXVIII. Location of the quarry not given. Workmen split off blocks and send them to the surface via hoists to be further split elsewhere. The building labeled 33 is a forge for repairing tools; 34 is a "retreat for the labourers" also used for mending machines and tools.

THOMAS MYERS & SON,
Tin, Slate, and Corrugated Iron
ROOFERS,
DEALERS IN
Gutters, Conductors, Roofing Cement, &c.
SLATE AND METAL CHIMNEY TOPS,
No. 38 Chapel Street, Albany.

Slate Roofs Repaired ; also Tin Roofs painted and
paired at short notice, and warranted not to leak.

THOMAS MYERS,   JOHN MYERS

Fig. 8.4 Combined use of different shapes of slate are evident in this illustration which shows a slate roof being installed on a gothic cottage, adding to its picturesque appearance. From Albany city directory, 1873, p. 360.

was it heavy on the roof, it was heavy to ship. It was also difficult to ship because there were no efficient ways of transporting the material.

But the situation soon changed, for during the mid-nineteenth century several sets of circumstances coalesced to increase the demand for slate and to make quarrying in America profitable. One of these was the rapid increase in America's population, which more than tripled between 1830 and 1860. The demand for all building materials, including roofing, was correspondingly large. Secondly, by mid-century, canals had been opened and railroads built. These networks allowed the slate to be shipped to the cities cheaply and from there to other, more widespread markets. Another important factor was that the roof became an increasingly important, more visible part of many architectural designs. Rural and suburban villas in the pointed or gothic styles filled the pages of the new architectural pattern books, and the complexity of their roof forms became their hallmark. One writer spoke of this quality as the "picturesqueness of roof."

Slate was a natural material for this type of roof. It could be readily fitted to the complex roof forms. Furthermore, slate lent itself to the fanciful patterns that were important to creating a visually interesting texture on the roof surface (Fig. 8.5). In fact, several patterns would often be combined on one roof by laying strips of different patterns one above another.[32]

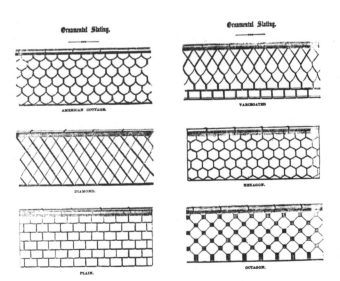

Fig. 8.5   Patterns of slate for roofing that were available in 1857 from the Eagle Slate Company, located near Hydeville, Vermont. Illustration from H. N. Stafford, *The Ready Calculator* (N.Y.: John F. Trow, 1857).

140

TINPLATE

During the nineteenth century, tinplate, like wood shingles, became truly an all-American roofing material.[33] In 1846 one English manufacturer wrote to an American customer in amazement, saying, "We do not see what you do with all the tinplate you Americans import."[34] But if he had come to this country he quite literally could have seen where it was used—on the roofs of American buildings.[35] It was used on roofs with a low pitch, on mansard roofs, and the many different shapes of roofs dictated by architectural fashion. It was relatively reasonable in cost.

Another important factor in the popularity of tin was durability. If properly maintained, primarily a matter of regular painting, a tin roof could last for a very long time. Thomas Jefferson, writing in 1821, called it the most desirable cover in the world. "We know that it will last 100 years," he wrote, "and how much more we do not know." He also recommended it because of another important characteristic—its light weight.[36]

Charles Peterson has noted that one of the first buildings in the United States to be covered with tinplate was the Exchange Coffee House in Boston, shown in Figure 8.6, completed about 1808.[37] The sheets used on this building and others up through the mid 1830s

Fig. 8.6 One of the first buildings in the United States to be covered with tinplate was the Exchange Coffee House in Boston, built about 1808. From Snow's *History of Boston* (1825), reproduced in *Stark's Antique Views of the Towne of Boston* (Boston: James H. Stark, 1901) p. 121.

measured approximately 10 by 13¾ inches[38] Apparently during the 1830s some technological changes were made in tinplate manufacture that resulted first in that size being enlarged slightly to 10 by 14 inches, and in a multiple of that, 20 by 14 inches.

By the 1870s a third size, 20 by 28 inches was being widely used. This size, like the 14 by 20 inch size, was frequently used for standing seam roofing. For this type of roof the sheets were often joined together with flat seams to form long strips of the metal while still on the ground. The rolled strips of metal were then lifted up to the roof, where they were joined together with standing seams.

In nineteenth-century documents it is apparent that there was considerable disagreement among roofers until at least the 1870s over whether standing seams were better than flat and whether one size of tinplate was better than another. But one thing that they did agree on was that it was important to keep a tin roof well painted. It was recommended that the underside of the sheets be painted before being laid to prevent their rusting out from underneath. Every few years it was necessary to paint the exterior surface.

The color of the paint used on tin roofs was commonly brown or red, made economically from iron oxide and linseed oil, both readily available from domestic sources. Other tin roofs were painted to look like slate. In a plate in the rare publication *Rural Residences*, Alexander Jackson Davis specified that the tin roof was to be painted a bronze green, apparently to imitate copper (Fig. 8.7).[39]

ZINC

Although tinplate was the foremost metal roofing material in nineteenth century America, it was by no means the only one. Another whose history covers about the same time span is zinc.[40] Although it was never as widely used, the history of zinc is equally interesting and perhaps even more intriguing.

Zinc was first rolled in about 1805 in Sheffield, England, but it was in Belgium that the

Fig. 8.7   Specifications called for the tin roof to be painted bronze green, apparently to imitate copper, in a plate from *Rural Residences* by Alexander Jackson Davis (New York, 1837). (*Courtesy New York State Library, Albany*)

industry was subsequently developed; it was largely Belgian sheet zinc that was used in America. Once the technique of rolling zinc was developed, a market had to be created. Toward this end the manufacturers arranged to have a house, a church, and a cathedral in Belgium roofed with sheet zinc in 1811 and 1812 as an experiment. They were successful and zinc quickly became a very popular roofing material on the Continent, especially in Germany and France.

Zinc was used in America for roofing soon after these experiments, at least as early as the 1820s, perhaps even before. In 1824, Peter Augustus Jay wrote his father, John Jay, that he would "like to try" it.[41] All the roofs "which have been covered with it heretofore," he wrote, "answer completely." By 1837 somewhere between seventy and one hundred buildings in New York had been covered with zinc.[42] Nor was interest confined to New York. The famed St. Charles Hotel in New Orleans, for example, was also covered with it.[43]

It appears that American interest in zinc roofing during the nineteenth century ran in cycles of approximately 25 years. About every quarter century, the possiblity of its use was raised and discussed. Each new generation apparently became interested, tried it, found it wanting, and then rejected it. The first of these periods, as we have seen, began in the 1820s. Zinc was enthusiastically received but apparently had largely fallen out of use by the 1840s.[44]

However, in the 1850s a new wave of enthusiasm swept the country, evidently largely the result of the exceptional sales techniques of the Vieille Montagne Zinc Mining Company, the leading Belgian firm which took full advantage of that newly popular advertising device—the trade catalog.[45] Within their pages are impressive lists of the many buildings in Europe that had been covered with the company's zinc; mention is also made of a few American buildings, such as the roof and dome of St. Vincent de Paul Church in New York (Fig. 8.8), illustrated in their 1851 cata-

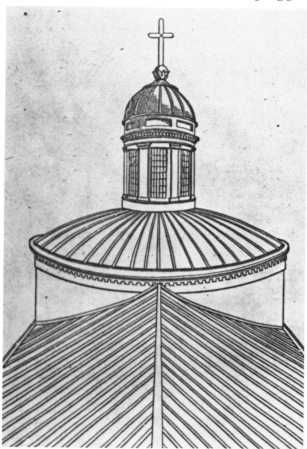

Fig. 8.8　Roof and dormers of the Saint Vincent de Paul Church in New York, which were covered with sheet zinc, laid with standing seams. Illustration from "Zinc Roofing" (New York: The Vieille Montagne Zinc Mining Company, 1851) p. 3.

log. The catalog also includes pages of written directions and beautifully detailed drawings showing how to install the roofing in virtually any situation. It was also suggested that under some circumstances the sheets be corrugated.

But it was not long before the tide of opinion once more began to turn against zinc roofing. In 1857 Calvert Vaux dismissed it completely in three words: "Zinc," he wrote, "is worthless."[46] An idea of the problems encountered in using zinc during this period was provided by those parties who in the 1870s were trying to promote its use once again. They claimed it had been utilized "in the wrong place and in a wrong manner" and that the gauge used was too thin.[47]

GALVANIZED IRON

If not too widely used by itself, zinc, when combined with iron in the form of galvanized iron, did prove to be a universally popular roof covering.[48] Coating iron with zinc had been suggested during the mid-eighteenth century, but galvanizing was not developed into a practical industrial process until 1836 by the French chemist I. M. Sorel. In this process, an alloy of zinc and iron is formed at the surface of the iron sheet; it is this oxide that gives the galvanized iron its important property of resisting rust, not simply the zinc coating as is the case with the tin coating of tinplate.

One of the important American firms that dealt in galvanized iron was Marshall Lefferts and Brother, a New York City Company. Like the Belgians, Lefferts used the trade catalog to promote his wares. His catalogs explained in words and pictures how to use the product. His catalog of 1854 showed the recommended way of laying flat sheets, with their edges joined over wooden rolls(Fig. 8.9). These galvanized sheets measured 24 by 72 inches and were thus much larger than contemporary tinplates. Lefferts also sold corrugated galvanized iron sheets, which were stiff enough that roof boarding was not necessary.

Galvanized iron was used on many fine structures. Apparently one of the first was the Merchant's Exchange in New York, where it was used for both the roof and the gutters.[49] In the Church of Annunciation (Fig. 8.10), also in New York City, architect Owen G. Warren reported that he had specified it for the entire roof, the spire, the cross, and even the battlements of the front gable where, he felt, it would equal granite in its durability.[50] Galvanized iron also was very widely used on the roofs of industrial complexes. Another use, increasingly popular as the years passed, was in cornices, window lintels, and other ornate trimmings.

Other sheet metals used in early America were copper, the aristocrat of roofing materials, and sheet iron. Sheet copper, originally for sheathing the bottoms of wooden ships,

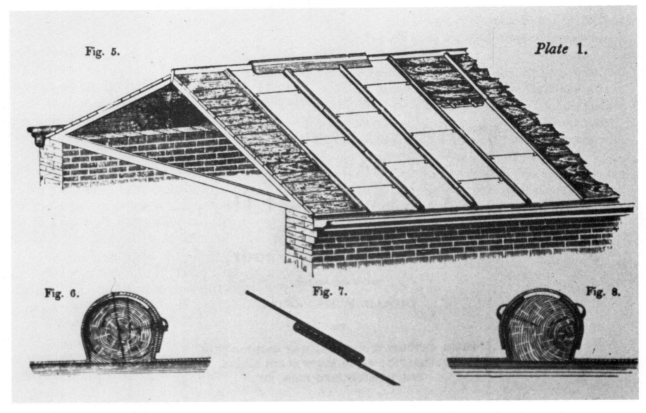

Fig. 8.9 Illustration from Marshall Lefferts and Brother catalog of 1854, showing roof of plain galvanized iron sheets laid with standing seams over wood rolls. From "Patent Galvanized Iron . . ." (New York: William C. Bryant and Co., 1854).

came into use as a roofing material just before the Revolution. The new roof of the New York City Hall, put on about 1763 or 1764, was covered with rolled copper ordered from Bristol, England. The lantern of the Sandy Hook lighthouse was soon roofed with the same material. About 1774 the Maryland State House in Annapolis was covered with copper. It was followed by the First Bank of the United States in Philadelphia (1796) and Bulfinch's State House dome in Boston (1802). But except for minor uses, such as flashings and downspouts, copper did not come into common architectural use until well into the nineteenth century with the development of the Lake Superior mines.[51]

COMPOSITION ROOFING

Along with the upsurge of interest in pitched roofs of complex forms during the mid-nineteenth century, there was increasingly widespread use of flat roofs, especially for commerical blocks, rows of townhouses, and industrial complexes. Board and shingles were used on roof flats but they warped and leaked. Later, various sheet metals were used. The problem inherent in both of these systems was the small size of the pieces or sheets, which resulted in a large number of cracks or seams where water could seep through. What was needed to alleviate this problem was a type of roofing that had fewer seams and would approach the ideal of a one-piece continuous membrane. These requirements led to the development of composition or built-up roofing.

Usually composition roofing, as built-up roofing was generally termed during the nineteenth century, consisted of pieces of cloth, felt, or paper that were first saturated with a tar-like substance and then nailed to the roof.

Fig. 8.10   Church of the Annunciation, New York City. Galvanized sheet metal was used for covering the entire roof, as well as for the spire, the cross and the battlements of the front gable. From *Putnam's Monthly*, Vol. 2 (September, 1853) p. 244.

then the excess tar pressed out. An important advance came when the felt became available in long rolls and could be impregnated by machines.

There was even more experimentation with the viscous or cementitious substances. In Europe, considerable interest developed in using natural rock asphalt for this purpose. This asphalt could also be used for paving. Another form—the type that oozes up through the earth in natural seepages—was used in Los Angeles for waterproofing the roofs of adobe houses during the early nineteenth century and perhaps earlier.[52] Pine pitch or tar and gravel had been used in Europe as early as 1772 and in America shortly after the Revolution. In Newark, New Jersey, during the 1840s, pine pitch was being used to coat roofs covered with canvas and it was reported that this technique had been used even earlier in Boston.[53] Canvas was used on the roof of the porch of the Commandant's house at Sackett's Harbor, New York, in the 1850s.[54]

What was to become one of the leading American companies dealing in composition roofing placed an advertisement in the Cincinnati city directory in 1853, mentioning an important material that had just recently appeared in the roofing picture—coal tar. It was a byproduct of the manufacture of illuminating gas, produced by burning coal. Before the discovery of aniline dyes in 1856, coal tar was considered a nuisance by the gas manufacturers and they were forced to pay to have it carted away. This the Warrens did, and found the coal tar more economical than the pine pitch used previously.

The Warrens were an interesting family: two brothers, Samuel M. and Cyrus M., began the business in Cincinnati (Fig. 8.11). Three other brothers and a sister's husband later joined the business, and taking advantage of their number, expanded their operations to many of the leading cities—New York, Brooklyn, Boston, Philadelphia, St. Louis, Louisville, and Buffalo. The Philadelphia firm of Drexel & Company had the flat roof of their new banking house (built in 1854) covered with Warren roofing (Fig. 8.12). Three years later,

The whole roof was then coated with more of the viscous substance. Sometimes it consisted of several of these layers and was usually finished off with a coat of sand or gravel.

During the nineteenth century, there was considerable experimentation with the two key elements of this roofing material—the fiber membrane and the cementitious substance. Ship sheathing paper was used, as was stout canvas, felt, burlap, pasteboard, and even woven strips of paper or twine. Initially these materials were applied as individual sheets; often they had first been dipped by hand into the tar or whatever viscous substance was being used, sheet by sheet, and

DIANA S. WAITE

S. M. & C. M. WARREN,

MANUFACTURERS AND DEALERS IN

COMPOSITION ROOFING

MATERIALS,

NORTH SIDE PEARL ST., BETWEEN VINE AND RACE,

CINCINNATI, O.

☞ **Tarred Paper and Felt,** (saturated either with Carolina or Coal Tar,) Composition in barrels, and all other articles used in manufacturing these roofs constantly on hand, and for sale at the **lowest prices.** All articles of best quality. Printed Directions for putting on roofs furnished with materials.

Fig. 8.11 Early advertisement for the S. M. and C. M. Warren firm, which manufactured composition roofing materials. From Cincinnati city directory, 1853, p. 350.

the firm endorsed the product, stating that they "so far have had no reason to doubt its efficacy and durability, nor regret having used it, and should in building hereafter prefer it to any other roofing."[55]

The Warrens were constantly on the lookout for ways to improve their product. An important advance was made by Cyrus M. Warren, who invented a process of fractional distillation that was applied to coal tar. Distilled coal tar was quickly recognized as being far superior to the crude coal tar previously used.

Of course the Warrens were not the only ones to be aware of coal tar and composition roofs. In fact, the American patent office was so flooded with applications for new roofing and paving compounds that finally a digest of those patents granted up through 1875 was published.[56]

The story of composition roofing is thus one of experimentation and of a constant search for new and better materials — the very qualities that are also characteristic of the history of American roofing as a whole. With virtually every material there was a constant attempt to improve or expedite its manufacture and its

Fig. 8.12 The Drexel Building, Philadelphia, built in 1854. The flat roof of the building was covered with the Warrens' "Fire and Water-Proof Roofing." *(Artist's copy from the Baxter street directory in the Dreer Collection: n.d., Historical Society of Pennsylvania)*

installation; there was also a constant search for new materials. As much experimentation was occurring with roofing as with any other aspect of building construction, perhaps even more.

Especially important were the technological advances made in shingle manufacture, in sheet metal fabrication, and in composition roofings. With many of these materials, there was also a slow but steady movement away from imported materials. But while these technological changes may also have helped create products that were cheaper in price and more readily available, quality was sometimes compromised. As early as 1847, for instance, warnings were published about the deficiencies of sawn shingles. The tinplate industry underwent may changes and so-called refinements, but the late nineteenth-century literature is filled with references to the good, "old-time" tinplate.

But no matter what technological advances were made with the roofing materials, the success of any one material rested largely upon the integrity of the roofer himself: the best material was only as good as the workmanship that went into the roof. The craftsman in every age has played a critical role.

NOTES

1. Because of space limitations, it is not possible to discuss the many materials in much detail. For general discussions of roofing materials, *see* Ernest G. Blake, *Roof Coverings, Their Manufacture and Application* (New York, 1925) and the *Bulletin of the Association for Preservation Technology,* Vol II (Nos. 1–2, 1970), a special edition devoted to historic types of roofing.
2. E. B. O'Callaghan, comp. and trans., *Laws and Ordinances of New Netherland, 1638–1674* (Albany, 1868), p. 82.
3. Fiske Kimball, *Domestic Architecture of the American Colonies and the Early Republic* (New York, 1966) pp. 24–25.
4. In 1631, for example, Governor Thomas Dudley of Massachusetts wrote that "noe man shall build his chimney with wood, nor cover his house with thatch" ("Boston Building Ordinances, 1631–1714", *Journal of the Society of Architectural Historians,* Vol XX [May, 1961] p. 90). Similarly, in New Amsterdam in 1656, Director General Peter Stuyvesant and the council decreed that no houses were henceforth to "be roofed with straw or reeds" (I. N. Phelps Stokes, *The Iconography of Manhattan Island,* Vol IV [New York, 1922] p. 163).
5. As late as 1873 John Bullock in *The American Cottage Builder* (Philadelphia, 1873, p. 51 ff) remarked that thatch was "admirably adapted . . . for the humble kind of dwellings" but noted that it was then falling into disuse.
6. In January, 1630, Kiliaen Van Rensselaer expressed his intent to build a tile and brick yard in Rensselaerwyck. From A. J. F. Van Laer (ed.), *Van Rensselaer Bowier Manuscripts,* (Albany: 1908), p. 160. Paul R. Huey supplied this reference.
7. The provincial council, for example, extended the time for covering William Pietersen de Groot's house in New Amsterdam until tiles could be received from Holland or Fort Orange. Quoted in I. N. P. Stokes, *The Iconography of Manhattan Island,* Vol IV (New York, 1922) p. 188.
8. While it has not been definitely established that the tiles excavated at Fort Orange were actually made near Albany, they are composed of the red clay which is characteristic of that area and which is very similar to the clay used in the bricks of the house, which may have been burned locally. A neutron activation analysis of clay samples is scheduled to be carried out shortly to determine the source of the clay used in the bricks and the tile. Excavations carried out at Fort Orange and other Dutch sites in the Albany area have yielded only pan tile and no plain—or flat—tile.
9. Rita Susswein Gottesman, *The Arts and Crafts in New York, 1726–1776* (New York, 1938) pp. 84, 187–8, 191.
10. Rita Susswein Gottesman, *The Arts and Crafts in New York 1800–1804,* (New York, 1965) p. 201.
11. Jean Carl Harrington, "Early Brickmaking, Roanoke Island, North Carolina, 1585," *Journal of the Society of Architectural Historians,* Vol XXV, No. 4 (December, 1966) pp. 301–302. Henry Chandlee Forman, *The Architecture of the Old South,* (Cambridge, 1948). Examples recorded at Jamestown during the excavations of the 1930s may be seen in the HABS records at the Library of Congress. — Ed.
12. "On the Comparative Expense of Covering the Roofs of Houses with Different Materials," *Journal of the Franklin Institute,* Vol. X (October, 1830) p. 256.
13. Ironically, the break was so complete that by the 1850s the designers of the popular Italian villas found themselves in an odd predicament. While they were able to accommodate most other parts of this type of structure to the American situation, the need for a richly textured, high relief roof covering in a non-tile producing country presented problems. The answer for many such villas was tinplate roofing, laid on with boldly rendered standing seams. In the late nineteenth and early twentieth centuries the manufacture of roofing tile in America was revived, and tile was often used for its striking architectural effect.
14. Horatio Gates Spafford, *A Gazeteer of the State of New-York* (Albany, 1813) p. 119.
15. The roofs of both houses were later covered with sheet metal.
16. Marcus Whiffen, *The Eighteenth Century Houses of Williamsburg* (New York, 1970) pp. 68–70.
17. Since the house was not being well maintained at the time of this view (1834), it is believed that the roofing dates from the eighteenth century.
18. A. J. Downing, for example, noted that "some character is given to the roof . . . by employing shingles of a uniform size, and rounding the lower ends before laying them on the roof", in *Cottage Residences, Rural Architecture & Landscape Gardening* (Watkins Glen, N. Y., Library of Victorian Culture), a reprint of the 1842 edition.

19. Sidney I. Pomerantz, *New York, An American City 1783–1803* (New York, 1938); Studies in History, Economics and Public Law, ed. by the Faculty of Political Science of Columbia University, No. 442, p. 232.

20. Charles E. Peterson, "California Brea Roofs", *The Masterkey*, Los Angeles, Vol. 39 (July–September, 1965) p. 114.

21. Whiffen, *op. cit.*, p. 212, and Nina Fletcher Little, *American Decorative Wall Painting 1700–1850* (Sturbridge, Mass., 1952) p. 2.

22. George H. Adams & Sons, *New Columbian Rail Road Atlas and Pictorial Album of American Industry* (New York, 1874) p. 109.

23. "On the Comparative Expense of Covering the Roofs of Houses with Different Materials," p. 256.

24. Hugh Morrison, *Early American Architecture*, (New York, 1952) p. 36. *See also* Edward Rowe Snow, *The Romance of Boston Bay*, (Boston, 1944).

25. The slate came from inland Lancaster where the quarries were worked until about 1825 or 1830 and then closed—to be reopened later. George P. Merrill, *Stones for Building and Decoration*, (New York, 1891) has a great deal of historical value. *See also* Peter Whitney, *The History of Worcester County*, (Worcester, 1793).

26. By 1806 the Pennsylvania Slate Company could claim that "a great number of the best and most elegant Buildings in the city of Philadelphia, likewise the Trenton Bank, in New Jersey" were using their Northampton County slate. The company furnished the mechanics and installed roofs for $15.00 per square. *The Trenton Federalist*, April 14, 1806.

27. Charles Scarlett, Jr., "Governor Horatio Sharpe's Whitehall," *Maryland Historical Magazine*, XLVI, No. 1 (March, 1951) p. 15. The slates were hung with small wooden pegs and buttered with mortar to make the roof tight.

28. Gottesman, *The Arts and Crafts in New York 1777–1799*, p. 204.

29. Gottesman, *The Arts and Crafts in New York, 1800–1804*, p. 194.

30. A near disaster occurred in Philipse Manor in Yonkers, New York, where this fact was not taken into account. During the early twentieth century, a wood shingle roof was removed and replaced with one of extremely heavy slates. The brick walls of the building could not bear this added weight and slowly began shifting. Once the situation was analyzed, emergency tension cables were installed, the slate removed, and new wood shingles installed. The building is now stable again.

31. For a listing of early slate quarries opened in the United States and a discussion of quarrying methods, see Harley J. McKee, "Slate Roofing," *Bulletin of the Association for Preservation Technology*, Vol II (Nos. 1–2, 1970) pp. 77–84.

32. Another factor that very rarely concerns us today, but was of considerable importance to many of our forefathers, was the likelihood of the surface to taint the rain water which ran off the roof to the cistern. Slate also contributed to that desirable quality of "picturesqueness" through the color it lent to a roof. The fact that there was a wider range of colors available from American quarries than from Wales also helped stimulate the demand to open American quarries. The newly popular mansard roof was another effective showplace for slate and its colors and shapes. Throughout the mid-nineteenth century slate was used on many other buildings, such as industrial structures, where its ornamental qualities were secondary.

33. For a more detailed discussion of tin roofing, *see* Diana S. Waite, *Nineteenth Century Tin Roofing and Its Use at Hyde Hall*, 2nd ed. (Albany, N. Y.: New York State Parks and Recreation, 1974).

34. Quoted in Elva Tooker, *Nathan Trotter, Philadelphia Merchant 1787–1853* (Cambridge, 1955) p. 126.

35. During the nineteenth century, nearly all tinplate in America was imported from Britain. There were a few abortive attempts to establish tinplate works in the United States during the latter half of the century, but they could not compete with the imported product and they failed. The only apparent exception to this pattern was the N. and G. Taylor Company of Philadelphia. Since 1830, and perhaps earlier, they had been producing "leaded" roofing plates. These were not terne plates in the pure sense of the word, for they were simply imported English tinplates that had been dipped into a bath of molten lead. The McKinley Tariff on imported tinplate put into effect in 1891 had been devised to stimulate establishment of the American tinplate industry, and it proved very successful. Three years later virtually no tinplate was imported to the United States.

36. *Thomas Jefferson, Architect* (New York, 1968) pp. 193–4.

37. Charles E. Peterson, "Iron in Early American Roofs," *The Smithsonian Journal of History*, Vol 3, No. 3 (1968) p. 42.

38. The history of tin roofing is complicated by a matter of nomenclature, due to the fact that a similar product called terneplate was also manufactured. It was made very much like tinplate, except for the fact that the iron sheets were dipped in a bath of an alloy of lead and tin, rather than just tin. Sometimes these other plates were called terneplates, sometimes leaded plates, or roofing plates, but often they were simply called tinplates.

39. Alexander Jackson Davis, *Rural Residences*, (New York, 1837).

40. The zinc mentioned here refers to rolled sheets of zinc, not galvanized iron, which is described below.

41. Letter from Peter Augustus Jay to John Jay, written in New York, March 5, 1824, *John Jay Papers* (MS) Columbia University. Lewis C. Rubenstein provided this reference.

42. L. D. Gale, "On Zinc Roofing," *American Journal of Science*, Vol 32 (no. 2, 1837) p. 316.

43. "Zinc Roofing" (New York: Vieille Montagne Zinc Mining Company, 1850) p. 7.

44. Part of the responsibility for its demise probably rests with L. D. Gale, a professor of geology and mineralogy at New York University, who in 1836 publicly urged that its use be "entirely abandoned", because of its unsatisfactory mechanical properties and also because of the deleterious effect that it had upon drinking water. Gale based his remarks upon his professional expertise and also on his experience, as he put it, of "having unfortunately resided under a zinc roof." Some builders continued to use it, but within the next decade it was largely abandoned as a roofing material. (See L. D. Gale, "On Zinc Roofing," *American Journal of Science* Vol 32 [no. 2, 1837] p. 316).

45. Apparently the authors of pattern books were also influenced by these promotional devices, for at about this same time—the 1850s—they were also recommending zinc for use. Downing, for instance, recommended the roof of a "villa in the pointed style" be "covered with zinc, laid on a ribbed sheathing without soldering." The diagonal effect is reminiscent of medieval English leadwork. (See A. J. Downing, *The Architecture of Country Houses*, [New York, 1969], reprint of 1850 edition, p. 342.)

46. Calvert Vaux, *Villas and Cottages* (New York, 1968), reprint of 1857 edition, p. 60.

47. "Zinc", *The Sheet-Metal Builder*, Vol I (October, 1874) p. 102. A quarter century later, in 1896, another attempt was made to promote zinc, this time in the form of a series of well-illustrated articles. (W. H. Seamon, "The Application of Sheet Zinc for Roofing and Other Purposes, *The*

Engineering and Mining Journal, Vol LXII [October 24, October 31, November 7, November 14, 1896].) The editors were promoting zinc roofing because once again a new market had to be created for the material; at that time the productive capacity of American zinc mines and smelting works was far beyond the demands of the market.

48. For a more detailed discussion of the use of galvanized iron for roofing and related uses, see Diana S. Waite, *Architectural Elements, The Technological Revolution* (Princeton, 1972).

49. "Patent Galvanized Iron," (New York: Marshall Lefferts & Brother, 1854) pp. 9, 11.

50. *Ibid.,* p. 12.

51. Charles E. Peterson, "Notes on Copper roofing in America to 1802," *Journal of the Society of Architectural Historians*, Vol. XXIX, No. 4, (December, 1965) pp. 313–318. Elva Tooker also discussed the use of copper for roofing purposes in *Nathan Trotter, Philadelphia Merchant, 1787–1853* (Cambridge, 1955). Sheet lead, commonly used in Britain on first class structures, never found much favor in the American colonies, where it did not hold up under the hot sun. Nevertheless, it was tried on the flats of major buildings such as the Governor's Palace in Williamsburg in 1709 (Marcus Whiffen, *The Public Buildings of Williamsburg*, [Williamsburg, 1958] pp. 57, 58 and 214). Lead was more widely used for flashings and for lining gutters. The stripping of lead on buildings during the Revolutionary War was widely reported.

Rolled sheet iron from the Delaware Works at Morrisville was used at Philadelphia as early as 1795. Benjamin Henry Latrobe, with Nicholas Roosevelt, was rolling iron sheets on the Schuylkill just after the year 1800. As architect, he used it at Nassau Hall, Princeton, the Capitol and White House in Washington and other buildings without permanent success. See Charles E. Peterson, *Iron Roofs,* pp. 43–45.—Ed.

52. Charles E. Peterson, "California Brea Roofs," p. 114.

53. *Ibid.*

54. Betsey Warren Davis, *The Warren, Jackson and Allied Families* (Philadelphia, 1903) p. 26.

An intriguing thread that runs through the history of many of the different roofing materials is its relationship to the history of materials used for ships. Copper, galvanized iron, and zinc, for instance, were all used for ship sheathing. The close relationship of pitch and canvas to naval interests is also hard to ignore.

55. George Copp Warren, *The Part of the Warrens in the Development of Coal Tar, Petroleum Oil and Asphalt* (New York: George Copp Warren, 1928?) p. 25.

56. L. W. Sinsabaugh, *Digest of United States Patents for Paving and Roofing Compositions* (Washington: United States Patent Office ?, 1875). One formula, for instance, contained coal tar, asphalt, gum shellac, plaster of Paris, soap stone, raw rubber, turpentine, alcohol, and boiled linseed oil. Other patentees suggested even more unlikely ingredients, such as eggs, rye flour, boiled fish oil, beeswax, and one even specified three gallons of blood as a drier. Most of these were disasters, failing shortly after they were applied.

Much interest was also evidenced in the use of natural asphalt for paving, especially in Paris. In 1876 on Pennsylvania Avenue in Washington, an important experiment took place in which natural asphalt from the Trinidad Pitch Lake was successfully used for paving. This marked the beginning of the widespread use of Trinidad asphalt paving in America, and that development spilled over to the roofing industry. Once again, the Warrens were in the forefront; one of the employees of the firm later recalled that Samuel Warren, its founder, had made a practice at this time of visiting architects' offices with a cigar box containing an "alcohol cooking machine" to demonstrate the superior qualities of asphalt over coal tar.

CHAPTER
9

Window Glass in America

KENNETH M. WILSON

MR. WILSON, currently Director of Collections and Preservation at the Henry Ford Museum, Dearborn, and formerly Chief Curator of Old Sturbridge Village and The Corning Museum of Glass, has published many articles on Early American glass.

EARLY TECHNIQUES

The history of window glassmaking is not quite so long as the general history of glassmaking, but it does date back to the days of the Roman Empire and American window glassmaking has its roots there. To England, source of our first glass, the techniques of manufacture had come from France and the Rhineland by the fourteenth century.[1]

The two principal ways of making our first window glass were the *cylinder glass* method and the *crown glass* method. While the former offered the advantage of providing larger sheets, and for that reason its product was also referred to as broad glass or sheet glass, it was inferior in quality to well-made crown glass because of imperfections inherent in the process.

The cylinder glass method consisted basical-

ly of blowing and forming a cylindrical tube, cracking it off the blowpipe, scoring and removing the ends after it was cool, and then slitting the cylinder lengthwise. The cylinder was then placed in a flattening oven, reheated and flattened with the aid of a wooden block. In the eighteenth century, as illustrated by Diderot, the French encyclopedist, the cylinders formed, often called "rolls" or "muffs," were small, perhaps 8 or 10 inches in diameter by about 24 or 30 inches long. The process also differed somewhat from what developed later (Fig. 9.1), in that the lower end was cut and part of the length of the cylinder slit while the glass was hot. The cylinder was transferred to a special holder and the other end reheated and cut off, then the remainder of the length of the cylinder was slit by cutting while hot.

In the nineteenth century cylinders were blown and formed by swinging to much larger sizes—up to 16 or 18 inches in diameter and 7 feet in length. To accomplish this swinging, which extended the cylinder by means of centrifugal force, the furnace was constructed so that the glassblower stood on platforms elevated 7 or 8 feet above the floor opposite the mouth of the furnace, or worked on the floor of the glasshouse beside "swing pits" 7 or 8 feet deep dug into the floor. The flattening

Fig. 9.1 Making cylinder or broad glass. Wood engraving from "Scenes in a Glass Foundry," by Theo R. Davies, *Harpers Weekly,* January, 1884. *(The Corning Museum of Glass.)*

ovens into which the slit cylinders were placed were of several different designs, sometimes arranged in a circular pattern, sometimes in a longitudinal one. As the cylinders were prepared for reheating and flattening, they were placed on the tables, one after another, and pushed into the flattening and annealing ovens. In either form, the tables bearing the cylinders were of stone covered with a smooth coating of plaster of Paris. Despite efforts to achieve a perfectly smooth surface by this means, undulations and other imperfections persisted.[2]

The second means of making window glass was the crown glass method: a good quality product could be obtained but in limited sizes. Basically, crown glass was made by blowing a large bubble, attaching a pontil rod opposite the blowpipe, then cracking off the blowpipe, which left a hole in the bubble. This hole was enlarged by a paddle and by successively re-

heating and twirling the piece until centrifugal force caused it to open onto a flat disc or crown (Fig. 9.2).[3]

In England in the eighteenth and nineteenth centuries, the glasshouse took the form of a large cone (Fig. 9.3) with the furnace set inside. The structure of the furnace was slightly different for producing crown glass than it was for the manufacture of cylinder glass—or bottle or flint glass—and a special furnace for reheating the partially formed crowns of glass (called a flashing furnace) was a distinguishing feature of a crown glass manufactory (Fig. 9.4).[4] A part of it was a stepped wall which served as a rest for the blowpipe and pontil rod while the glass was twirled or "flashed" to expand it into a disc, or crown, with the aid of the heat from the furnace and centrifugal force (Fig. 9.5).

The layout and cutting of these crowns, or tables, was a very carefully considered pro-

Fig. 9.2 Glassblower with a completed crown or table of glass attached to the pontil rod. The crown would then be cracked off and placed in an annealing furnace and slowly cooled to remove the strain. Woodcut from Rev. George Richardson Porter, "Treatise on the . . . Manufacture of Porcelain and Glass," 16, quoted in Dionysus Lardner's *Cabinet Cyclopedia*, London, 1832. *(The Corning Museum of Glass.)*

cess (Fig. 9.6). It is described and illustrated in detail by William Cooper in his *Crown Glass Cutter and Glazier's Manual*, published by Oliver and Boyd in Edinburgh in 1835. The bull's-eye, or the center of the crown remaining after it had been cut up, is usually thrown back into the batch for cullet at the glass factory. However, in many instances, particularly in this country, carpenters or builders would often order the complete tables of glass and then cut them to required size on the job. The bull's-eyes remaining were sometimes used in transoms or other areas where light was needed but where clear visibility made no difference. In some instances a complete crown or

half of a crown was set in a gable end of a house or building as a window.[5]

SOURCES AND USES

Window glass, first brought across the Atlantic in the 1620s, continued to be imported in ever larger quantities throughout the seventeenth, eighteenth and nineteenth centuries. But as time went on, more and more of this demand was satisfied by domestic glasshouses.

The first attempts to make glass here were at Jamestown, Virginia, in 1608 and 1621: both failed. We do know that from the first attempt, a "tryal of glasse" was sent back by Captain John Smith to England. Archaeological excavations by the National Park Service

Fig. 9.3 Typical English glasshouse cone of the eighteenth and early nineteenth centuries. Another characteristic form of this type of glasshouse was the bottle-shaped cone. From William Cooper, *The Crown Glass Cutter and Glazier's Manual*, Edinburgh, 1835. *(Henry Ford Museum.)*

Fig. 9.4   Interior of an English Crown-Glass House. The melting furnace is in the center, flashing furnaces around the periphery on the left and annealing ovens on the right and center beyond the melting furnace. From Cooper. *(Henry Ford Museum.)*

Fig. 9.5   Glassblower at the flashing furnace reheating a large portion of glass during one of the stages of producing a crown. Woodcut illustration from Cooper.

have turned up small quantities of green glass, the largest pieces about one inch square. A lot of this may have been from bottles; some of it *may* have been window glass. However, the Jamestown glass was not intended primarily for use of the colonists, but rather for commercial exportation to England.

J. Paul Hudson, veteran National Park curator at Jamestown, writes from the scant documents of early Virginia and the overwhelming collection of artifacts excavated during four decades of archaeological investigations. He states that the earliest houses were built without glazed windows. Wall openings for windows in the very early structures are believed to have been closed by simple batten shutters, often operated on hinges of wood and fitted with wooden fastening devices. A little later, some of the less humble houses were glazed with small diamond-shaped panes known as quarries or quarrels. They were held in place by means of slotted lead strips known today as cames.[6] Square or rectangular glass panes—

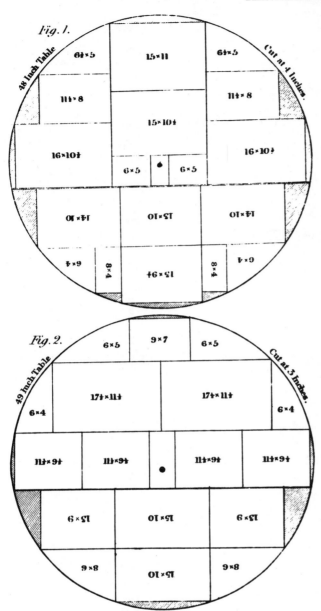

Fig. 9.6 Layouts for cutting up 48 and 49 inch tables into standard size panes. From Cooper (*Henry Ford Museum.*)

the later rectangular panes cut from the crowns of glass, were systematized as indicated in Figure 9.8. This is explained in Richard Neve, *The City and Country Purchaser and Builders' Dictionary,* the second edition of which was published in London in 1726. It is an alphabetized book for builders. Under *Glass* and *Glazier's Work* is a great deal about current practices. Neve describes the sizes and methods of cutting glass and the construction of sash. Among other things, the use of the square quarrel, ten of which make up a square foot, and the long quarrel, in which there is a lesser angle at the top and bottom points, are discussed. It was the long quarrel which was primarily used in the American colonies

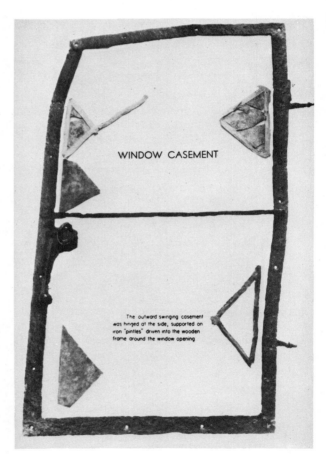

Fig. 9.7 Iron window casement with fragments of turned lead, or cames as they are called today, and diamond-shaped window panes. Height 2 feet, ½ inch. Excavated at Jamestown, Virginia. From Isabel Davies, "Window Glass in Eighteenth-Century Williamsburg," *Five Artifact Studies,* Vol 1, Williamsburg, Virginia (1973) p. 81.

somewhat thicker than the diamond shaped ones—were probably first used in the Virginia colony toward the end of the seventeenth century. Wrought iron casement frames (Fig. 9.7), once containing diamond-shaped quarrels, have been excavated at Jamestown. These frames vary in size from 2 feet, 1½ inches to 2 feet, 4½ inches high, by 1 foot, 5¾ inches wide.[7]

The sizes and shapes of such quarrels, like

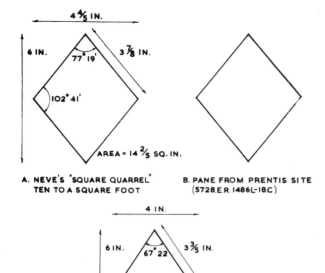

A. NEVE'S "SQUARE QUARREL"
TEN TO A SQUARE FOOT

B. PANE FROM PRENTIS SITE
(5728.E.R.1486L-18C)

C. NEVE'S "LONG QUARREL"   TWELVE TO A SQUARE FOOT
NO PANES OF THIS SHAPE HAVE BEEN FOUND IN
WILLIAMSBURG

Fig. 9.8 Forms and sizes of diamond-shaped panes as dimensioned by Richard Neve, 1726. Illustration from Isabel Davies, "Window Glass in Eighteenth-Century Williamsburg," p. 79.

throughout the seventeenth century, although toward the end some square ones were also used. A number, mostly of the long type, have been excavated in Jamestown and Williamsburg. The remains of at least two casement windows, excavated from beneath the present Nassau Street in Williamsburg, were from houses built in the late seventeenth or very early eighteenth century, prior to the establishment of the present street plan for the city.[8]

In New England, glass had come into use at an early date. We know that in 1629 Higginson, writing back to his friends in England as to what should bring with them, advised: "Be sure to furnish yourselves with glass for windows,"[9] and in 1634 William Wood wrote in his *New England Prospect*, urging immigrants: "glasse ought not to be forgotten of any that desire to benefit themselves, or the Countrey: if it be well leaded, and carefully

pak't up, I know of no other commodity better for portage or sayle."[10]

Fragments, and occasionally whole units, of casement sash have been found in New England. Examples have been described and illustrated by architectural historians, notably J. Frederick Kelly in *The Domestic Architecture of Connecticut* (New Haven, 1924) pp. 87–90.

The Hudson River community very early produced some elaborate figured glasswork, including stained glass in the Dutch manner. In 1638 Evert Duyckink came over to New Amsterdam, establishing a family line of painters, glassmakers and master glaziers. The late R. G. W. Vail in an essay, "Storied Windows Richly Dight," illustrates a number of elaborate and competent heraldic and "proverb" windows installed in the churches of Manhattan and Albany. These were unique in the New World.[11] Some elaborate lead camework recently excavated from the Schuyler Flatts site near Albany corresponds roughly to that delineated in the paintings of Johannes Vermeer (1632–1675).[12]

English Pennsylvania was founded too late for us to expect any complete examples of the early type. But architect John D. Milner, while

Fig. 9.9 Thomas Massey House, Delaware County, Pa. Fragments of window glass and lead cames excavated at site. The panes were 4 by 6 inches and of a thin, greenish glass. (*Eisenmann photograph, courtesy John D. Milner AIA.*)

Fig. 9.10 Massey House fixed window restored. The original head and jamb frame members were present until restoration began (only the former could be used) and clearly revealed the conformation of the camework. The half panes were presumably cut on the job.
(Eisenmann photograph, courtesy John D. Milner AIA.)

recently restoring the little brick Thomas Massey House (built ca. 1696) in Delaware County,[13] found in the soil fragments of both small, thin rectangular glass panes and of lead cames (Fig. 9.9). These provided models for new windows, which were fitted into some of the original frames (Fig. 9.10).

SASH WINDOWS

At the turn of the century, especially with the coming of the Georgian style, there was a general change from the medieval type of window to those with vertically sliding *sash.* Many of these sash were fixed—that is, only one of the pair could be pushed up and down. It was secured by a stick or some other simple means.

Professor Charles F. Montgomery ran across some interesting mentions in the manuscript letterbooks of Thomas Bannister, a merchant of Newport, Rhode Island. In a letter dated April 29, 1705, written during a visit to Boston, he noted that "Sash windows are the newest Fashion." Ten years later in a gesture to the new technology, Bannister "open'd his sashes to whom the Govr. made a deep bow" as he rode down the street.[14]

Later in the eighteenth century, as window glass became much more common, it also became somewhat less costly and there was a growing demand for larger panes. In 1737 merchant Thomas Hancock of Boston sent for "380 squares of best London crown glass, all Cutt Exactly 18 Inches long and 11 Inches wide of a Suitable Thickness to the Largeness of the Glass, free from Blisters and by all means be careful it don't wind or worp." He also ordered "100 Squares Ditto, 12 inches Long, 8-1/2 wide, of the Same Goodness as above."[15] As the available supplies of window glass increased, the fenestration of numerous earlier houses was revised. In 1737, the roof of the first parish meeting house in Lynn, for example, was altered and the windows changed from leaded to wooden sashes.[16]

Some fifty Boston newspaper advertisements offering window glass were collected by George Francis Dow for *The Arts and Crafts in New England.*[17] London, Bristol and Newcastle seem to have supplied most of it, shipped over in boxes and cases. Sheet and crown glass are mentioned, the shapes both diamond and rectangular. The earlier glass was often associated with "window lead" in the notices. The later glass ranged from 4 by 6 inches up to (in 1772) large ones, 19 by 24 inches in size,

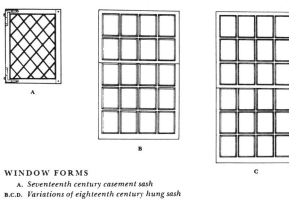

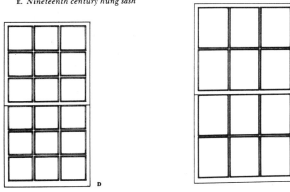

WINDOW FORMS
  A. *Seventeenth century casement sash*
  B.C.D. *Variations of eighteenth century hung sash*
  E. *Nineteenth century hung sash*

Fig. 9.11 Window forms in America, seventeenth through nineteenth centuries. From drawings by K. M. Wilson in Abbott Lowell Cummings, *New England Architecture*, Old Sturbridge Village booklet.

though the latter were perhaps intended for framing pictures. It is noteworthy that "Diamond Quarreys by the Cribb" were still being sold up until the Revolution, presumably for the repair of old windows. Ready made glazed sash "suitable for hot beds" were also coming in at that time.

Rita Susswein Gottesman did the same service for us in searching the early newspapers of New York City. Richard Wistar of Philadelphia on August 17, 1769, advertised four hundred boxes of his window glass in New York consisting of "the common sizes": 6″×8″, 7″×9″, 8″×10″, 9″×11″, and 10″×12″.[18] By that time these sizes were more or less standard everywhere in the colonies. But they were increasing. Thirty years later Matthew Hawkins, who also sold ready-glazed sash, had in stock panes 8″ × 10″, 9″ × 11″, 10″ × 12″ and 10″ × 14″.[19]

DOMESTIC SOURCES

Some twenty attempts were made to establish glasshouses here in the eighteenth century, mostly to produce window glass and/or bottles.[20] Only four or five of them were even moderately successful. Of them, three are well remembered today. The first was Caspar Wistar, who established a small glasshouse at Alloway in South Jersey in 1739. To do so he imported workmen from Rotterdam and agreed to share profits with them. The factory was carried on by his son until 1780. The Wistars sold not only window glass they produced, but also that imported from England.[21] William Henry Stiegel made some window glass from 1765–1774 in his Elizabeth Furnace and at his two Manheim factories near Lancaster, Pennsylvania, as did the factory established in Kensington, Philadelphia, in 1772.[22] The third of the best-known glassmakers of the eighteenth century was John Frederick Amelung, a practical glassmaker. He landed in Baltimore in August of 1784 with sixty-eight workmen and equipment to build three "glass-ovens" (probably three different sets of glass furnaces). The remains of his factory in New Bremen, about 9 miles south of Frederick, Maryland, were excavated in 1962–63 by The Corning Museum of Glass and the Smithsonian Institution under the archaeological director, Ivor Noël Hume, of Colonial Williamsburg. Within the foundations of Amelung's extensive factory were found a total of fourteen furnaces, including a melting furnace where bottles and window glass were produced, fritting ovens for pre-preparation of the batch material, annealing ovens and rooms for pot making and other activities associated with glassmaking. Amelung advertised in the *Maryland Journal and Baltimore Advertiser* of February 11, 1785, ". . . that said Glass Manufactory will consist in making all kinds of Glass-Ware, viz., window glass from the lowest to the finest sorts, white & green Bottles, Wine & other Drinking Glasses, . . ."[23]

Another establishment—not so successful as the builders of the new State House in Hartford would have liked—was the Albany Glass

Works. Unfortunately, the committee found that when its glass was delivered, there were two problems: 1) approximately one third of the glass was broken and 2) the glass was so dark that a wag indicated that if a nocturnal meeting place were needed, they had with the use of this glass, one on a 24-hour basis. Some of the original panes still to be seen in this building attest to that problem and also to excessive irregularities such as striations and "cords."[24]

One of the most successful window glass factories was the Boston Crown Glass Manufactory. The company was organized in 1787 but did not begin producing glass until 1792, when a complement of workmen was recruited in Germany. The original furnace was torn down and one of a different construction erected. This company developed a very fine reputation for their crown glass, which appeared in a number of buildings. They advertised, for example, in 1799, "The Boston Crown Glass Manufactory offers the best window glass at their manufactory on Essex Street. You may look at the new State House and the new Branch Bank in Boston, both of which are glazed with the glass from this particular factory."[25] There are numerous other instances of the use of their glass in public buildings. According to a glassblower, John Tates, who worked there in his earlier years: "the company made good and brilliant glass of a light, bluish white color that was quite thick and strong. The pots contained enough glass to make 80 or 90 crowns or tables, each 48" to 50" in diameter. The worker took 25 pounds of metal on the blow pipe, which weighed 18 to 20 pounds."[26] This statement is corroborated by that of Richard Bentley, who began to work at the Company's Essex Street factory in 1815 and continued there until 1824. He said in an interview some years later:

At Essex Street blowers made pots and bricks for furnaces of six pots each; each pot made about 75 sheets to the pot, 48" in diameter. It took six months to make glass for the Capitol at Washington after it was burned. They made two melts, Monday and Thursday. The government could have got English plate at half the price but would not. The company never made any money on the lot, as it was double thick and was badly broken in blowing . . . thirty-two pounds would be taken on the pipes to make a sheet 54" in diameter.[27]

Despite these and other attempts to produce window glass here, imported products fulfilled most of the growing demand in the eighteenth century, as already noted. There is evidence that the importations of window glass came from many sources after the Revolution; many advertisements attest to this. In the *New Hampshire Gazette* of June 1, 1782, J. M. Sparhawk advertised among other things, "German steel, nails and glass." William Allen on April 10, 1777, in Boston advertised glass for windows imported from France.[28]

Thomas Jefferson wrote the merchant Joseph Donath in Philadelphia on October 9, 1807, ordering 250 panes 12 by 18 inches and 150 panes 12 inches square, of Hamburg or Bohemian glass "of the middle thickness." Earlier, on September 16, 1795, he had written to Donath from Monticello ordering "350 panes of Bohemian glass 18 inches square" and "25 panes do 18 by 24 inches to be of 1½ thickness."[29] Later, after the War of 1812, there are also references to glass having been imported from Sweden and even from Russia.[30]

SOME FANCY TYPES

In addition to the decorative use of glass in fanlights and sidelights in the late eighteenth and early nineteenth centuries, other types of windows came into use in decorative ways during the 1830s. An advertisement of the Pittsburgh Flint Glass Manufactory of Bakewell & Company, about 1840, offers for sale, in addition to numerous other types, window glass and "cut and pressed Panes for steamboats."[31] One of these, marked Bakewell, is shown in Figure 9.12. In addition to apparently having been customarily used on steamboats, probably in partitions, these panes are known to

Fig. 9.12 Lacy pressed glass window pane marked "Bakewell." Made at Bakewell, Page & Bakewell's Pittsburgh Flint Glass Manufactory, Pittsburgh, Pennsylvania, about 1835–1850. Height 6¾ by 4¾ inches wide. *(Photograph, collection of the Henry Ford Museum.)*

Fig. 9.13 Lacy pressed glass window pane made at the Boston & Sandwich Glass Works, probably between about 1835 and 1865. Height 13⅓ by 9½ inches wide. *(Photograph, Collection of Henry Ford Museum.)*

have been used in secretary desks. A competitor in Wheeling, West Virginia, the J. C. Ritchie Company, also produced similar pressed glass panes; one, at least, bears a steamboat as part of its design, probably indicating its association with such boats. Decorative pressed glass panes were produced by other companies as well and sometimes used for other purposes, such as lanterns. The Boston and Sandwich Glass Works also produced these pressed glass window panes (Figure 9.13) in varying designs up to about the 1870s. In the late 1800s and early 1900s, colored pressed glass panes of an entirely different character bearing geometric, floral and bird designs were produced by several factories, including one in Addison, New York.[32]

Pressed glass cup plates, which were in high fashion and in widespread use in America

between about 1825 and 1850 were, at least in several instances, also used in conjunction with plain window glass to create effective and very decorative fanlights during that era. One such fanlight still exists in a house in Westford, Connecticut; two others are in the collection of the Henry Ford Museum (Figure 9.14).

The Gothic revival style of architecture provided another type of decorative fenestration in the form of pointed Gothic windows including, in some cases, leaded glass panes. Lyndhurst on the Hudson in Westchester County, designed by A. J. Davis in 1838 and enlarged by him in 1861, well illustrates one of the more pretentious houses in this style of architecture. Its Gothic-shaped windows are divided in some cases into lancets and, in others, are in the form of roundels with trefoil decoration. Both are part of the fenestration

Fig. 9.14 Fanlight, composed in part of thirteen Henry Clay cup plates dating from about 1830–1850 and attributed to the Boston and Sandwich Glass Works. One of several known decorative fanlights using such plates. Length: 52 inches, height, 21 inches. *(Henry Ford Museum.)*

characteristic of this style. In smaller houses, the Gothic revival style is best epitomized by A. J. Downing's books illustrating cottage architecture (Fig. 9.15).[33]

STAINED GLASS

Stained glass was a more common decorative type of window glass revived for use in the

1830s. It is mentioned in advertisements during that period by several glass companies. On July 19, 1831, the *Boston Daily Evening Transcript* carried an article about the Boston & Sandwich Glass Works and described in glowing terms a window stained and painted after the manner of the ancients which had been executed at that flourishing establishment. The reporter noted that:

the coloring is magnificent, the design chaste and the whole beautiful. In the center is the head of Christ after Guido and the coloring of which is, he thinks, equal to any oil painting of that master. This is high praise, and if it emanates from one who is familiar with the works of Guido, and believes that he asserts, we have made unparrallelled progress in an art exceedingly difficult in acquisition, which has been, although erroneously, believed to have been lost. We have no doubt, however, that the window spoken of merits commendation. We have seen some specimens of stained glass from this manufactory which were truly elegant and convinced us that the power to execute an elaborate work was not wanting.

Fig. 9.15 Rural residence in the Gothic Revival style, from Andrew Jackson Downing, *Cottage Residences . . .* 3rd ed., New York and London, 1847, fig. 61. *(Robert Hudson Tannahill Memorial Research Library, Henry Ford Museum.)*

In 1832, an article in the *New Bedford Mercury* noted that the New England Glass Company was also making stained glass.[34] By the mid-nineteenth century, there were a number of firms producing decorative stained glass for over door panels, side lights and partitions. Numerous examples of stained glass made or decorated by American firms were exhibited at the New York Crystal Palace Exhibition in 1853.[35] Illustrated catalogs also show a wide variety of cut, engraved and obscured window glass. From about 1850 onward, many of these cut and engraved decorations were done on "cased" glass—a sheet of plain glass covered by a thin layer of colored glass, usually red or blue. The design thus appears as colorless against the red or blue background.

In Philadelphia, St. Stephen's Episcopal Church, designed in the Gothic style by William Strickland and built in 1822–23, had some stained glass imported from England.[36] Local manufacture was perhaps not tried until the 1830s, as the accounts above indicate, and even by the mid-nineteenth century the art of making stained glass in the country was still considered novel, as the following suggests. William D. Jones of Philadelphia produced a specimen described by the *Public Ledger* for January 24, 1850, as:

> a landscape, preserved with remarkable force and brilliancy. The difficulty of doing this may be inferred from the fact, that the glass has to be heated so hot as to incorporate the colors with it, so that they cannot be rubbed off by any process. This is a beautiful art, and may be employed in decorating churches and other public and private buildings in an elegant and tasteful manner.

The next year Charles P. Fox's new five-story cast iron front Cornucopia Hotel on Third Street had "stained glass partly dimond shape" across its large front windows, as well as in two small windows with "dimond shaped glass figured" over the bar.[37] By 1859 Philadelphia had two large manufacturers of stained glass. One of them was the J. & G. H. Gibson Company, which produced for Thomas U. Walter the great glass ceilings of the House of Representatives and the Senate Chamber in Washington "done in rich colors, which give the effect of Mosaics set in silver."[38]

Through the years many special types of decorative glass were evolved. In 1843, Matthew Lane at 97 State Street, Boston, sole agent for Redford and Saranac Crown Glass, carried double-thickness glass recommended for skylights, steamboats and lighthouses (Fig. 9.16). He also sold *bent* glass for bow-windows (just give "the sweep of pattern on paper, by mail or otherwise") and *ground* glass for artists' studio windows.[39] And in the city directory for Newark, New Jersey, for 1859–60 we find Cooper & Belcher "Inventors and Manufacturers of New Engine Engraved, Flocked, and Vitrified Enamelled Glass for Churches and Dwellings, Bent Glass, &c."[40]

PLATE GLASS

A demand for *plate* glass to be used in shop fronts developed in the nineteenth century—first abroad, then in America—as interest in larger display windows grew. Since there were limits to the sizes that glass made by the cylinder method could achieve (36 by 48 inches being about the largest in view of its thickness), plate glass was required. A shop front in London supported by narrow iron columns was glazed with plate glass in 1834.[41]

Plate glass came into use in America in the late 1830s. The *New Bedford Daily Mercury* for April 10, 1838, carried the following announcement. "The window glass in the New Custom House in New York is to be all of the first quality plate glass. There are to be 1200 panes of various sizes from 15″ by 22″ to 38″ × 42″. The smaller panes will be a quarter inch thick and the larger ones one-half inch thick." To its Honolulu readers, *The Polynesian* reported on December 21, 1844, that "Petit's Shawl store in Boston has plate glass windows of but one pane each containing 48 square feet, while H. P. Stewart's store located on Broadway, New York between Chambers and Reed had French plate glass seven feet wide by eleven feet, two inches high."

REDFORD GLASS COMPANY'S
PRICES CURRENT OF CROWN WINDOW GLASS,
MANUFACTURED AT REDFORD, CLINTON CO., N. Y.

REDFORD

SIZE.	Wholesale Price per Light.	SIZE.	Wholesale Price per Light.	SIZE.	Wholesale Price per Light.	SIZE.	Wholesale Price per Light.	SIZE.	Wholesale Price per 100 ft.
22 by 17	$1 25	19 by 15	$0 74	16 by 14	$0 54	13 by 13	$0 36	10 by 8	$16 00
22 .. 16	1 04	19 .. 14	0 69	16 .. 13	0 50	13 .. 12	0 33	9 .. 9	16 00
22 .. 15	0 95	19 .. 13	0 63	16 .. 12	0 45	13 .. 11	0 30	9 .. 8	16 00
22 .. 14	0 87	19 .. 12	0 57	16 .. 11	0 40	13 .. 10	0 26	9 .. 7	15 00
22 .. 13	0 80	19 .. 11	0 53	16 .. 10	0 36	13 .. 9	0 23	8 .. 8	16 00
22 .. 12	0 73	18 .. 15	0 69	16 .. 9	0 33	13 .. 8	0 19	8 .. 7	14 00
21 .. 16	0 96	18 .. 14	0 63	15 .. 13	0 44	12 .. 12	0 30	8 .. 6	13 00
21 .. 15	0 87	18 .. 13	0 59	15 .. 12	0 40	12 .. 11	0 26	7 .. 6	10 00
21 .. 14	0 80	18 .. 12	0 55	15 .. 11	0 36	12 .. 10	0 23	7 .. 5	10 00
21 .. 13	0 73	18 .. 11	0 50	15 .. 10	0 33	12 .. 9	0 18	7 .. 4	9 00
21 .. 12	0 67	18 .. 10	0 45	15 .. 9	0 30	12 .. 8	0 15	6 .. 4	9 00
20 .. 16	0 87	17 .. 15	0 63	15 .. 8	0 27	11 .. 11	0 23		
20 .. 15	0 80	17 .. 14	0 58	14 .. 13	0 40	11 .. 10	0 18		
20 .. 14	0 73	17 .. 13	0 54	14 .. 12	9 36	11 .. 9	0 15		
20 .. 13	0 67	17 .. 12	0 50	14 .. 11	0 33	11 .. 8	0 12		
20 .. 12	0 60	17 .. 11	0 45	14 .. 10	0 30	10 .. 10	0 14		
20 .. 11	0 57	17 . 10	0 40	14 .. 9	0 26	10 .. 9	0 12		

Terms.—Under $100, cash: $100, 60 days: $200, 90 days: $250 and upwards, 4 months.

SARANAC.

SIZE.	Wholesale Price per Light.	SIZE.	Wholesale Price per Light.	SIZE.	Wholesale Price per Light.	SIZE.	Wholesale Price per Light.	SIZE.	Wholesale Price per 100 ft.
22 by 17	94 cts.	19 by 15	55 cts.	16 by 14	37 cts.	13 by 13	27 cts.	10 by 8	$14 00
22 .. 16	78 ..	19 .. 14	50 ..	16 .. 13	34 ..	13 .. 12	24 ..	9 .. 9	14 00
22 .. 15	71 ..	19 .. 13	45 ..	16 .. 12	31 ..	13 .. 11	21 ..	9 .. 8	14 00
22 .. 14	65 ..	19 .. 12	41 ..	16 .. 11	29 ..	13 .. 10	20 ..	9 .. 7	13 00
22 .. 13	60 ..	19 .. 11	37 ..	16 .. 10	26 ..	13 .. 9	17 ..	8 .. 8	14 00
22 .. 12	55 ..	18 .. 15	50 ..	16 .. 9	24 ..	13 .. 8	14 ..	8 .. 7	12 00
21 .. 16	72 ..	18 .. 14	45 ..	15 .. 13	31 ..	12 .. 12	21 ..	8 .. 6	11 00
21 .. 15	65 ..	18 .. 13	41 ..	15 .. 12	28 ..	12 .. 11	20 ..	7 .. 6	9 00
21 .. 14	60 ..	18 .. 12	37 ..	15 .. 11	26 ..	12 .. 10	17 ..	7 .. 5	9 00
21 .. 13	55 ..	18 .. 11	34 ..	15 .. 10	24 ..	12 .. 9	14 ..	7 .. 4	8 00
21 .. 12	50 ..	18 .. 10	30 ..	15 .. 9	21 ..	12 .. 8	12 ..	6 .. 4	8 00
20 .. 16	65 ..	17 .. 15	45 ..	15 .. 8	18 ..	11 .. 11	17 ..		
20 .. 15	60 ..	17 .. 14	41 ..	14 .. 13	28 ..	11 .. 10	14 ..		
20 .. 14	55 ..	17 .. 13	37 ..	14 .. 12	26 ..	11 .. 9	11 .		
20 .. 13	50 ..	17 .. 12	34 ..	14 .. 11	24 ..	11 .. 8	9 ..		
20 .. 12	45 ..	17 .. 11	31 ..	14 .. 10	21 ..	10 .. 10	11 ..		
20 .. 11	41 ..	17 .. 10	28 ..	14 .. 9	20 ..	10 .. 9	9 ..		

Terms.—Under $100, cash: $100, 60 days: $200, 90 days: $250 and upwards, 4 months.

SIZES.—10 by 8 and under, are packed in boxes of 50 feet each. 10 by 9, to 14 by 13, are packed in boxes of 50 lights each. 15 by 8, to 18 by 15, are packed in boxes of 30 lights each. 19 by 11 and upwards, are packed in boxes of 25 lights each. All sizes cut to order, not named in the Price Current, will be at the same rate as regular sizes. Fractional parts of inches will be charged as whole inches

The **Redford Crown Glass** is made from WHITE FLINT SAND, obtained in the vicinity of the Works, and is the only Crown Glass made from *that species of Sand* in this country. It is capable of standing every change of climate—nor will it lose its lustre by age.

The REDFORD CROWN GLASS is distinguished from ordinary Crown Glass, by its uncommon evenness and beauty of surface—by its superior transparency and lightness of color—by its great thickness, and the general excellence of the materials which compose it. This Glass possesses a remarkable lustre, which causes it to reflect like the purest specimens of Plate Glass. Its surface not being polished after being blown, retains its enamel, brilliancy and hardness, and is not liable to the objection which applies to Plate Glass, of being easily and permanently bedimmed by dust; being all made of *double* thickness, will be found on comparison cheaper and stronger than Foreign or Cylinder Glass. The subscribers have spared no pains to make this one of the best establishments in the United States. Orders addressed to them or their Agents will be executed with care and prompt attention.

COOK, LANE, CORNING & CO.

Agent.
TROY, N. Y., January 9, 1837.

N. Tuttle, printer, Troy—Sign of the Printing-Press.

Fig. 9.16   Broadside for Redford Crown Glass, 1837. *(Courtesy Prof. Harley J. McKee.)*

Notable were the windows in a Mr. Taylor's saloon-restaurant, on the corner of Broadway and Franklin Streets, New York City. *The Syracuse Standard* reported on July 12, 1853, that "the windows are most noticeable, aside from the display of glass shelving, ornamental vessels, & etc., for the immense size of the window panes. The original, however, which was sixteen and one-half feet high by seven feet wide cost $3000.000, was unfortunately broken in setting." All of this plate glass was imported. The first attempt to make it here was in 1853 or 1854 when the Lenox Glass Works was established in the Berkshires of Massachusetts. Rough and polished plate glass were made under several managements until the company failed in 1872.[42]

Among the other plate glass specialties were the sidewalk lights used to illuminate vaults and rooms below. Thaddeus Wyatt, an American, was awarded a patent for them in 1856. They were made by "inserting small glass discs in perforated iron plates."[43] The Philadelphia Glass Company at the end of our period was making "Rough Plate Glass particularly rolled or hammered Glass for green-houses &c and flooring Glass."[44]

The production of American window glass increased steadily in the nineteenth century but large quantities were still imported. In general, too, it became less expensive as time went on. Despite these advantages, the use of window glass depended to a large extent upon one's socioeconomic circumstances. As late as 1856, Frederick Law Olmstead in traveling through the South spoke of the wretched slave cabins divided into two sections, each section having but one wooden window in which was set a single pane of glass.[45]

NOTES

1. Not too much window glass was made in the Roman Empire, but some was used in Pompeii, where fragments of glass in a bronze frame have been found. Other fragments have been excavated from the remains of Roman villas in England and from buildings in Wales. Those windows were apparently small panes made by a casting process, one of the three methods of making window glass. George C. Boon, "Roman Window Glass from Wales," *Journal of Glass Studies*, vol. VIII, 1966, Corning, New York, pp. 41–45.

   For a sketch of European developments see Norman Davey, *A History of Building Materials*, London, 1961, pp. 189–194. For early English documentation, L. F. Salzman, *Building in England down to 1540*, Oxford, 1967, pp. 173–186.

2. Later, in 1903, the cylinder glass method was perfected by J. H. Lubbers, who developed a mechanical means for blowing and drawing the cylinders approximately 40 feet long by 30 inches in diameter. This is called the "Lubbers' method."

3. n.a., *Conversations on the Art of Glassblowing*, New York, 1836. A juvenile book of twenty-three pages.

4. *Ibid.*

5. A half-crown window still exists in the gable end of a barn in Wiscasset, Maine.

6. Diderot's "Vitrier," plates III and IV, and William Cooper, *Glazier's Manual*, plate 18, illustrate the machines and describe the process used for casting and forming the H-shaped cames which held the diamond-shaped quarrels, and also the rectangular panes, in place within either iron or wooden casements.

7. J. Paul Hudson, "Glassmaking at Jamestown 1608–09 and 1621–24", *The Ironworker* (Lynchburg, Virginia) Vol. XXVII, No. 1 (Winter 1962–63) pp. 1–7, 22–25, cites J. C. Harrington, *Glassmaking at Jamestown, America's First Industry*, Richmond, 1952 and Charles E. Hatch, Jr., "Glassmaking in Virginia, 1607–1625", *William and Mary College Quarterly*, April and July, 1941. Mr. Hudson is currently completing for publication an essay on Jamestown houses and buildings. Examples of Jamestown casement windows are illustrated in Henry Chandlee Forman, *Virginia Architecture in the Seventeenth Century*, Williamsburg, 1957, plate J, and John L. Cotter, *Archaeological Excavations at Jamestown*, Washington, 1958, p. 170.

   That window glass was prized and considered valuable is indicated by the fact that when pirates took over a Thomas Cornwall's house in Tidewater, Virginia, in 1652, after living there for awhile, they stole not only the locks when they left but also all the windows. (Esther Singleton, *The Furniture of our Forefathers*, New York, 1901, pp. 27–29.)

8. Isabel Davies, "Window Glass in Eighteenth-Century Williamsburg," *Five Artifact Studies*, Colonial Williamsburg, Vol. I, pp. 78–99.

9. The Rev. Francis Higginson, *New-Englands Plantation or, a Short and True Description of the Commodities and Discommodities of That Countrey* (London, 1630). Reprinted by the Essex Book and Print Club, Salem, Mass., 1908.

10. Other accounts confirm the high value placed on window glass in the seventeenth and eighteenth centuries. For example, town meeting records of Dedham, Massachusetts, in 1672 note that "Hen. Wight is deputed to take down so much of the glasse in the meetinghouse windows as he shall see cause in the summertime and carefully lay it by safely and in due time set it up again when the hott season is past for this yeare."

11. *The New-York Historical Society Quarterly*, Vol. XXXVI, No. 2 (April, 1952) pp. 149–159.

12. E. g. "Het Straatje" (Street Scene) in the Rijksmuseum, Amsterdam.

13. On Lawrence Road, Marple Township, near Broomall.

14. "Thomas Bannister on the New Sash Windows, Boston, 1701," *Journal of the Society of Architectureal Historians*,

Vol. XXIV, No. 2 (May, 1965) pp. 169–170. Bannister requested the glass to be "ready fitted" because "few of our workmen know how to doe it."

15. Singleton, Vol. 2, p. 374.

16. Alonzo Lewis, *History of Lynn*, Boston, 1829, p. 159.

17. New York, 1967, pp. 244–251.

18. *The New York Journal or General Advertiser* (Supplement).

19. *The Time Piece*, March 26, 1798 and again in the following year. Peter LaCoer, a glazier and house painter at his glass store on Cedar Street had glass from 7″ × 9″ to 18″ × 24″. (*The French and American Gazette*, January 27, 1796.) William Cowley on Queen Street carried 6″ × 8″, 7″ × 9″, 8″ × 10″, 9″ × 11″, 10″ × 12″, 10″ × 14″, 11″ × 15″, 12″ × 16″ and 13″ × 18″. (*The Daily Advertiser*, April 5, 1791.)

20 George S. and Helen McKearin, *American Glass*, (New York, Crown Publishers, 1941) pp. 75–131.

21. Milo M. Naeve, "Richard Wistar, His Glasshouse and Country Store," *Winterthur Newsletter*, Vol. V, No. 4, April 28, 1959 and Arlene Palmer, "Benjamin Franklin and the Wistarburg Glassworks," *Antiques Magazine*, January, 1974, p. 207.

22. Frederick William Hunter, *Stiegel Glass*, (New York, Dover Publications, 1950) pp. 180, 181; George S. and Helen McKearin, *American Glass*, (New York, Crown Publishers, 1941) pp. 81–86.

23. Rhea Mansfield Knittle, *Early American Glass*, (New York, The Century Co., 1927) p. 177.

24. Newton C. Brainard, *The Hartford State House of 1796*, (The Connecticut Historical Society, 1964) pp. 13–15; ALS, Joseph Woodbridge to Elijah Hubbard, agent for the Albany Glass Works, December 6, 1795; *Connecticut Courant*, December 7, 1795.

25. *Boston Gazette and Weekly Republican Journal*, February 21, 1799. For additional testimonials and references to the high quality of this firm's product see Kenneth M. Wilson, *New England Glass and Glassmaking*, (New York, Thomas Y. Crowell, 1972) pp. 82–84.

26. MS, *Glass Journal*, Thomas Gaffield, Vol. II, pp. 205–208. Gaffield Collection, Arts and Humanities Library, Massachusetts Institute of Technology.

27. Gaffield, Vol. I, p. 12, 1862.

28. *Continental Journal and Weekly Advertiser*, Boston, April 10, 1777.

29. Edwin Benjamin Betts, *Thomas Jefferson's Farm Book*, Princeton, 1953, p. 340; A.L.S., Thomas Jefferson, Monticello, Sep 16 '95, to Josiah Donath, Philadelphia, in the possession of Mrs. Robert Hopkins.

30. *Columbian Centinel*, Boston, May 19, 1821.

31. Knittle, *op. cit.*, p. 309, advertisement reprinted from a Pittsburgh directory ca. 1840.

32. The Addison Glass Works, Addison, New York, operated from about 1890–1906. Among fragments excavated at the site of the factory were quantities of pressed colored glass panes about 5 inches square and about 5 by 10 inches, of which numerous counterparts were found still extant in local homes.

33. Andrew Jackson Downing, *The Architecture of Country Houses* . . . (New York: D. Appleton and Company; Philadelphia: G. S. Appleton, 1850); *Cottage Residences* . . . (New York and London: Wiley & Putnam, 1842); and *A Treatise on the Theory and Practice of Landscape Gardening* . . . *With Remarks on Rural Architecture* (New York and London: Wiley & Putnam; Boston: C. C. Little & Co.; 1841).

34. *New Bedford Mercury*, September 28, 1832, an article describing a specimen of STAINED GLASS had been left for inspection with their agent in New Bedford, Mr. S. Babcock.

35. C. R. Goodrich, ed., *Science and Mechanics:* illustrated by *Examples in the New York Exhibition*, 1853–54, (New York, G. P. Putnam, 1854).

36. Theo B. White, ed. *Philadelphia Architecture in the Nineteenth Century*, Philadelphia, 1953, p. 25.

37. Philadelphia Contributionship Fire Insurance Survey (MS) Policy No. 8689, October 22, 1851. Courtesy Mrs. Earle R. Kirkbride.

38. Edwin T. Freedley, *Philadelphia and its Manufactures. . .in 1867*, (Philadelphia, Edward Young & Co., 1867) p. 302.

39. Janet R. MacFarlane, "A History of the Redford Crown Glass Works at Redford, Clinton County, N. Y." *New York History*, Vol XXVI, No. 3 (July, 1945), p. 373.

40. B. T. Pierson, *Directory of the City of Newark for 1859–60*, New York, 1959, p. 56.

41. Raymond McGrath and A. C. Frost, *Glass In Architecture and Decoration*. With a section on the Nature and Properties of Glass by H. E. Beckett (London, 1937), Fig. 101, p. 235.

42. Wilson, p. 108.

43. *A History of Real Estate, Building, and Architecture in New York City during the Last Quarter of a Century*, n.a.

44. Freedley, p. 277. These were, he continued, "articles which previously had not been made in this country. The excellence of their produce so effectually alarmed foreign manufacturers, and they reduced the price at once, from $2.25 per square foot to 75 cents, and are now actually losing money on their sales, in order to crush an American competitor."

45. Warren S. Tryon, *A Mirror for Americans, Life and Manners in the United States, 1790–1870, by American Travelers*. (Chicago, University of Chicago Press, 1952) p. 353.

CHAPTER
10

An Historical Sketch of Central Heating: 1800-1860

EUGENE S. FERGUSON

MR. FERGUSON, author of the *Bibliography of the History of Technology* (1968), is Professor of History (Technology) at the University of Delaware and Curator of Technology at the Hagley Museum.

Charles Peterson, who invented and organized the symposium and edited this volume, once wrote that he did not believe that Philadelphia is the only place in the world worth writing about.[1] Much as I may agree with Mr. Peterson, I nevertheless find myself being led back to Philadelphia, far more frequently than even the piety of the native would require, whenever I look for origins of some aspect of American technology. For example, my story of central heating begins a few blocks from The Carpenters' Company, in that remarkable section of early nineteenth-century Philadelphia that was leveled a few years ago by the landscapers of Independence Hall.

One branch of central heating in the United States started, of course, with Benjamin Franklin's Pennsylvania fireplace in 1744. That is the branch that uses a fire to heat air directly, now used largely in domestic heating systems. The other branch of central heating, which uses steam or hot water as an intermediary between the fire and the space being heated, springs in a curious way from a Philadelphia shop near Market and Seventh Streets. Very briefly, in 1815 Jacob Perkins came to Philadelphia from Newburyport, Massachusetts. Perkins, who was then nearly fifty years old, already had established his reputation as a prolific inventor of nail-making machines and fire-engine pumps, and he was now working on important innovations in printing and engraving. In Philadelphia he joined a community of active and innovative mechanics that included Oliver Evans, Charles Willson Peale, Pat Lyon, Matthias Baldwin, and many others. Perkins was easily the most active of the lot: he was a man obsessed where inventions were concerned. Visitors to his little shop were treated to a succession of novel ideas delivered in his quick, enthusiastic manner; he could be aroused instantly by the excitement and challenge of a new technical problem.

When one of his regular visitors suggested to Perkins the notion of making steam at very high pressures (far higher than those used by

Note: I am indebted to Robert Bruegmann for guiding me to a number of sources that he had located earlier in his work on eighteenth and nineteenth century prisons and hospitals. His suggestions and comments have been most valuable.

Oliver Evans in his high-pressure steam engine), he seized the suggestion and developed it in several directions over the next two decades, particularly after he emigrated to England in 1819.[2] One of those directions was to a very high-temperature hot water heating system. It was this system, developed in England and brought back to the United States around 1840, that was installed in the White House in Washington and the Custom House in New York.[3] Through Joseph Nason, an American who learned heating and ventilating practice from Perkins and his son in London, the Perkins influence extended to the heating and ventilating system of the new wings of the United States Capitol building, one of the most extensive and advanced systems of its time.

It is hardly necessary to say that the history of technical details of heating systems is a complex affair, with many strands intertwined and overlapping. Most of what I have learned of such details I have placed in footnotes, not because the details are not interesting or exciting—at least *I* can get excited over finding a perfectly practicable steam trap in a book published in 1824 (Fig. 10.1)—but because I want to emphasize two things that I didn't expect to find in the history of central heating before 1860: the first was the vigorous if confused response of the technical community to new scientific information on human respiration; the second, the unconscious subordination of human criteria to the requirements and limitations of mechanical equipment. Finally, I shall want also to suggest a tendency at work in heating systems as in all other complex technical systems: a tendency, over a considerable period of time, of technical virtuosity to defeat itself and collapse. You may recall the Tacoma Narrows bridge of 1940, which exhibited the most advanced design in the world just before it fell into Puget Sound. This is the phenomenon I shall call the Vogel effect after Robert Vogel, the author of Chapter 7, because a chance remark of his brought home to me the fact that the tendency is a universal one.

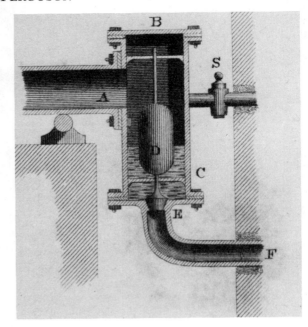

Fig. 10.1 Steam trap, 1824. The purpose of a steam trap is to remove, automatically, water from a steam space. Increments of condensate from pipe A collect in reservoir C. When sufficient water accumulates, float D rises and opens conical valve E. Internal pressure blows water out of trap (through drain pipe F); the valve E recloses and the trap cycle is repeated. Air is bled as necessary through cock S. Thomas Tredgold, *Principles of Warming and Ventilating Public Buildings* (3rd edition, London, 1836) plate III.

Before proceeding to my arguments, I will sketch in, with some names, dates and places, a minimum context for the thesis that follows.[4]

The English have for centuries favored fireplaces for heating.[5] The cheerful blaze and ventilation of an open fire have been frequently extolled by English writers. Thomas Tredgold, an early nineteenth century English authority on heating, pointed out the advantage of a "common fire"—that is, an open fireplace: it provided a "genial warmth" for one's body and at the same time permitted one to breathe cool air. The radiant heat from the fire provided a feeling of warmth even though the air of the room might be in the fifties (as Tredgold and others thought it should be).[6]

Austrians, Russians and Scandinavians have used stoves for space heating at least since the fifteenth century, although most surviving examples are of the eighteenth century and later.[7] Freestanding wood or coal-burning stoves of cast iron, which once again became popular a few years ago as insurance against massive power failures and even more recently to supplement uncertain oil-fired heating systems, were very widely used in the United States from around the time of the Revolution until well into the twentieth century.[8]

Benjamin Franklin's response to the English fireplace was to improve it without shutting up the cheerful blaze. By conducting fire and smoke over and around an air box before discharging them into the chimney, his Pennsylvania stove of the 1740s introduced fresh, heated air into the room, which the unaided common fireplace was unable to do (Fig. 10.2). Franklin thus reduced substantially the general complaint regarding fireplaces that he had put in the words of Everyman: "A man is scorch'd before," he wrote, "while he's froze behind."[9] The Pennsylvania stove was superseded before 1800 in the United States by freestanding cast iron stoves, which permitted the heat source to be placed well out into the room, requiring only a simple stovepipe connection to a chimney.[10]

Before 1820, numerous examples of hot air and steam central heating systems were to be found both in the United States and in England. Hot water systems came a little later, the ubiquitous rotating fan to propel air through ducts later still, although of course centrifugal fans were well known in the eighteenth century.[11]

Between 1802 and 1810, the historic Birmingham firm of steam engine builders, Boulton and Watt, installed fifteen extensive steam heating systems in cotton and paper mills, a sugar house, two warehouses, Lowther Castle in Westmorland and Covent Garden Theatre in London. A typical early mill installation, shown in Figure 10.3, carried low-pressure steam from a boiler throughout the building in cast iron pipes about 4 inches in diameter, which served as virtually a continuous radiator.[12] Probably the first system in the world to

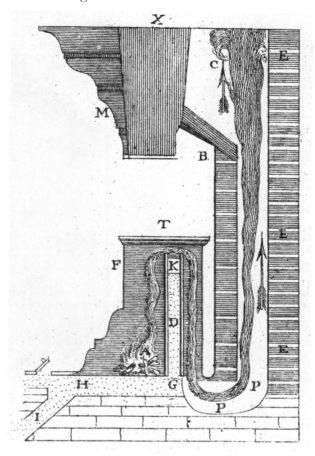

Fig. 10.2 Pennsylvania Fireplace, 1744. Designed by Benjamin Franklin, the Pennsylvania fireplace permitted the discharge of fresh heated air into the room through opening K in the fireplace casing. Fresh air, brought in through passage IHG, was warmed in box D by the hot products of combustion as they made their way toward the chimney. The flap valve at the left admitted cold outside air to the hearth. Benjamin Franklin. *Descrizione della Stufa di Pensilvania* (Venezia, 1791) fig. X.

use exhaust steam from a high-pressure steam engine (there was no usable exhaust from the condenser of a Watt engine) was to heat a factory in Middletown, Connecticut, in 1811. The engine was one of Oliver Evans's. Another mill, near Baltimore, was similarly heated a year later.[13]

Central hot-air systems were being used in English mills by 1792. William Strutt's installation in his Belper Mill was copied soon after in the Derby Infirmary and in the Wakefield Insane Asylum.[14] In both England and the

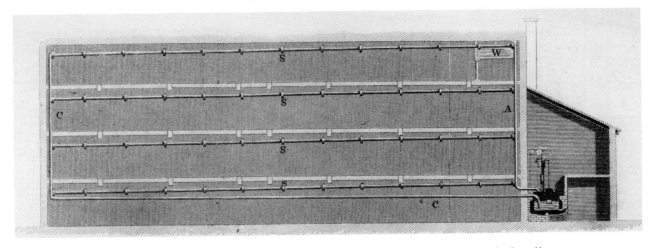

Fig. 10.3   Steam heating system in silk mill, 1817. Cast iron pipes, about 4 inches diameter, conducted steam through mill rooms just below the ceiling. Condensate was returned to the boiler through pipe C. This mill, in Watford, Herts., formerly was heated by thirteen individual iron stoves. Thomas Tredgold, *Principles of Warming and Ventilating Public Buildings* (3rd ed., London, 1836) plate VIII.

United States, a number of independent stoves, located throughout a mill, were used long after the idea of central hot-air heating was well known. A long horizontal run of stovepipe from stove to chimney helped to distribute the heat more evenly, as, for example, in the Duplanty and McCall textile mill, now the Hagley Museum, near Wilmington, Delaware.[15] A number of central hot-air heating systems, in hospital, almshouse, insane asylum, factory and medical college, had been installed in the United States before 1820; all were similar in principle to the system built by Oliver Evans and illustrated in 1795 in the first edition of his *Young Mill-Wright and Miller's Guide* (Fig. 10.4).[16] A closed stove or furnace, located in the cellar of a building, was surrounded by a brick or metal chamber (Fig. 10.5). Heated air, issuing from the top of the heating chamber, was led through air ducts to the respective rooms to be heated.

American stoves and hot-air central heating systems were perennial subjects for indignant comments by nineteenth-century British itinerants. Swedish visitors might complain that American houses were generally too cool, with bedrooms often not heated at all,[17] but the British could agree that an American heating system was "a terrible grievance to persons not accustomed to it, and a fatal misfortune to

those who are. Casual visitors are nearly suffocated, and constant occupiers killed." And so forth.[18]

The idea of the hot-water heating system was quite as obvious as steam and hot-air systems to the early English and American writers on heating, but it is difficult to point to even one large hot-water system anywhere before the 1830s. The Marquis de Chabannes, in his book of 1818, illustrated a hot-water heating system for a hot house (Fig. 10.6) but not until 1836 do we find published a drawing of an actual installation, when the orangery of the royal palace at Windsor was so heated.[19] Tredgold, in his standard work on warming and ventilating, first published in 1824, said that water could not be circulated through pipes unless it was converted to steam, because the difference in densities of hot and cold water was insufficient to induce circulation.[20] Tredgold was wrong, of course, but domestic heating systems in the United States were generally steam systems until about 1880, when a massive shift to hot water occurred.[21] I suspect that a vast amount of tinkering with heating systems that should have worked but didn't was required to prove Tredgold wrong. A sampling of early radiator types is shown in Figures 10.7 – 10.9.

This brings us back to Jacob Perkins, who by

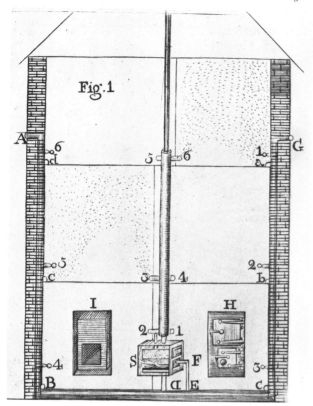

Fig. 10.4 Central hot air heating system, 1795. Hot gases, produced by combustion of fuel in furnace S, are conducted to chimney through the smoke pipe, shown extending into the attic. Heated fresh air is carried from top of furnace casing through T-shaped pipe 2–1 to annular passage surrounding the vertical smoke pipe. Warm air flows into rooms at 1, 2, 3, 4, 5, 6, near the central trunk. Exhausted air enters vent pipes which discharge outside at A and G. Oliver Evans, *Young Millwright and Miller's Guide* (Philadelphia, 1795) plate X.

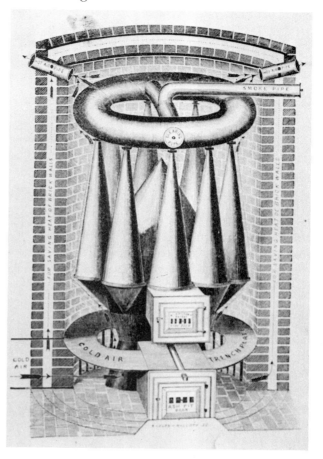

Fig. 10.5 Central hot-air heating furnace, 1857. This coal-burning furnace was to be located in a basement in a brick chamber. Fresh cold air enters at the bottom, flows over the conical heating surfaces of the furnace, and is conducted through "hot air pipes" *(upper left and right)* to rooms of the house. The annular space between inner and outer brick walls permits recovery of heat otherwise lost through the walls. David Bigelow, ed., *History of Prominent Mercantile and Manufacturing Firms in the United States* (Boston, 1857) p. 15.

the 1830s was the impressario of the Adelaide Gallery in London, a gallery of popular science and museum of "objects blending instruction with amusement." The Perkins steam gun, periodically discharging a terrifying fusillade of lead down the length of the "Long Room," was one outgrowth of Perkins' obsession with high pressure steam that had its origins in Philadelphia. Perkins, you will recall, had been in London since 1819. In 1831, his son, Angier Marsh Perkins, took out an English patent for a high-pressure hot-water heating system; in 1835 the younger Perkins' apparatus was heating the Long Room and its gallery in the Perkins emporium in Adelaide Street.[22]

The novelty of the Perkins system lay in the way water was used to convey heat from fuel in the furnace to the space to be warmed. A continuous, relatively small-bore pipe—about ¾-inch inside diameter, as compared to the usual 3½-inch inside diameter of other hot-water systems—formed a coil in the fire box of the furnace and a coil, not unlike a modern radiator, in the rooms to be heated; these coils were connected by supply and return lines. As just mentioned, the pipe was continuous throughout the system: no valves or other constrictions interrupted a free

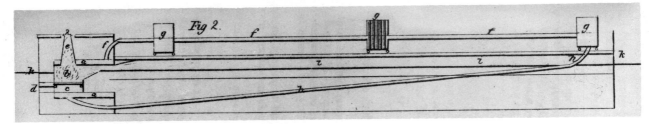

Fig. 10.6 Gravity-type hot water heating system, 1818. The furnace at left (e, b, c) supplies warmed water to large-diameter pipe f, f, f. Air is heated by the pipe and by "patent cylinders" g, g, g a rudimentary form of convector. Return pipe is marked h, h. Marquis de Chabannes, *On Conduction of Air by Forced Ventilation* (London, 1818) plate XIII.

gravity flow and the entire length of pipe formed a continuous, closed loop. Because the loop was closed and full of water, the water temperature could be raised well above 212°F, the boiling point of water at atmospheric pressure. When the water reached 212°F, there was no space for steam, so both the pressure and temperature in the system increased as heat was added. Perkins intended that the normal water temperature would be 350°F; the pressure in the system would then be about 125 pounds per square inch above atmospheric or, stated another way, a pressure of nearly ten atmospheres. Having no temper-

ature controls, both temperature and pressure could rise precipitously; at 500°F, for example, the pressure was close to 50 atmospheres; at 636°F, 135 atmospheres or 2,000 pounds per square inch.[23] Even today, these figures will cause a normally conservative engineer

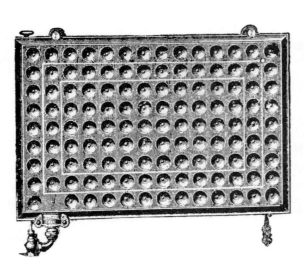

Fig. 10.7 "Mattress type" radiator, patented in 1854 by Stephen J. Gold, of Connecticut, inventor of a very popular home heating boiler. This radiator, in which two embossed iron sheets were fastened together by rivets, was similar to James Watt's steam radiator of 1784. Susan R. Stifler, *The Beginnings of a Century of Steam and Water Heating* (Westfield, Mass.: 1960) ff p. 64.

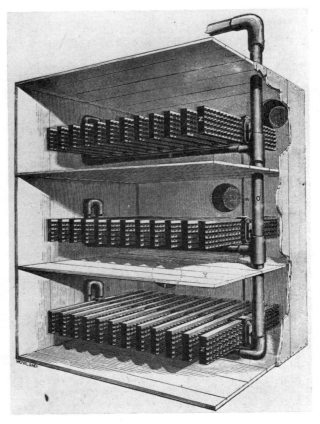

Fig. 10.8 "Pin-type" radiators (actually convectors), patented by Samuel Gold, Stephen's son, in 1862. Air was heated by the steam radiator sections as it passed through the wooden chambers adjacent to the heating boiler. Massachusetts Steam & Water Heating Co., trade catalogue, 1861. (*Eleutherian Mills Historical Library.*)

NASON'S
"STANDARD" WROUGHT IRON TUBE RADIATOR

Fig. 10.9 Joseph Nason's steam radiators, ca. 1870, employed wrought iron return tubes fastened with ferrules into a cast iron base. The iron casting at the top of the radiator acted as a decorative diffuser and protective guard against contact with hot tubes. (*Broadside, n.d. Smithsonian Institution, Warshaw Collection, National Museum of History and Technology.*)

to turn slightly green, and not with envy.

While the low-pressure hot-water system was only slowly being adopted, the Perkins high-pressure system won astonishingly prompt acceptance. I can only guess that promotion rather than intrinsic merit was the secret of its commercial success. In November, 1835, two Perkins systems were installed in the British Museum for the architect Sir Robert Smirke: one system to heat the reading rooms and one to heat the Bird and Print Rooms (Fig. 10.10).[24] Before 1839, three Perkins high-pressure hot-water heating systems had been installed in Edinburgh; others were in the Duke of Wellington's country seat "Strathfieldsaye," the London Patent Office and the Atlas Insurance Office, Cheapside, Charles Babbage, designer of the noted calculating engine, heated his residence on Dorset Street with a Perkins system, placing one coil in an upstairs cistern that held enough water for two and one-fourth hot baths.[25] In the New York Custom House, a Perkins system was installed in 1841 despite the determined opposition of the Building Commissioner. The system was successful; the architect was delighted; eventually, even the Comissioner pronounced it satisfactory.[26]

A disturbing characteristic of the Perkins system was the tendency of the hot water pipes to char and sometimes ignite adjacent wood. In some tests made by the Manchester Assurance Company, blocks of wood in contact with Perkins pipes were charred, matches were lighted, gunpowder fired, and lead melted (which required a temperature of 612° F). Sir

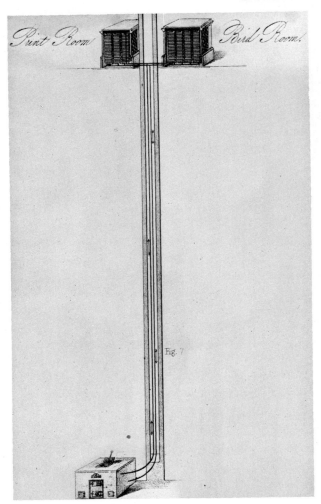

Fig. 10.10 Perkins high-pressure hot-water heating system in the British Museum. The furnace, at bottom of illustration, was about 50 feet below the heating coils, at top. Charles J. Richardson, *A Popular Treatise on the Warming and Ventilation of Buildings* (2nd ed., London, 1839) plate 2.

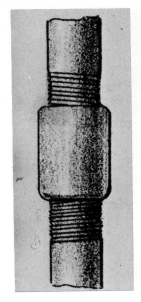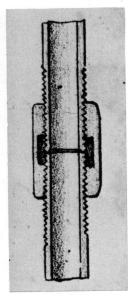

Fig. 10.11 Pipe coupling, used in the Perkins system. The end of one tube is sharpened, the other flattened. When the coupling, with left- and right-hand threads, is pulled up tight, the sharpened pipe-end is indented in the flattened surface of the other. Charles J. Richardson, *A Popular Treatise on the Warming and Ventilation of Buildings* (2nd ed., London, 1839) plate 1.

Robert Smirke reported that he had seen a red-hot water pipe some twelve feet from the furnace and that he would not consider installing a Perkins furnace in any but a fire-proof room. Being less dogmatic than Perkins, Sir Robert also placed a safety valve near the furnace on the systems he had installed, thus limiting both pressure and temperature.[27]

Sharing the excitement of the Perkins invasion of Great Britain was Joseph Nason, a native of Boston; in 1837, when he went to work for Angier Perkins in London, Nason was 22 years old. He returned to America about 4 years later, bringing with him the knowledge and experience gained from his association with Perkins. By the time he was 40, Nason was easily the most knowledgeable person in the United States on matters pertaining to heating and ventilating.[28] In that year — 1855 — he was the designer of the system for heating and ventilating the House and Senate wings of the Capitol in Washington, which I shall describe a little later. First, it is necessary to turn from developments in systems to a revolutionary incursion of science into traditional ways of thinking about the rooms we inhabit and the air we breathe.

Building on the work of a group of experts on ventilation who were active from about 1830 onward, a pure air movement rose to a crescendo right after the Civil War, when the Philadelphia lecturer and self-styled "heating and ventilating engineer," Lewis W. Leeds, coined and gave currency to the slogan, "Man's own breath is his greatest enemy."[29]

The facts of human respiration had been established in the late eighteenth century by Priestley, Lavoisier and others. Priestley demonstrated that in the lungs dephlogisticated

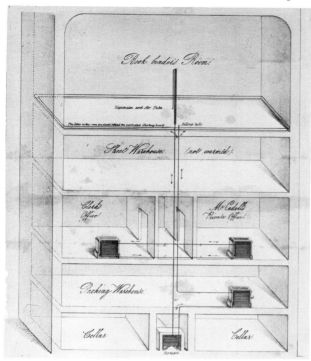

Fig. 10.12 Perkins high-pressure hot-water heating system, 1833. This installation was in the bookbinding establishment of Robert Caddell in Edinburgh. Charles J. Richardson, *A Popular Treatise on the Warming and Ventilation of Buildings* (2nd ed., London, 1839) plate 2.

air (Lavoisier would call it oxygen) changed blue blood to red while the air picked up phlogiston; part of the exhalation from the lungs was thus fixed air, now known as carbon dioxide, in the early nineteenth century also known as carbonic acid gas. Carbonic acid gas would not support life; oxygen would.[30] When, in the nineteenth century, the facts of respiration were translated into a need to have sufficient oxygen in the air one breathes and to avoid breathing the carbonic acid gas that was exhaled by oneself or by others, an embargo was placed on folk wisdom and common sense.

The entrepreneurial experts who jumped into the imaginary void mapped out a mechanism of hygienic ventilation which in the next generation became a powerful dogma. An adult inhales and exhales a third of a cubic foot of air per minute; but Dr. Arnott says that "air expelled from the lungs is found to vitiate, so as to render unfit for respiration,

twelve times its own bulk of pure air." ($1/3 \times 12 = 4$ cu. ft. per min.) Not less dangerous than the carbonic acid of exhalation are the miasma, or effluvia, and vapor of animal matter, exuded from the whole surface of the body. This will require replacement by another 3½ cubic feet per minute of fresh air. Added to this is the vitiation of air by gas used for lighting.[31] Dr. Reid, an Edinburgh doctor of medicine who turned to the theory and practice of ventilation, made a series of tests, as careful as most, on human subjects. His conclusion, "given with much diffidence," was that 10 cubic feet per minute of fresh air should be supplied for every person in a given space.[32] Needless to say, other practical men were not so diffident and prescriptions of requisite fresh air varied widely. More importantly, a general underlying assumption was that the fresh air must be swept up through a room in order to avoid any possibility of having the same air "breathed over" or breathed twice; vitiated air was a terrible menace that must be avoided at all costs.[33]

In 1850, the American architect A. J. Downing, in his *Journal of Rural Art and Rural Taste*, deplored the prevalence of what he named "the national poison," which was "nothing less than the vitiated air of *close stoves*, and the unventilated apartments which accompany them."[34] In 1866, Lewis Leeds told an audience in Franklin Institute that although Philadelphia was one of the healthiest cities in the United States, he was "forced to the conclusion that about forty percent of all deaths that are constantly occurring are due to the influence of foul air."[35]

Leeds illustrated his lectures, using a magic lantern to show the way air currents behave with various arrangements of heating apparatus. In the little domestic scenes on the screen, the audience could observe how the people sat in clouds of purple air that had been vitiated while robust, pink, fresh warm air clung to the ceiling because of faulty design of the heating system (Figs. 10.13 through 10.15). The pictures are delightfully amusing, but two aspects of them are worth noting: first, they provide the graphic foundation for the fresh-air fads

Fig. 10.13 Popular exposition of the principles of warming and ventilating. These illustrations were projected onto a screen by Lewis W. Leeds during his 1866–67 lecture series in the Franklin Institute, Philadelphia. They appear in his *Treatise on Ventilation* (New York, 1871). The clouds represent air exhaled and thus vitiated. The straight lines represent radiated heat. *Fig. 10* shows how feet are chilled by cold current from the closed window while reader breathes warm but vitiated air. *Fig. 11* shows how, with the given heating system, one may "keep the feet warmer than the head, and the back warmer than the face." *Fig. 12* notes the radiation from the man's back to the cold window, chilling his back and legs. The room's air is totally vitiated. *Fig. 13* depicts an optimum condition. Some fresh air is admitted to the room just above the radiator at right. Vitiated air is removed at floor level *(left)* and from the gas lighting fixture near the ceiling.

Fig. 10.14 Fresh, heated air, admitted at lower left, rises and flows across ceiling to the exhaust register at upper right. The clouds of heavily vitiated air, purple and unpleasant, which surround the man and woman, are nearly undisturbed by the badly designed heating system. Lewis W. Leeds, *Treatise on Ventilation* (New York, 1871) fig. 2.

that were so numerous and influential in the early twentieth century; second, they are technically quite accurate. Leeds understood very well the mechanisms of heat transfer, recognizing, for example, that the cold walls of a room can make the inhabitants uncomfortable even though the air temperature is at a comfortable level. He was also full of slogans. If you would be healthy, he admonished, "always keep your feet warmer than your head, and your back warmer than your face."[36]

Fig. 10.15 The open window, at left, permits cold air to enter, but the air flows across the floor and up the chimney while the sleeping victims are surrounded by the air that they have vitiated. Lewis W. Leeds, *Treatise on Ventilation* (New York, 1871) fig. 6.

In 1869, Catharine Beecher and her sister, Harriet Beecher Stowe, told the numerous readers of their *American Woman's Home: or, Principles of Domestic Science,* that "tight sleeping-rooms, and close, air-tight stoves, are now starving and poisoning more than one half this nation." Carbonic acid, they wrote, is a fatal poison; even a small amount, when mixed with pure air, is "a slow poison, which imperceptibly undermines the constitution."[37]

The rapid advances in ventilating practice, tried out in England and brought to fruition in the Capitol building in Washington, should thus be seen against the rapid rise of a shrill dogma of ventilation. The tendencies of heating and ventilating practice in England in the first wave of new knowledge are epitomized in the system installed in 1836 by Dr. Reid in the House of Commons. Reid's guiding principle was to control the flow of air. In his words, "The movement of air, from its ingress to its egress, was regulated as in a pneumatic machine, the house, in this respect, being treated as a piece of apparatus."[38] The entire base-

ment was devoted to air filters and sprays, heating coils and air ducts. After being drawn in from the Old Palace yard, the air passed through a "suspended fibrous veil," 42 by 18 feet, for excluding mechanical impurities, through steam coils (which might also in summer contain cold water to cool the air),[39] and through ducts that spread the air evenly under the floor. The floor was pierced by nearly a million holes, drilled with gimlets, and the entire floor was covered by a loosely woven hair carpet. Reid expected that the air would be perfectly diffused and rise imperceptibly but uniformly throughout the House, making impossible the "breathing over" of vitiated air (Fig. 10.16). Spent air was removed from the top of the hall, being directed into a large duct which descended, outside the building, to the base of a huge chimney. In the center of the chimney, a brisk coke fire was maintained to produce a constant draft and, hopefully under full control, conduct the air through Reid's huge pneumatic machine.

Seldom is any complex machine under full

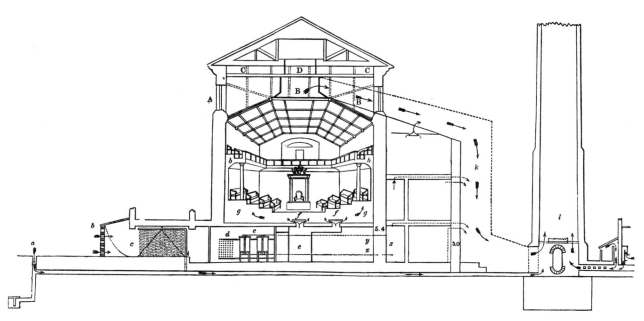

Fig. 10.16 Warming and ventilating system, House of Commons, 1836. Outside air was drawn in through wall b, at left. Screen c removed large dust particles; the air, heated as it flowed through coil D, was distributed under the House floor and balcony in chambers f, g, and b before it entered the chamber through a million small holes in the floor. Spent air was drawn through vent B and descended to the base of large chimney at right. A coke fire at l maintained a brisk draft in the chimney. David B. Reid, *Illustrations of the Theory and Practice of Ventilation* (London, 1844) p. 283.

control; this heating and ventilating system of the House of Commons was no exception. External conditions sometimes intruded. The smell from gasworks on both sides of the river occasionally reached the air inlets; "emanations" from the graveyard of St. Margaret's Church were occasionally very offensive; a barge laden with manure produced "extreme complaints"; in calm weather, even the smell of a crowd of coachmen smoking their pipes might enter the system. Internal conditions were controlled to the best of the ability of operators on duty. Every complaint was acted upon and in a single sitting of the House from fifty to a hundred changes might be made in quantity or quality of the air supplied.[40] Reid opined that if Members would dress more uniformly, particularly around the ankles and feet, "a greater unity would prevail as to the state of the air demanded."[41] Nor could much be done by Reid about the complaint that dirt in the carpet was blown into Members' eyes and noses, except to install boot scrapers at the entrances.[42] It was reported years later, perhaps with some exaggeration, that the Members one by one took matters into their own hands, placing sheets of lead beneath the carpet in their own vicinity to keep the dirt from blowing up into their faces. When nine tenths of the million holes in the floor were thus covered, Goldsworth Gurney, the expert of the next wave of change, reversed the air flow, supplying air from above and removing it below.[43]

This brings us at last to the United States Capitol building, where in 1855 the problem of heating and ventilating the new House and Senate wings presented Captain Montgomery C. Meigs, U.S.A.—the able, if occasionally pompous, superintendent of construction— with "the most difficult piece of engineering and construction that I have yet to undertake."[44]

The present Senate and House wings were added to the Capitol Building in 1855–61. Because the older parts of the building were inadequately warmed by sixteen hot-air furnaces,[45] it was clear to Captain Meigs that the haphazard approach to heating and ventilat-

ing which had gotten the Capitol through its first fifty years was inadequate for the future, particularly in view of the heightened importance of ventilation and the accepted need to avoid vitiated air in any form. The tide of best practice was setting hard toward complete control of air flow. The zeal with which Meigs approached the problem of complete and continuous control of the atmosphere hastened the day when architectural textbooks would leave unmentioned the principles or even the possibility of natural ventilation.[46]

In 1853, after visiting public buildings in Boston, New York, Brooklyn and Philadelphia, in company with Secretary Joseph Henry of the Smithsonian Institution and Alexander Dallas Bache, chief of the Coast Survey, Meigs modified plans of the new wings that had been made earlier. In the plans, the hall of the House of Representatives was located at the western end of the south wing—that is, overlooking the present Mall. Windows were to be provided on three sides of the room. Meigs reasoned that the primary purpose of the hall was to provide an effective setting for debate. He worried that noise, coming through the windows, might interfere with debate. When the weather was cold, the expanse of glass (a modest expanse, remember, by twentieth century standards) would produce unpleasant air currents inside the hall. And the light from the windows might be disagreeable in a speaker's eyes. Mostly, though, he believed that he "could secure a greater uniformity of temperature by placing the room in the center of the building," as well as better control of public access to Congressmen entering and leaving the hall. Accordingly, taking into account the factors that seemed relevant to him, Meigs moved the hall to the center of the building, making for posterity the choice that a controlled situation was more important than, say, a glimpse of the western sky as gas lamps were being lit overhead. Testifying in 1865, twelve years after he had made the decision to move the hall to the center of the building and to eliminate all windows, Meigs said, "It seems to me that members, occupied in the business of legislation, did not need, and would not have time to enjoy, any external prospect."[47]

After the basic decisions as to arrangement of rooms in the new wings of the Capitol had been made, Constructor Meigs called Joseph Nason to Washington to discuss the requirements of the new buildings. Meigs told Nason that he wanted "something more than a mechanic—one who has some scientific knowledge, so as to be able to appreciate the scientific discussions of the subject which have been published abroad, and to apply the principles they unfold with mechanical skill and practical knowledge, enough to lay down his system of pipes without error."[48] Nason proposed to supply all necessary pipe and fittings at fifteen per cent discount from printed list prices, to furnish an engineer and draftsman for six dollars a day, and first and second class workmen at proportionate rates. Nason's services, as consultant, were supplied at no extra charge to the purchasers of his pipe. Meigs promptly accepted Nason's proposal; design of the intricate system of steam boilers, steam coils, brick-masonry ducts and huge centrifugal fans began in the fall of 1855. Because no reliable data were available to determine power requirements of the fans, Meigs ordered immediately the making of a model fan, on a scale of 10 to 1, to be driven by a weight on a cord, falling through a distance of 50 feet.[49]

Nason's general approach to the heating of the new wings was to blow air across nests, or coils, of pipe containing steam at low pressure (Fig. 10.17). The heated air was then carried through ducts, divided and subdivided to reach every part of the rooms to be heated and ventilated. Pipe for the coils was wrought iron, slightly larger in size but otherwise the same as that used in the Perkins high pressure system.[50] In the House of Representatives wing, twenty-two heating coils were provided, each comprising over a thousand lineal feet of pipe. Four boilers supplied steam to the coils and steam to drive the two fan engines; two enormous fans—the larger some 16 feet in diameter—were required, one for the House chamber and one for all the other rooms in the wing (Fig. 10.18). An equivalent system was installed in the north, or Senate, wing.[51]

The detailed design of the fans was based on a study by Robert Briggs, Jr., of the data collected from the model fan and dynamometer, mentioned earlier.[52] Briggs was the "engineer and draftsman," furnished as part of the piping contract, who was to become a recognized authority on heating and ventilating, reinforcing in the next generation the Nason approach.[53]

The apparatus of the House wing was thoroughly tested during the winter of 1857–58, while the House was in session; the Senate system was completed soon after. In the Senate chamber, warm air was initially introduced through registers in risers just behind the Senators' chairs. Complaints by the Senators forced Meigs to change the duct locations immediately, and he also added deflectors to wall registers because drafts annoyed visitors sitting on sofas nearby.[54]

In 1861, nevertheless, Meigs noted that the new wings, having been in use for some time, had "realized all that I undertook to accomplish in regard to light, warmth, ventilation, and fitness for debate and legislation. The health of the legislative bodies has been better, [he asserted, and] more business has been accomplished in the same time than in the old halls."[55]

The inhabitants of the new halls would not have consented unanimously to Meigs' own appraisal of his success. In 1865 an inquiry, no doubt politically inspired (for there is a tangled history of wrangling amongst appointed and disappointed architects and engineers), led at length to a general diatribe by a number of Senators on the miserable conditions under which they worked. Suffice it to say that some Senators could see advantages in natural ventilation, perched on a hill as they were; Senator Charles Sumner of Massachusetts observed that "there is no public edifice in the world which enjoys the advantages of sight equal to that of this Capitol. . . . Now, Sir," he continued, "when we voluntarily shut ourselves up in this stone cage with glass above, we renounce all the advantages and opportun-

Fig. 10.17   Heating coil, U.S. Capitol, 1858. Wrought iron pipe, assembled with return bands, formed a continuous steam coil to heat air that was blown across it. Note the large pot-type steam trap at lower left. *(Photograph, kindness of the Architect of the Capitol.)*

ities of this unparalleled situation. Unless changed by legislature," he concluded, "this room will continue uncomfortable as it now is for centuries."[56] The charges and countercharges of the inquiry served chiefly to highlight the differing perceptions of technical triumphs by those who build them and by those who merely use them.

Lewis Leeds, the popular lecturer, expressed in his discussion of the defects of the House and Senate heating and ventilating system a bit of folk wisdom that was, I presume,

drowned out by the whoosh! of the great fans. Leeds said:

It appears to have been originally designed to exclude the main halls as much as possible from all external influences, and to have all the currents, the heating and the lighting, under perfect artificial control.

But if the whole nation could be taught the valuable lesson, of the great folly of attempting to produce artificial light, artificial heat, and artificially mixed air, that shall

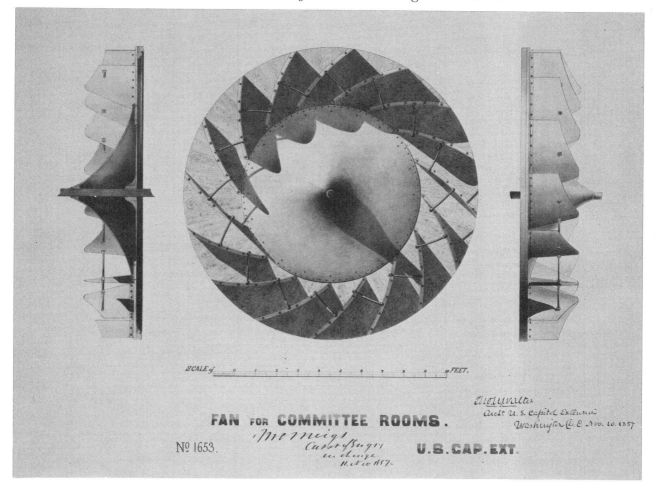

SCALE of   0   1   2   3   4   5   6   7   8   9   10 FEET.

FAN FOR COMMITTEE ROOMS.

Nº 1653.

U.S.CAP.EXT.

Fig. 10.18 Fan rotor, U.S. Capitol, 1857. The passages in which were located the large centrifugal fan rotors, designed by Robert Briggs, were so shaped that no fan casing was required. The central cone of the fan was of cast iron; the vanes were of wood, stiffened and fastened with metal angles and rods. This rotor was 14 feet in diameter. *(Photograph, kindness of the Architect of the Capitol.)*

equal to that which our Creator has provided for us, that knowledge would be cheaply bought at the great price paid for these buildings.[57]

I have no wish to use Montgomery Meigs as whipping boy, but the enormity of his decision, as viewed from my perspective (I hasten to say), not his, should, I think, give us pause. Here was an important, perhaps crucial, step in the direction of designing buildings whose internal conditions depend entirely on machines. Over the years I have been grateful to the maverick architects who put windows in the committee rooms in several places (not in the Capitol) in which I've spent uncounted hours. I did need, and I did have time to enjoy, the external prospect. My point is simply that Meigs' choices, made from the perspective of a man impelled to avoid imponderables and to embrace only the predictable and controllable, are the kinds of decisions that appear to a conditioned mind so right and natural, but which make a perceptible difference in the tone of a civilization. To complete my little soliloquy, I say simply that we would do well to hire poets as building commissioners in order to face the large issues before they are decided by small choices.

Finally, the Vogel effect was bound to show

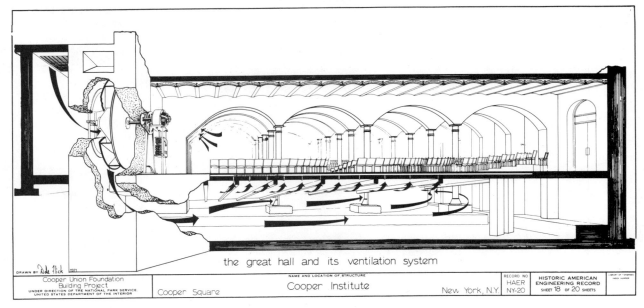

the great hall and its ventilation system

| | NAME AND LOCATION OF STRUCTURE | | | RECORD NO | HISTORIC AMERICAN | LIBRARY OF CONGRESS |
| Cooper Union Foundation Building Project UNDER DIRECTION OF THE NATIONAL PARK SERVICE UNITED STATES DEPARTMENT OF THE INTERIOR | Cooper Square | Cooper Institute | New York, N.Y. | HAER NY-20 | ENGINEERING RECORD SHEET 18 OF 20 SHEETS | |

DRAWN BY Dale Flick 1971

Fig. 10.19   The ventilation system of the Great Hall of Cooper Institute, New York City.
Measured drawing by Dale Flick for Historic American Engineering Record, 1971.

up in central heating and ventilating systems, as it must eventually in any other progressive, equipment-oriented technology.

Near the end of 1973, it began to dawn on a few architects and engineers that the acute stage had been reached in a technically impressive progression that has led to overlighted, overheated and overcooled airtight boxes used for offices, stores and schools. The notion of the completely controlled environment, even if the systems actually operate as they promise to (and who does not have his own story of capricious failure of such systems?), requires acceptance of a cooling system that does battle with a lighting system, declares obsolete the use of spring breezes to produce a spring-like interior climate, and frustrates anyone who would turn off a light or otherwise upset the desperate balance of mechanism. Feature articles, appearing recently in newspapers and technical journals, explain that such buildings require at least twice as much energy—some closer to four times—as those making adequate but moderate use of heating, cooling and ventilating systems.[58]

Robert Vogel has told me that the way to overcome the aftermath of the Vogel effect

(there seems to be no way to avoid it in the first place) is to back down a little, to retain the useful parts of the system that has just thrashed itself to pieces, and to discard the rest. To be carefully avoided is the seductive notion that the defects of a complex technical system can be permanently overcome by making it still more complex—by adding another layer of controls or other mechanism—rather than reducing its complexity. Yet it is this very notion that has propelled the winds of change since Reid and Meigs were at work.[59]

In closing, let me return to the beginning. We started with Jacob Perkins in early nineteenth century Philadelphia, thoroughly engrossed in the technical problems that he and his colleagues found interesting and solvable. We have seen how his enthusiasm and that of his son for technically advanced heating systems blossomed in England, was brought back to the United States by Joseph Nason and placed in the mainstream of American development through Nason's undoubted ability and professional connections. We have seen also how the new facts of science were given a twist in their interpretation that led to a new doctrine of ventilation as it affected public health. In the heating and ventilating systems of the House of Commons and the new wings of the United States Capitol, we found what can and will be done by the leading technical experts in the virtual absence of economic restraints. These great works provided the vision as well as the problems for following generations of experts, and those experts in turn advanced and modified their systems to incorporate new ideas and new enthusiasms.

In heating and ventilating, as in any technically sophisticated enterprise, the view from outside is likely to come to rest on an aspect of the subject that the technical community has considered neither interesting nor relevant. As I see it, that is precisely the justification for an historical review of a progressive, successful field of technology.

NOTES

1. In his piece on early house-warming; (*see* note 4).

2. The Philadelphia sojourn of Jacob Perkins (1766–1849) is described in Greville and Dorothy Bathe, *Jacob Perkins* (Philadelphia, 1943), pp 59–77; and in George Escol Sellers, *Early Engineering Reminiscences (1815–1840)*, Eugene S. Ferguson, ed. (Washington, 1965), pp. 12–28.

3. U.S., Serial Set no. 1031 (see note 60) p. 8. The White House appears to have been warmed by hot air which was heated by twelve high-pressure hot water coils, operating above 212°F, in the basement. Louis Torres, "John Frazee and the New York Custom House," Society of Architectual Historians, *Journal*, XXIII, 3 (October 1964) 143–150.

4. A brief but comprehensive historical sketch is in John S. Billings, *Ventilation and Heating* (New York, 1893) pp. 26–41. A bibliography of 32 titles includes most of the significant historical and early technical works. More recent but much less satisfactory is Neville S. Billington, "A Historical Review of the Art of Heating and Ventilating," *Architectural Science Review*, (November 1959), 118–130; the references are cryptic, the information frequently wrong. A.F. Dufton, "Early Application of Engineering to the Warming of Buildings," Newcomen Society, *Transactions*, XXI (1940–1941) 99–107, pls. XVII–XX, written when libraries were being blitzed, is understandably very thin. Charles Peterson, in "American Notes," Society of Architectural Historians, *Journal*, IX, 4 (December 1950) 21–24, has supplied an astonishing amount of fresh information on "Early House-Warming by Coal-Fires." Benjamin L. Walbert, III, "The Infancy of Central Heating in the United States: 1803 to 1845," Association for Preservation Technology, *Bulletin*, III, 4 (1971) 76–87, another fresh article, demonstrates the existence before 1820 of a substantial number of central hot-air systems.

Maurice Daumas, ed., *Histoire Générale des Techniques*, vol. 3 (Paris: Presses Universitaire de France, 1968) pp. 516–523, has a general review of French heating experience before 1850, following and borrowing from British practice, including the Perkins high-pressure system. Tredgold's 1824 book was translated into French in 1825.

A review that tells how the earlier notions were revised after 1900 is George T. Palmer, "What Fifty Years Have Done for Ventilation," in American Public Health Association, *A Half Century of Public Health*, M. P. Ravenel, ed. (New York, 1921; reprinted, Arno, 1970) pp. 335–360. The manuscript history of heating by the stove designer and builder William John Keep (1842–1918), which was deposited in Baker Library by his daughter, is concerned chiefly with the internal details of stove design; it was of marginal interest to me in writing this paper. However, a section on the making of iron suitable for stoves refers to a series of technical papers by Keep and others, 1888–1905. Wiley appears to have published a book by Keep on cast iron. I am indebted to Mrs. Diana Waite for telling me of the existence of Keep's manuscript.

Marshall B. Davidson, "American House-Warming," Metropolitan Museum of Art, *Bulletin*, N. S. III (March 1945), pp. 176–184, has an 1882 cartoon of Oscar Wilde on a red-hot American stove.

5. Fireplaces were present in castles in the fourteenth and fifteenth centuries. Viollet-le-Duc, *Dictionnaire raisonné de l'architecture française du XI. au XVI. siècle* (10 vols., Paris, 1854–68) vol. 3, p. 206, illustrates a spectacular fifteenth-century triple fireplace in Poitiers. He asserts (p. 207) that many small fireplaces were used in private rooms in castles. Lynn White,

Jr., in "Technology Assessment from the Stance of a Medieval Historian," *American Historical Review*, LXXIX, 1 (Feb. 1974) 1–13, notes the tendency toward privacy and individualism that chimneys and fireplaces in northern Europe encouraged in the eleventh and twelfth centuries. He refers to a graduate dissertation on an aspect of the subject (pp. 8–9).

6. Thomas Tregold, *Principles of Warming and Ventilating Public Buildings*, (3rd ed., London. 1836) pp. 2, 4. The notion that cool air is healthier to breathe than warm air is examined by Billings, *op. cit.* (note 4), p. 213. Billings rejects the notion, but it turns up after 1900, intertwined now with comfort and industrial efficiency. Cf. Palmer, "What Fifty Years Have Done," *op. cit.* (note 4), pp. 344–346.

7. The great ceramic stove ca. 1500 in Hohensalzburg, Austria, is beautifully illustrated in William Anderson, *Castles of Europe* (New York: Random House, 1970) p. 223. A number of ceramic stoves of the eighteenth century are illustrated in Josephine H. Peirce, *Fire on the Hearth. The Evolution and Romance of the Heating-Stove* (Springfield, Mass., 1951) pp. 78–86.

8. Peirce, *op. cit.* (note 7), claims an earlier date for American stoves, but her data hardly support the claim. However, her book, a standard work, illustrates dozens of nineteenth century stoves and lists several hundred patents before 1860. See also the important article by Samuel Y. Edgerton, Jr., "Heating Stoves in Eighteenth Century Philadelphia," Association for Preservation Technology, *Bulletin*, III, 2–3 (1971) pp. 15–104. Many stoves of the mid-nineteenth century are illustrated and a checklist of makers is in John G. and Diana S. Waite, "Stovemakers of Troy, New York," *Antiques*, CIII, 1 (January 1973) 134–144. See also Marcel Moussette, "Quatre fabricants d'appariels de chauffage du Bas-Canada," Association for Preservation

Technology, *Bulletin*, V, 4 (1973) pp. 50–64, which notes a steam system in 1815 for house warming, cooking and washing.

9. *The Papers of Benjamin Franklin*, Leonard W. Labaree, ed. vol. 2 (New Haven: Yale University Press, 1960) pp. 419–446. The common fireplace is purely a radiant heater; it heats no air that will be discharged into the room. In the Franklin system, cold air, from outside the room that contained the Pennsylvania stove, was heated in the stove's air box and discharged into the room. In order to reduce the amount of warm air drawn out of the room merely to supply the needs of the fire, unheated air could be introduced through a flap in the hearth just in front of the fire. Franklin mentioned his debt to Nicholas Gauger, whose *Mécanique de Feu* (1713) described a similar scheme of warming air in an air box set in a fireplace. An illustration of Gauger's fireplace is in A. F. Dufton, *op. cit.* (note 4), pl. XVII. A substantial analysis of Gauger's text is in John S. Billings, *op. cit.* (note 4) pp. 27–29. Franklin's later thoughts on heating and ventilation and his description of a new stove are in American Philosophical Society, *Transactions*, II (1786), 1–36, 57–74. See also J. Pickering Putnam, *The Open Fireplace in All Ages* (Boston, 1882), in which all the variations of "ventilating fireplaces" that one could wish for are illustrated (pp. 32–88).

10. "Franklin stoves," which may still be bought from Sears, Roebuck & Co., are simply cast iron fireplaces; the air box (the essential feature of the Pennsylvania stove) has been omitted. Substantial changes in the masonry of a fireplace were required to install the original Pennsylvania fireplaces.

11. A rotating positive-displacement fan, not centrifugal in its action, is described in Georgius Agricola, *De Re Metallica* (1556), Herbert and Lou Henry Hoover, eds. and trans. (London 1912, New York, 1950) pp. 203–207. A centrifugal fan is described and illustrated in J. T. Desaguliers, *A Course of Experimental Philosophy* (2 vols, London, 1734–44), II, pp. 563–568, pl. XLVI. Desaguliers's fan, rotated by a man he called a "Ventilator," blew fresh air into and sucked foul air out of the House of Commons from 1736 until 1791, perhaps as late as 1819. Cf. Charles Tomlinson, *Cyclopaedia of Useful Arts and Manufactures* (2 vols, London, 1854), II, 931. Incidentally, the Speaker of the House was the control device, giving orders to suck or blow, depending on conditions within the House.

12. Jennifer Tann, *The Development of the Factory* (London: Cornmarket Press, 1970) pp. 109–122. An 1816 installation in Samuel Oldknow's six-story Meller Mill (pp. 115–116), 28 feet wide by over 300 feet long, required over 1500 feet of pipe, whose surface area was about 1500 square feet. The cubic content of the mill was about 500,000 cubic feet. Essentially, the arrangement gave each running foot of each story ($28 \times 10 \times 1 = 280$ cu. ft.) a square foot of heating surface. Neil Snodgrass, whose system won a prize of the Society of Arts, in 1799 installed perhaps the earliest successful British steam heating system. *Ibid.*, p. 111; *Philosophical Magazine*, XXVII (Feb.–May 1807) 172–181.

A paper by Mårten Triewald, of steam engine fame, on heating hotbeds with steam, published in 1739 in the *Transactions* of the Swedish Academy of Sciences, is listed in Mårten Triewald, *Short Description of the Atmospheric Engine* (Cambridge: Heffer, 1928) p. 56.

13. Greville and Dorothy Bathe, *Oliver Evans* (Philadelphia, 1935) pp. 184, 193. I am indebted to Ben Walbert III for these references. In his article, *op. cit.* (note 4), Walbert notices Benjamin Latrobe's claim that in 1807 he recommended steam heat for the hall of the House of Representatives in the Capitol.

14. Jennifer Tann, *op. cit.* (note 12), p. 109.

15. *Ibid.*, p. 109; Acc. 500 Add., Papers of Duplanty, McCall & Co., General file, Inventory, Sept 26, 1820, Eleutherian Mills Historical Library, Greenville, Delaware. An invoice for two stoves, 120 feet of stovepipe, and 6 elbows is in Correspondence, In-file, Nov. 29, 1817.

16. Oliver Evans, *Young Mill-wright and Miller's Guide* (Philadelphia, 1795), Appendix, pp. 5–7, pl. X. In the second edition, 1807, the text is unchanged, but appears on pp. 358–359. A footnote in the first edition (1795) indicates the actual construction of an example by Evans. The plate was reengraved without alteration and the description of the "philosophical and ventilating stove" was modified only slightly when Thomas P. Jones took over editorship of the Guide at its fifth edition in 1826; Walbert, *op. cit.* (note 4), identified central hot air installations in a hospital, an almshouse, an insane asylum, jail, and medical college. U.S., Congress, House Committee on Public Buildings and Grounds, *Report . . . relative to Daniel Pettibone's Petition*, Feb. 19, 1817, mentions the installation of Pettibone's system in 1815 in the Patapsco cotton mill: safe, economical, providing "an agreeable and pleasant heat," an improvement on the principle of the Russian stove.

There are several contenders for the earliest U.S. central heating system. Zachariah Allen, in his manuscript autobiography, p. 29, Rhode Island Historical Society, believed his system of 1820 to be first in New England; I am indebted to Ted Z. Penn for this reference. Claims for Jacob Perkins and Solomon Willard are noted in Society of Architectual Historians, *Journal*, XII, 3(Oct. 1962), p. 145.

17. *Baron Klinkowström's America 1818–1820*, Franklin D. Scott, trans. (Evanston: Northwestern University Press, 1952) pp. 66, 68; Fredrika Bremer, *The Homes of the New World: Impressions of America* (2 vols., New York, 1853), I, 66, cited in Association for Preservation Technology, *Bulletin*, IV, 3–4 (1972) p. 60, n. 139.

18. Thomas C. Grattan, *Civilized America* (2 vols., London, 1859), quoted in Allan Nevins, ed., *American Social History as Recorded by British Travellers* (New York, 1923) pp. 254–255. There is also a minor professional literature on the health hazards of overheated stoves. Tredgold, *op. cit.* (note 6), pp. 2–3, thought that when metal surfaces exceeded 212°F, the air was burned, and "burnt air is neither healthful nor agreeable." Tomlinson, *op. cit.* (note 11), p. 909, said burnt air had a "sulphurous smell." No, Benjamin Franklin had said, *op. cit.* (note 9), p. 439, iron is a "sweet" metal; what you smell is evaporated grease and spittle. The *Scientific American*, n.s. XXIII (Dec. 3, 1870) 359, warned that overheated iron stove plates were "permeable to carbonic oxide, generated in the combustion of coal, a very poisonous gas." The sad case of an English cat, reduced to convulsions by air that had passed over a hot stove, was duly reported in the *Philosophical Transactions* and repeated in U. S., Serial set no. 545 (see note 60) p. 9.

19. Marquis de Chabannes, *On Conduction of Air by Forced Ventilation* (London, 1818), described hot air, hot water, and steam heating systems. His plate XIII shows details of a hot water system for a hot house. Daniel Pettibone, *Pettibone's Economy of Fuel; or Description of His improvements of the Rarefying Air-stoves . . . for Warming and Ventilating . . . with or Without the Application of Steam* (2nd ed., Philadelphia, 1812; 62 pp.) pp. 38–40. The Windsor Palace orangery installation is illustrated in the appendix (by T. Bramah), of Tredgold, *op. cit.* (note 6). Charles Hood, *A Practical Treatise on Warming Buildings by Hot Water*

(2nd ed., London 1844), p. 3, notes that in 1822 Atkinson designed what became the conventional hot-water apparatus. Cf. David B. Reid, *Illustrations of the Theory and Practice of Ventilation* (London, 1844) p. 244. I have not identified Atkinson or the other men mentioned there.

In the "number of recent conversations at the Institution of Civil Engineers," reported in Franklin Institute, *Journal*, XIV (Sept. 1832) 211-212, three or four hot water installations were mentioned by the members.

20. Tredgold, *op. cit.* (note 6), p. 12.

21. Susan Reed Stifler, *The Beginnings of a Century of Steam and Water Heating by the H. B. Smith Company* (Westfield, Mass, 1960) pp. 34, 68, 100-102. There were two crucial inventions in American domestic hot water heating systems. The first was the sectional boiler, invented in 1859 by Samuel F. Gold, of Connecticut, which simplified boiler construction and permitted a new boiler to be carried literally in pieces down narrow cellar stairs of old or new houses. The second was the cored, cast iron, sectional radiator, attempted unsuccessfully in the early 1870s and perfected by several makers in the 1880s. Billings, *op. cit.* (note 4), p. 238, in 1893 advised hot water radiator pipes no smaller than 3 inches in diameter, because "friction increases rapidly with the reduction in the diameter of the pipe, and thus impedes the circulation."

22. *Society for the Illustration and Encouragement of Practical Science, Gallery for the Exhibition of Objects Blending Instruction with Amusement, Adelaide Street and Lowther Arcade, West Strand. Catalogue for 1835* (11th ed., London, 1835;48 pp.) p. 14. Angier's patent, dated July 30, 1831, was noticed in Franklin Institute, *Journal*, XIV (July 1832) 45-49. A biographical sketch of Angier Marsh Perkins (1799?-1881) appears in the *Dictionary of National Biography*.

23. The best description of the Perkins system is in the promotional book by Charles J. Richardson, *A Popular Treatise on the Warming and Ventilating of Buildings* (2nd ed., London, 1839), *passim*.

24. J. Mordaunt Crook, "Sir Robert Smirke: A Pioneer of Concrete Construction," Newcomen Society, *Transactions*, XXVIII (1965-66), 5-22; see p. 5. Smirke employed other heating systems in the British Museum and elsewhere, but he heated his office with a Perkins system; he put one in the Custom House and one in the temporary houses of Parliament, following the fire of 1834.

25. Richardson, *op. cit.* (note 23), pp. 38-

39, 42, 44, 45, 47, 119, 121. A Perkins system, installed in Kew Gardens, is illustrated in Derek Linstrum, *Sir Jeffrey Wyattville, Architect to the King* (Oxford: Clarendon Press, 1972) p. 208.

26. Torres, *op. cit.* (note 3).

27. John Davies and George V. Ryder, *Report on Perkins's System of Warming Buildings by Hot Water* (Manchester, 1841; 15 pp.). This report was the subject of a "conversazione" in the Royal Victoria Gallery of Practical Science, Manchester; the report and the ensuing discussion were published in Sturgeon's *Annals of Electricity*, VI (1841), 475-499.

28. *National Cyclopaedia of American Biography*, vol. 24 (1935) p. 150; portrait opp. The judgment on Nason's eminence is mine, hopefully supported in the present paper. Nason died in 1872.

29. Lewis W. Leeds, *Treatise on Ventilation: comprising seven lectures delivered before the Franklin Institute, Philadelphia, 1866-68* (New York: John Wiley & Son, 1871). His home in 1867 was in Germantown; he was addressed as "heating and ventilating engineer" in one of the solicited testimonials (p. 218); Leeds's slogan is displayed on the title page.

30. J. R. Partington, *A History of Chemistry*, vol. 3 (London 1962) pp. 196, 253, 284-286 (Priestley on respiration), 471-479 (Lavoisier on respiration). Early and influential surveys of hospitals by Lavoisier are reported in Denis I. Duveen and Herbert S. Klickstein, "Antoine Laurent Lavoisier's Contributions to Medicine and Public Health," *Bulletin of the History of Medicine*, XXIX, 2 (March-April 1955) pp. 164-179.

The importance of pure air was well known to miners and sailors. Agricola, *op. cit.* (note 11), pp. 200-212, has a section on mine ventilation. Stephen Hales, the Englishman whose influence upon Lavoisier was traced in Henry Guerlac, "The Continental Reputation of Stephen Hales," *Archives internationales d'histoire des sciences*, IV (April 1951) pp. 393-404, was concerned with ventilation of ships. I was introduced to the windsail in a troopship in the tropics in World War II; it was and probably still is the most effective ventilator of troop spaces in a ship's hold.

Original documents of the sixteenth and seventeeth centuries illustrating discoveries that led to our current understanding of respiration are given and sensibly explained in Mark Grabaud, *Circulation and Respiration: The Evolution of an Idea* (New York, 1964).

31. T. Antisell, *Hand-Book of the Useful*

Arts [Putnam's Home Cyclopedia] (New York, 1852) pp. 663-664. Calculations and conclusions are to be found in many places, e.g., Byrne and Spon, eds., *Spon's Dictionary of Engineering* (London, 1874) pp. 3024-3025, which summarized the findings of Péclet, Morin, Arnott, and others. For long term changes in quantities of air considered necessary, see Palmer, *op. cit.* (note 4), p. 344. A chemical solution to the problem of vitiated air was suggested in the 1820s by a chemist in Manchester. Where work rooms were too small and too crowded to be adequately ventilated, he wrote, one might obtain a "supply of oxygen from manganese." His method was not further elaborated. See C. Turner Thackrah, *The Effects of the Principal Arts, Trades and Professions . . . on Health and Longevity* (Philadelphia, 1831) p. 33.

32. Reid, *op. cit* (note 19), p. 176. Reid is noticed in *Dictionary of American Biography*.

33. *Spon's Dictionary, op. cit.* (note 31), pp. 3024-3025. The term "breathed over" appears in *The Horticulturist and Journal of Rural Art and Rural Taste*, V (Nov. 1850) p. 205.

It is easy to dismiss the fresh-air advocates as fools or charlatans, but the matter is not as simple as that. Epidemic diseases, such as yellow fever and cholera, were thought to be caused by a "miasma" that arose from swamps or other putrid sources. By 1849, the nature of the miasma was being debated—carbonic acid gas, electricity, ozone, fungi or animalculi—but of 146 physicians sampled in that year only four considered microscopic germs to be a possible cause of disease. Cf. Charles Rosenberg, *The Cholera Years* (Chicago, 1962) pp. 75n22, 164n31, 165-166. Andrew Ure, in his *Dictionary of Arts, Manufactures, and Mines* (4th ed., 2 vols., London, 1853) II, 890, assumed that stoves in the Custom House were the cause of "the malaria which prevailed there."

34. *The Horticulturist, op. cit.* (note 33), V (Nov. 1850) pp. 201-202.

35. Leeds, *op. cit.* (note 29), p. 13

36. Leeds, *op. cit.* (note 29), pp. 98, 155, 187.

37. Catharine E. Beecher and Harriet Beecher Stowe, *The American Woman's Home: or, Principles of Domestic Science* (New York, 1869) pp. 47-49.

38. Reid, *op. cit.* (note 19), p. 274

39. Precursors of cooling and "air-conditioning" will be found throughout the nineteenth century; that is a subject whose early history has not yet been traced. For a framework, see Margaret Ingels, *Willis Haviland Carrier: Father of Air Conditioning* (Garden City,

1952) pp. 107–170, a "Chronological Table of Events Which Led to Modern Air Conditioning 1500–1952."

40. Reid, *op. cit.* (note 19), pp. 270–299.

41. Reid, *op. cit.* (note 19), p. 297.

42. Tomlinson, *op. cit.* (note 11), p. 930.

43. U.S., Serial set no. 1211 (see note 60), p. 31.

44. U.S., Serial set no. 994 (see note 60), p. 96. Meigs was referring specifically to the heating and ventilating system.

45. U.S., Serial set no. 545 (see note 60).

46. Billington, *op. cit.* (note 4), p. 126. An earlier proposal for heating the House wing was made in 1843 by A. A. Humphreys, first lieutenant, Topographical Engineers. Humphreys reported that, until a few weeks earlier, heating and ventilating was a "subject upon which I have never bestowed any attention." Leaning heavily on Tredgold and Ure, he recommended steam coils and a "spiral fan," similar to one giving good service, according to Ure, in the Reform Club House, London. See U.S. Serial set no. 441 (see note 60); Ure, *op. cit.* (note 33), p. 890.

47. U.S., Serial set no. 1211 (see note 60), p. 50.

48. U.S., Serial set no. 1031 (see note 60), pp. 1–2.

49. *Ibid.*, pp. 3, 6, 7, 8. Although the services of consultants are frequently obtained in this way, at no extra charge, the implications regarding the kind of advice thus made available are seldom thought much about. It was not to be expected, for example, that Nason, committed as he was to a system employing pipe coils, would recommend an alternative system that would not require large quantities of the pipe he had for sale.

Before the design was fairly under way, Nason's firm Nason and Dodge, in what appears to me to be a promotional puff, wrote a letter to Meigs informing him that Péclet, the French author of a standard work on heat and heating, had just published a new book, *Nouveaux documents relatif au Chauffage* (1854), which gave more fully than his earlier volumes the results of recent French experience. Furthermore, the letter continued, Mr. Nason had "added to his experience with further investigations at Utica," where the heating system in the Insane Asylum had proved more than adequate during the previous winter.

50. Wrought-iron pipe for the Perkins heating systems was supplied in England by the Russells of Wednesbury, Staffordshire, who, before 1835, were making lap-welded pipe by running formed skelp between grooved rollers; a supported spherical mandrel, inside the pipe, provided the pressure to complete the weld. The Philadelphia firm of Tasker, Morris, and Tasker, later Pascal Iron Works, made the first wrought iron pipe in the United States. In 1848–49, James Walworth (Nason's first partner), after visiting the Russells in England, organized and operated a tube mill near Boston. The consulting engineer to the tube mill was Nason. Five brief but useful articles on wrought-iron pipe making and early steam heating in U.S. are in *Valve World*, I, 1 (Jan. 16, 1905) 7; I, 2 (Feb. 1905), 5, 7; I, 9 (Oct. 1905), 5; I, 11 (Dec. 1905), 7; II, 1 (Jan. 1906), 7. Additional details are in Richardson, *op. cit.* (note 23), and in the biographies of James Walworth in *National Cyclopaedia of American Biography,* XVIII (1922) pp. 374–375, and Joseph Nason in *ibid.*, XXIV (1935) p. 150. Nason is credited here with the invention of the term "radiator." R. T. Crane, in the *Valve World*, I, 11 (Dec. 1905), 7, credits Nason with the present design of the globe valve, and also a most elegant invention: taper threads for pipe and fittings. The standard pipe thread assumed Briggs's name. See references in C. P. Auger, *Engineering Eponyms* (London: The Library Association, 1965), p. 18.

51. U.S., Serial set no. 1031 (see note 60), pp. 9–11; U.S. Serial set no. 1263 (see note 60), pp. 2–3. An apparently similar system was installed in 1859 to ventilate the Cooper Union building in New York City. The following description is quoted from *The Cooper Survey,* November 1859, in the Historic American Engineering Record (National Park Service) report of "The Cooper Union for the Advancement of Science and Art, HAER NY-20," page 59:

"The basement contains a large hall which will seat 2500 persons. A steam engine in an adjoining room drives a fan 14 feet in diameter, by which warm or cold air may be thrown into the several halls at pleasure; the air being admitted to the large hall through small holes under each seat so that the most perfect ventilation is secured without disagreeable currents of air."

Who designed this heating and ventilating system I do not know. The HAER report, p. 22, attributes it to Peter Cooper's son Edward; the source cited does not support the attribution.

52. U. S. Serial set no. 1263 (see note 60) pp. 15, 23. Briggs based his design upon methods suggested by E. Péclet, *Traité de la chaleur* (2nd. ed., 3 vols., Paris, 1843). He described and explained his design in Robert Briggs, "On the Conditions and Limits Which Govern the Proportions of Rotary Fans," Institution of Civil Engineers, *Minutes of Proceedings,* XXX (Session 1869–70), pt. II, pp. 278–308. Comments by prominent English engineers in his audience made clear their disdain for the rational approach to design employed by Briggs. Briggs's understanding of several aspects of vitiation of air and conditions for comfort are given at some length in his paper "On the Ventilation of Halls of Audience," American Society of Civil Engineers, *Transactions,* X (1881) pp. 53–106. He refers to other sources of further details of the Capitol heating and ventilating system. His obituary notice is in *loc. cit.* XXVI (1896) pp. 542–545. He died in 1882, aet. 50.

53. Billings, *op. cit.* (note 4), pp. 38–39, described Briggs as "a well-known American engineer, who died in 1882." Through his numerous papers, given before engineering societies, "as well as through those trained in the shops under his direction, he has exercised a very considerable influence on . . . heating and ventilating during recent years."

54. U.S. Serial set no 1211 (see note 60), p. 49. Billings, *op. cit.* (note 4), p. 198, noted the phenomenon as a sheet of air "swept across the ankles of the member just in front [of the register]. When it had passed his desk and reached the next riser, it was reinforced by a fresh stratum, and the honorable member next in front received the upper current on the calves of his honorable legs while the floor current swept his ankles, to his great discomfort and dissatisfaction."

55. U.S., Serial set no. 1118 (see note 60, below). Meigs, who was on the speakers' stand at the opening of the Philadelphia Centennial Exhibition, May 10, 1876, inserted a half-page "Note on the Direction of Currents in a Crowd in Open Air" in Franklin Institute, *Journal,* LXXII, 2 (Aug. 1876) p. 85. The key passage: "a striking example and illustration was observed of *the fact that a crowd lives only by the aid of the ascending current of the persons composing it.*" [italics added].

56. U.S., Serial set no. 4585 (see note 60), p. 847. See also later comments by Henry Dawes and Benjamin Harrison in David J. Rothman, *Politics and Power: The United States Senate, 1869–1901* (New York: Atheneum Reprint, 1969), p. 143.

57. Leeds, *op. cit.* (note 29), p. 115. Discussions of proposed modifications of the Capitol heating and ventilating system were of widespread interest. Besides Leeds's study, pp. 114–127, I have noted an article in Edward H. Knight, *Knight's American Mechanical*

Dictionary, vol. 3 (Boston, 1876) pp. 2704–2705. Cf. also note 56, above.

58. Paul Goldberger, "Energy Crisis May Doom Era of Glass Towers," *New York Times,* December 6, 1973, pp. 45, 85; a similar article appeared in *VDI Nachrichten* [organ of the German Engineers' Society], XXVIII, 2 (Jan. 11, 1974) pp. 1–2.

59. It is well to read and ponder Reyner Banham, *The Architecture of the Well-tempered Environment* (Chicago, 1969), and the extended review of the book by John Kouwenhoven in *Technology and Culture,* XI, 1 (Jan. 1970) pp. 85–93. Banham notes (p. 17), "The history of the mechanization of environmental management is a history of extremists, otherwise most of it would never have happened." Amen. In another place he writes, the "arts of both ventilation and heating really waited upon the development of effective blowing fans."

60. Titles in the U.S. Serial set are arranged below by volume serial number. They are as listed in John R. Kerwood, *The United States Capitol;* *an Annotated Bibliography* (Norman, Okla., 1973).

Serial 441. Kerwood 1279.
U.S., Colonel of the Corps of Topographical Engineers, *Report,* Nov. 30, 1843 (28th. Cong., 1st. sess., H. Doc. 51, pp. 1–20).

Serial 545. Kerwood 1552.
U.S., Congress. House, Committee on the Public Buildings, *Report,* July 26, 1848 (30th Cong., 2d. sess., H. Rept. 90).

Serial 994. Kerwood 1682.
U.S., Congress. Senate, Committee on the Public Buildings, *Report,* Feb. 21, 1859 (35th Cong., 2d. sess., S. Rept. 388).

Serial 1031. Kerwood 1851.
U.S. President, *Message . . . Communicating . . . Information in Relation to the Heating and Ventilating of the Capitol Extension,* 1860 (36th Cong., 1st. sess., S. Ex. Doc. 20).

Serial 1118. Kerwood 1110.
U.S., Captain of Engineers in Charge of the U.S. Capitol Extension, "Report of Operations," Nov. 26, 1861 (37th Cong. 2d sess., S. Ex. Doc. 1, pp. 77–83).

Serial 1211. Kerwood 1641.
U.S., Congress, Joint Select Committee to Examine into the Present Condition of the Senate Chamber and the Hall of the House. . . ,*Report,* Feb. 20, 1865 (38th Cong., 2d sess., S. Rept. 128).

Serial 1263. Kerwood 930.
U.S., Architect of the U.S. Capitol. To the Secretary of the Interior, 4 May 1866. *On Warming and Ventilating the Capitol: Letter from the Secretary of the Interior in Answer to a Resolution of the House of the 4th. Instant, Transmitting a Report of T. U. Walter Relative to Warming and Ventilating Both Houses of Congress* (39th Cong., 1st. sess., H. Ex. Doc. 100) 1866 95 pp.

Serial 4585. Kerwood 274.
Documentary History of the Construction and Development of the United States Capitol Building and Grounds, 1904 (58th Cong., 2d. sess., H. Rept. 646).

CHAPTER
11

Early Nineteenth-Century Lighting

LORIS S. RUSSELL

Dr. Russell is Curator Emeritus of the Royal Ontario Museum, Toronto, and formerly the Director of the National Museum of Canada. He is the author of *A Heritage of Light* (1967), *Everyday Life in Colonial Canada* (1973), and other works.

Artificial lighting for interiors is a subject that differs from most other aspects of Building Early America in that the devices used were, for the most part, brought into the house and were not part of the architectural structure. Only when gas lighting, and much later electric lighting, arrived was the lighting system built in. If an excuse were needed for including the subject of domestic lighting in this book, one might be that the art provides a chronological framework on which one can hang many other developments in the building of domestic and public structures. Just as geologists use fossils as a means of dating rocks, so the antiquarian can date interiors and architectural designs by the lighting devices involved. This works both ways, because the nature and the date of the architecture stipulate the kind of lighting that would have been in use at that particular time and place.

Modern reconstructions, I am sorry to say, create many anachronisms in the field of lighting. I shall mention only one. The famous motion picture *Gone with the Wind,* in an effort to depict the opulence of ante bellum society in the southern states, portrayed lighting devices that were not known until the 1870s and did not come into general use until the 1890s. Hence the horrible designation "Gone-with-the-Wind lamp" for something that was characteristic of the late Victorian parlor.

Lighting for interiors goes back to the fireplace that the cave man built for heat, to warm himself and cook his food, but which also provided artificial light. When the Europeans came to North America and set up domestic establishments, they brought with them the lineal descendant of this fireplace, the domestic hearth. This was the source of comfort and of cooked food, and also the main illumination of the household (Fig. 11.1). Almost all systems of domestic lighting can be looked upon as extensions of the fireplace. They were devices for taking some of that fire and moving it elsewhere to provide an independent source of light. We can paraphrase the old saying and, with the exceptions to be noted later, pronounce that where there is light there is fire.[1]

As a source of light the fireplace was very inadequate; long before the eighteenth centu-

Fig. 11.1 Reading by firelight. *The Boyhood of Lincoln* by Eastman Johnson, 1868. *(The University of Michigan Museum of Art, gift of Henry C. Lewis)*

Fig. 11.2 Candle holder, candle mold and candle extinguisher. *(National Museum of Man, Ottawa)*

ry, something better had come into use — the candle. It seems hardly necessary to describe a candle, but by definition it is a lighting device in which the wick is imbedded in the fuel, which is solid and self-supporting. The typical candle holder of the eighteenth and early nineteenth centuries had a long tube socket for the candle, with a little handle projecting from a slot in the side. This handle is connected to a sort of piston, by means of which the candle can be moved up or down. This kind of holder, shown in Figure 11.2, was especially designed for the tallow candle, made from the solidified fat of cattle or sheep. At ordinary temperatures tallow is a solid, but with warmth it tends to collapse. By having a long socket with most of the candle inside, one could push up a little at a time as it burned away and thus prevent the remainder from falling over.

Tallow candles were made by two methods, molding and dipping. The mold had six tubes which tapered slightly and ended below in an inverted cone with a perforated tip. The string-like wicks were threaded through these holes and tightened. Then the liquid tallow, melted in a pot with a little water in the bottom to prevent burning, was poured into the upper ends of the tubes. When the tallow cooled the candles could be withdrawn easily after dipping the mold in hot water.

The other method of making tallow candles was by dipping the wicks into melted tallow, then taking them out and hanging them up until the tallow hardened, repeating the process until there was a sufficient thickness of tallow built up around the wick. It was a slow, tedious operation but some people preferred it to the molding method. The great advantage of molding candles was that any amount of tallow could be used, large or small: as few as one candle could be made at a time. The dipping method, however, required a large amount of tallow to provide sufficient depth for the wicks.

A practical disadvantage of tallow candles was that they were edible. It is said that some Arctic expeditions in the early nineteenth century stipulated that their candles be tallow, something to fall back on in an emergency. But if humans could eat them, so could rats and mice, so it was customary to put the tallow candles into some kind of solid, tight container called a candle safe, of which the cylindrical type was the most characteristic; it could be hung on the wall, safe from the ravages of rodents.

Tallow candles had a third bad property. If the slightest bit of unburned carbon ("snuff") from the wick fell on the rim of the burning candle, it would cause a gutter, through which liquid tallow would run off. This could be remedied temporarily by using an instrument called a candle snuffer. It looked like a pair of scissors, but one blade had a box-like receptacle to catch the bits of partly burned wick as they were snipped off. Here is one of the curious anomalies of nomenclature: when people speak of snuffing a candle they probably mean putting it out, but the candle snuffer was designed to improve the burning of the candle. The true extinguisher was a small metal cone on the end of a rod-like handle.

Another type of candle holder had a convex rather than saucer-shaped base. This apparently gave greater stability although it would not catch the tallow drip as well. It was known as the hog scraper candle holder because it resembled the device used to dehair the carcasses of freshly slaughtered pigs. It is hinted that the farmer might use the candle holder if he could not find the real hog scraper.

These simple candle holders were not elegant enough for some people. There were the familiar brass and silver candlesticks, partly cast, partly turned, and still common as decorative pieces. Then there were the multiple candle holders. In the 1840s a very popular type was the girandole, which came in sets of three: a three-socket holder for the center of the mantle (Fig. 11.3) and single-socket versions for the ends. These elaborate devices got their name from a French word for fireworks and were characterized not only by branching, but also by brass decorations and glass pendants and perhaps a bas-relief placque at the base.

Another form of multiple candle holder was the chandelier, which hung from the ceiling and, like the girandole, could have elaborate

Fig. 11.3   Girandole. *(Author's collection, Royal Ontario Museum photograph)*

ornamentation and numerous prism-like pendants (Fig. 11.4). Such fixtures would have been used in public buildings or the mansions of the wealthy. For less opulent establishments a chandelier of sheet metal (tole) would be appropriate, consisting of four or more sockets in a circle, supported by straight or gracefully curved arms.

For those who could afford them, spermaceti candles, made from the wax in the head of the sperm whale, replaced tallow candles in the 1830s. Spermaceti candles gave a steady light and they did not collapse nor readily form gutters. But in a few years, as whale hunting became less successful, spermaceti candles reached a prohibitively high price. In the 1850s the stearine candle appeared, made from chemically purified animal or vegetable fat. Finally, about 1860, the paraffin candle became available, a by-product of the emerging petroleum industry. This is still with us,

Fig. 11.4   Glass chandelier, Chateau de Ramezay, Montreal.

Fig. 11.5   Double crusie, burning suet. *(Author's collection, Royal Ontario Museum photograph)*

mostly for decorative purposes, but useful occasionally when the electric power fails.

Another type of lighting device that the early settlers of this continent brought with them was the pan lamp. This has its origins in the piece of rock with a hollow scraped in it, filled with grease and a bit of plant fibre, that may have been man's first lamp. Unlike the candle, the lamp uses fuel that is not self-supporting, but has to be contained in some sort of vessel, called a font. One of the early forms of pan lamp is known to collectors by the Scottish name of crusie, although it may be of French origin. It was a small pan with the side opposite the handle drawn out to a point (Fig. 11.5). The fuel was animal fat of some sort. The wick, a strip of cotton rag, was draped over the tip and it had a habit of dripping grease while burning. To guard against this, a second pan was hung underneath to serve as a drip catcher. Even so, it was not a very clean type of

lighting device and was usually restricted to the vicinity of the fireplace. The peculiar handle, which looks like a boat hook, could be caught on the edge of the mantle or driven into a chink in the wall.

The problem of the dripping pan lamp was solved in another way — by an invention, characteristic of the Pennsylvania German communities, the betty lamp. There was a special trough for the wick, at about a 45° angle, inside the margin of the pan, so that any drip would fall within the lamp. This efficient device was brought by Pennsylvania emigrants to Upper Canada (Ontario), where it persisted in use until at least 1840. In New England a pan lamp with an internal wick trough was made of wrought iron rather than sheet metal.

If you had said "lamp" to a person in the early nineteenth century, he would probably have thought of the whale-oil lamp (Fig. 11.6). The characteristic burner with its two wick

Fig. 11.6 Whale-oil lamp burning whale oil. *(Author's collection, Royal Ontario Museum photograph)*

changed the whole direction of lamp development and began the history of scientific lighting. A frustrated Swiss chemist, Ami Argand, unable to get adequate light for his midnight studies on fermentation, developed the first truly scientific illuminating device, the Argand lamp shown in Figure 11.7. The principle of this lamp is that the wick is tubular, i.e., hose-shaped, and is supported both on the outside and the inside by metal tubes. The ingenious arrangement of tubes and font permits the fuel to reach the wick between the tubes, while air can enter the inner tube and flow up to provide oxygen to the inside of the circular flame at the same time that air around the burner is reaching the outside of the flame.

Fig. 11.7 Argand lamp. *(New Brunswick Museum, Saint John)*

tubes is based on a principle supposed to have been discovered by Benjamin Franklin: that two adjacent flames produce more light than the same two flames separated, due to mutual augmentation of the draft. This type of lamp with short wick tubes and low, heavy glass font is typical of the early nineteenth century. But more elaborate lamps were being produced by the blossoming glass industry of New England, many with the lower part of the font of molded glass and the upper part handblown. They had the same type of double-tube burner, which often was not screwed on but plugged in like a cork. Whale-oil lamps were also made of metal, such as bronze, brass and pewter.

Something happened in France in 1783 that

Argand also invented — it is said, by accident — the glass chimney which, by concentrating the rising air above the flame, increased the draft of air from below. He also incorporated a spiral device within the burner for raising and lowering the wick by turning the chimney support, and the reservoir supplied fuel to the font by the Cardan fountain or "bird-bath" principle.

Double-burner and multiple-burner Argand lamps were made, with two or more lamps fed from a common reservoir. There is a handsome double Argand lamp of silver in the Smithsonian Institution's Museum of History and Technology that was once the property of George Washington.

Argand found it impossible to get support in France for his invention, owing to the political unrest; he went to England, where he was enthusiastically received and took out a patent in 1784. This is the first patent for a scientifically designed lighting device. He was offered a pension in England but he chose to go back to France, where he found that other people had stolen his idea and were making Argand lamps without even the courtesy of naming them after him. But in England his fame was preserved and Argand lamps continued to be made there for many years, some for export to North America. On this side of the Atlantic, at least, these devices were called mantel lamps (Fig. 11.8) because they came in sets of three and were placed on the fireplace mantle, like the girandoles.

There were certain disadvantages to the ordinary Argand lamp, one of them being that the reservoir cast a shadow. In 1809 a French inventor, J. A. Bordier-Marcet, made a reservoir in the shape of a ring, with the font and burner in the center. Light from the burner could come down through the opening in the ring. His design was for a hanging lamp, providing light from above, so it was called *lampe astrale,* or astral lamp. Another Frenchman, J. F. Chopin, mounted the astral lamp on a table stand and produced the *lampe à coronne.* The climax of this line was the invention by George Phillips of London in 1812 of the sinumbra or shadowless lamp (Fig.

Fig. 11.8 Mantle lamp, double arm. Made by Messenger & Sons of London and Birmingham for Jones, Low & Ball, Boston ca. 1839. *(Author's collection, Royal Ontario Museum photograph)*

11.9). This was Chopin's crown lamp with the ring-shaped reservoir having an obliquely oval cross-section, which reduced the shadow to a narrow area next to the lamp. In order to get maximum benefit from the design the lamp had to be tall, with a glass shade curving up to a narrow neck. If a chimney was also used, it was short and entirely within the shade. The burner was mounted on the upper end of the supporting column and was of the later Argand type, with spiral wick raiser and slotted drip catcher.

While these elegant sinumbra lamps were coming into use in the 1840s the price of whale oil was going up; by 1845 the cost of the better grades had risen to about a dollar a gallon. The cheaper grades were too heavy to flow readily in the lamps. Also available were vegetable oils — such as colza oil made from rape seed — which were cheaper than whale oil, but much too viscid for lamps. So efforts were made to devise lamps that could use these cheaper but heavier fuels, and about 1850 a French inventor named Hadrot produced a lamp in which the flow of oil was stim-

and one of those tried was lard, which was readily available and burned nicely in pan lamps. But because lard is more or less solid at ordinary temperatures it was unsuitable for the conventional oil lamp. To make lard flow it was necessary either to melt it or to squeeze it. For this reason lard lamps were made of metal, which would transmit heat and withstand pressure.

A simple lard lamp depending on heat had two flat wicks with an opening for an air draft between, a simplification of the Argand principle (Fig. 11.10). Lard was melted and poured into the font, making sure that the cotton wicks, if new, were well saturated. To start the lamp after the lard had solidified, a little was spooned out and placed in the priming pan. Here it could be melted with a match flame, providing enough liquid fuel to start

Fig. 11.9   Sinumbra lamp. *(Author's collection, Royal Ontario Museum photograph),*

Fig. 11.10   Flat-wick center-draft lard lamp burning lard, purchased Windsor, New York. *(Author's collection, Royal Ontario Museum photograph)*

ulated by means of a spring. This was called the moderator lamp. The lower, cylindrical part contained a spring-loaded piston. In the long stem that projects upward from the cylinder there was an inner tube to feed the oil to the burner, containing a steel rod which kept the oil from gushing out under the pressure of the spring. The burner was of the Argand type, with the wick height controlled by a rack and pinion. There would have been a tubular glass chimney. Many moderator lamps had the basic parts enclosed in a handsome ceramic or brass casing.

Other substitutes for whale oil were sought

flames on the wicks. Heat from these was conducted down the metal wick tubes to the lard in the font, melting it and providing liquid fuel to maintain the flames of the burner. Lard lamps working on this principle appeared in 1842, and one with a true Argand tubular-wick burner in 1856.

A popular type of lard lamp using heat to make the lard fluid was the Kinnear lamp, patented in 1851. This had a very broad, flat wick, the flame of which provided plenty of heat to melt the lard in the hatchet-shaped font. There was also an auxiliary wick tube, which functioned as a night light and kept a little lard melted for ready starting of the main wick.

Another lard lamp, the fuel flow, was patented in 1842 by B. K. Maltby and J. Neal. The font was in the form of a hollow piston, with the wick tubes inside, and was pushed down into the reservoir as the lard was consumed. As the font by this time must have been hot, it would have required a cloth pad to grasp it safely. When the piston reached bottom, the lamp would have to be opened and the reservoir refilled.

There was a lamp of German origin manufactured in Philadelphia that was characteristic of the lard era. This was the solar lamp. A variation on the Argand lamp, it was originally designed to burn oil. It differed from the typical Argand lamp in that there was no reservoir, the large metallic font around the burner containing all of the fuel. The burner had a low, dome-shaped deflector which concentrated the flame into a tall, narrow cone. It was found, soon after these lamps came into general use, that they worked very well with lard. Robert Cornelius of Philadelphia, who made many of these lamps, took out a patent for heat-conducting plates to increase melting of the fuel. It is said that solar lamps using lard had to be cleaned each time that they were used in order to retain a good flow of heat.

Most solar lamps now in collections have been converted at a later date to burn kerosene. The example shown in Figure 11.11 is original except for the chimney. The Argand burner has the spiral wick raiser, which is op-

Fig. 11.11   Solar lamp. *(Author's collection, Royal Ontario Museum photograph)*

erated by turning the rim on which the shade rests. The popularity of solar lamps in the 1840s is shown by the illustration reproduced in Figure 11.12: almost all of the lamps that can be seen through the window and the door are solar lamps.

Another substitute for whale oil was "burning fluid," popularly but erroneously called camphene. Burning fluid was a mixture of very strong alcohol with redistilled turpentine, the latter being the true camphene. Burning fluid was an excellent lamp fuel. It was light and clear, it flowed readily up the wick and it burned with a clean white flame. It had one small disadvantage: it exploded at the slightest increase in pressure. In fact, it blew up sometimes when the lamp wasn't even lit. The typical burning-fluid lamp (Fig. 11.13) had a thick

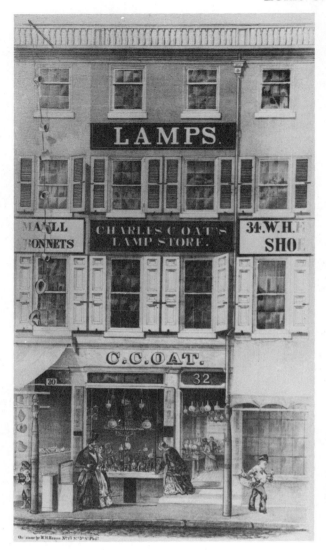

Fig. 11.12 Oat's lamp store, Philadelphia. Broadside lithograph, ca. 1848. *(Historical Society of Pennsylvania)*

Fig. 11.13 Burning-fluid lamp in operation, burning mixture of alcohol and rectified turpentine. *(Author's collection, Royal Ontario Museum photograph)*

glass font, relatively high and narrow. The burner was superficially like that of the whale-oil lamp, but differed in that the wick tubes were long and diverging, and were provided with little caps on chains. These caps served a double purpose: as extinguishers and to prevent evaporation of fluid when the lamp was not in use. Because the glass was thick it provided an opportunity for ornamentation in the mold; for years the fonts were collected as attractive glass objects and the burners discarded. Now we are searching for authentic burners to restore the lamps to their original state.

Burning fluid had a long history of fatal ac-

cidents that resulted from spillage or explosion. The *Scientific American* for 1853 reported thirty-three such occurrences in the previous year. An example from Canada was the burning of the St. Louis Theatre in Quebec City on June 12, 1846. This was the entertainment center for the garrison and on that occasion was offering a special display of what we would today call illuminated dioramas. At the close of the show somebody knocked over one of the burning fluid lamps and in a few minutes the whole place was afire. There was panic, with the inevitable congestion at the exit. Forty-five persons lost their lives as a result. This disaster, not so great as others from the

194

same cause, showed how rapidly a fire could spread when fed by burning fluid.

Many ingenious devices were tried in an attempt to reduce the danger from burning fluid lamps, but none was very effective. Perhaps the most drastic was the conversion of the conventional lamp into a vapor lamp. The wick tube was closed at the top except for six pinholes under the rim. In use the burning fluid would saturate the internal wick and would be vaporized by holding a burning match against the wick tube. The fumes emerged through the pinholes and were lit to produce a crown of flame. The same principle was used later with gasoline and survives today in the Coleman gasoline lantern.

Among the other lamp fuels tried was rosin oil, made by distilling pine rosin. A number of patents were issued in the 1850s for lamps burning this fuel, the most successful being the Blake lamp. Rosin oil is a heavy liquid and requires careful adjustment of the draft to produce a good flame, a feature provided for in the Blake lamp.

In 1846 the whole picture of domestic lighting changed. A physician in Nova Scotia, Abraham Gesner by name, had tried all his life to be something other than a doctor of medicine. In that year he discovered that by distilling oily coal at a lower temperature than that required to produce illuminating gas, he could get a liquid that would serve as an excellent lamp fuel. Later he christened it kerosene, which deviously is supposed to mean wax-oil. He preceded by two years the invention of paraffin oil, the same thing, by James Young in England.

Gesner set up a kerosene manufactury in Brooklyn in 1855, where the fuel was made from oil shale and coal, but he recognized from the first that petroleum would be the ideal source of the substance. In 1858 a man named J. M. Williams dug a hole in the ground at Oil Springs, Canada West (Ontario) and bailed out petroleum. In Canada we think that this was the world's first oil well. It depends on the definition, but if an oil well is a hole in the ground from which you get petroleum, we are right. The following year, on

August 27, 1858, "Colonel" Edwin L. Drake, at Titusville, Pennsylvania, brought in the first drilled well and in the process invented the technique of drilling through casing. Soon there were many producing wells in both Ontario and Pennsylvania, providing an abundant and cheap source of kerosene, so that almost everyone could afford the new lamp fuel (Fig. 11.14).

At first the burning-fluid and solar lamps were adapted to burn kerosene, but soon a characteristic type of lamp emerged, with a globular glass font, a cylindrical brass stem, and a square marble base which remained popular until the late 1860s. Most of them were made in the Boston area or at Pittsburgh. There was great variety in the patterns of the pressed-glass fonts; on some the double layer or "cased" technique was used to create designs in color. There were also more plebian

Fig. 11.14 Advertisement for kerosene. The new fuel quickly swept across the world in the middle nineteenth century. *New-Haven Daily Register*, December 31, 1859. *(Yale University Library)*

lamps, with blown fonts and fabricated handles. As for the burners, neither those for burning fluid nor whale oil were satisfactory for kerosene. When Gesner set up his factory in Brooklyn he engaged two brothers named Austen as public relations men. One of these searched Europe for a suitable burner, which he found in Austria and brought back to the United States as the Vienna burner. It was the granddaddy of all the kerosene burners with flat wicks. The most characteristic feature was the high, dome-shaped deflector to concentrate the draft on the flame. There was a toothed wheel for raising or lowering the wick, and a collar-like receptacle for the base of the glass chimney. The first burner for kerosene to be patented in the United States was the Jones burner of 1858 (Fig. 11.15). During the 1860s there were many designs for kerosene burners, but in 1873 Lewis J. Atwood of Waterbury, Connecticut, introduced a version with a hinged deflector and four springy prongs to hold the chimney. You can still buy the Atwood burner from hardware stores today, almost unchanged after more than a century.

During the years that lamp fuel was changing from whale oil to lard to burning fluid and, finally, to kerosene, a different method of interior illumination was being developed. This was based on the use of illuminating gas, produced by the distillation or incomplete combustion of coal. Here at last was a means of lighting that could be incorporated into the building (Fig. 11.16).

The system of gas lighting that came into general use, i.e., a central generating plant and a network of distributing pipes, was devised in England by John Murdoch in 1792 and elaborated by Samuel Clegg in the early 1800s. In 1799 Philippe Lebon of Paris anticipated the producer-gas technique, in which gas is made by incomplete combustion. The first American patent for gas lighting was issued to David Melville of Newport, Rhode Island, in 1810. Baltimore was the first American city to have gas lighting; this was in 1816.

Early gas burners, called cockspurs from the shape of the flame, were simple orifices at

Fig. 11.15 Composite table lamp, early 1860s, fitted with Jones burner, burning kerosene. Toronto area. *(Author's collection, Royal Ontario Museum photograph)*

the ends or sides of pipes, the emission of the fuel being controlled by a stop-cock. One of the first elaborations was a version of the Argand burner, in which a ring of openings produced a circular flame, with a central air draft. But the flame from gas came from incandescent particles of unburned carbon, so the greater the surface of the flame the brighter the light. A broad, flat flame was produced about 1810 by means of a burner with a slit-shaped opening. This was called the bat-wing burner, from the shape of the flame. A more effective method of making a wide, flat flame

Fig. 11.16 Gas entry lantern ca. 1840 with original fittings. From a house at 221 South Sixth Street, Philadelphia (now demolished). *(Courtesy The Athenaeum of Philadelphia)*

complex and ornate during the nineteenth century. The simple L-shaped pipe with stopcock, projecting from the wall, had by the 1850s incorporated two or more swivel joints so that the burner could be extended or folded back as desired. Other wall fixtures had the pipe bent into single or double curves, with extraneous ornaments of metal simulating leaves and flowers.

The portable table lamp using oil or kerosene could not be completely imitated by a gas burner, but the effort was made. By the 1850s

Fig. 11.17 Chandeliers converted, 1848. When gas was piped into the White House, Washington, the elegant East Room oil fixtures installed by President Jackson were refitted by Cornelius & Co. of Philadelphia. *Frank Lesley's Illustrated Newspaper, 1857. (Library Company of Philadelphia)*

was developed about 1820. In this the burner had two small holes slanting toward each other. The two fine jets of gas, emerging at high pressure, impinged and so produced a fan-shaped flame. This burner was called union-jet from the manner of operation, or fish-tail from the shape of the flame.

Much trouble was experienced in the early days of gas lighting due to the presence of moisture, especially if traces of ammonia were present. This produced corrosion of the tips of iron or brass burners. The difficulty was eliminated in 1858 when burners were made with the tips composed of the inert mineral steatite.

Gas fixtures became progressively more

Fig. 11.18 Gas chandelier. One of a pair shown in the Crystal Palace, London, by Cornelius & Baker of Philadelphia. Described as of lacquered brass about 15 feet high with 15 burners. *The Art Journal Illustrated Catalogue*, London, 1851, p. 212.

providing space for a glass shade. This forecast the "harp" support of late-Victorian hanging lamps. More remarkable were the vertical fixtures introduced in the 1850s, in which the burners could be raised or lowered. The movement was made possible by having the vertical component in the form of two telescoping pipes. To prevent the escape of gas between the two pipes, cork washers were used as the seal. A more elaborate version, shown in Figure 11.19, was called a water-sliding lamp: it had an ornate cup attached to the top or bottom of the outer pipe. Water in this receptacle provided the seal against gas escape. In addition to elaboration, this type of fixture had counterweights suspended on chains over pulleys, to keep the burners at the desired height.

In the parlors of the well-to-do, the ceiling fixtures consisted of multiple-burner devices with gracefully curved pipes, metal ornaments, glass pendants, and globular or vase-shaped cut glass shades. These were called

Fig. 11.19 Water-sliding triple burner gas ceiling fixture. From an 1861 advertisement. (*Courtesy Consumers Gas Company, Toronto*)

vulcanized rubber tubing was available and, with this as the connection to the gas outlet, table fixtures with a flat base and a column-like stem could be used with moderate portability.

It was in ceiling fixtures that gas lighting achieved its greatest elaboration, e.g., Figures 11.17 and 11.18. The simplest form, which persisted in the more plebian installations, was of an inverted T-shape with stop-cocks and burners at the ends of the two horizontal arms. More elaborate versions had curved arms and elaborate ornamentation. There were single-burner ceiling fixtures having a double feed pipe, which curved gracefully down and around on each side of the burner, gaseliers, by analogy with chandeliers. It was in public halls that such fixtures achieved their ultimate elaboration. In 1847 the original Senate Chamber in the United States Capitol had the multiple Argand-lamp "chandelier" replaced by a 24 burner gas fixture of highly ornate design. In Toronto the St. Lawrence Hall, where Jenny Lind sang in 1851, had a central gas fixture with 24 burners elaborately festooned with glass beads. Similarly ornate double-burner wall fixtures were mounted at intervals along the sides of the hall.

The gas flame could be extinguished by turning the stop-cock, but an open flame was required to light it. This presented a problem with ceiling fixtures that were mounted in a fixed position. A special tool called a gas lighter (Fig. 11.20) consisted of a double brass tube with a handle. One of the tubes terminated in a kind of wrench, which could be used to turn the stop-cock. The other tube carried a taper, a cord impregnated with wax, and its flame served to ignite the released gas.

Gas lighting, because of the elaborate instal-

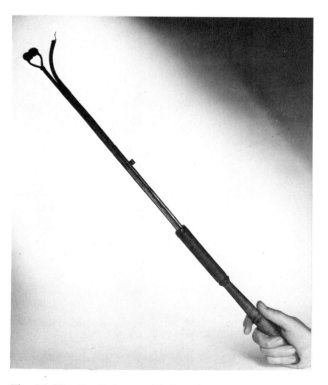

Fig. 11.20 Gas lighter, with burning taper. Purchased East Aurora, New York. (*Author's collection, Royal Ontario Museum photograph*)

lations required to manufacture and distribute the fuel, was largely restricted to cities. Hence its popularity in public buildings, where the many outlets justified the installation. Because of its convenience and efficiency, however, it came to be used even in homes of modest income. Most of the pipes within the house (fittings) or those connecting to the distributing system (service pipes) were of wrought iron. For those of smaller diameter near the burners, brass, tin or pewter might be used. Moisture in the gas was not only a source of corrosion in burners but also was capable of causing blockage in the pipes. For this reason the service pipes were laid with a slight slope away from the building. As an additional precaution, a U-shaped section of pipe might be incorporated, with a stop-cock at the bottom of the bend, through which condensed moisture might be drawn off.

The pipes used to distribute the gas to the individual installations were called mains. They were usually made of cast iron and varied in diameter from 2 inches to 2 feet. Each section, usually 9 feet in length, had an expanded socket at one end to receive the end of the next section. The joint was sealed with packed fibres and molten lead or, in some cases, with tar.

The only practical and fair way of computing the charge to the consumer was by measuring the gas that came into the house. The first gas meter was invented by Samuel Clegg in 1816 and progressively improved subsequently. Basically it consisted of a drum with chambers, partly immersed in water. The incoming gas filled each chamber in turn, causing the drum to revolve. When each chamber reached the upper position, out of the water, the gas escaped to the outlet and was replaced by the water. The revolution of the drum actuated the recording dials. A dry gas meter was introduced in the 1850s, in which the gas pressure caused an oscillatory motion of metal pistons, transmitted to the recorder as rotary motion. Gas meters of large size were also used to measure the output from the central plant.

Gas was stored after manufacture in large

containers called gas holders or, less aptly, gasometers. One consisted of a large tank inverted in a somewhat wider container of water. Gas was introduced from below, displacing the water and causing the inverted tank to rise, controlled by rollers in vertical tracks. In cases where the available area was restricted, a two-section or telescoping gas holder was used.

Gas pressure varied with the amount present in the gas holder, so a gas governor was used to produce a uniform flow. This was like a small gas holder, with a conical valve that rose or fell with the inverted tank. As the tank rose the valve progressively restricted the outlet, thus compensating for increased pressure.

Each innovation in household lighting improved the illumination, and people became concerned that the users, unaccustomed to such brilliance, might suffer eye damage. Colored shades and even colored chimneys were used to avoid this. In 1871 the comparative efficiency and economy of the then-available light sources were set forth in precise terms by Dr. Stevenson Macadam, Lecturer in Chemistry at the University of Edinburgh.[2] His standard of illumination was the light emitted by a spermaceti candle burning at the rate of 120 grains (140 grams) per hour. With this as unity (1 candle-power), the following relative intensities were determined by Dr. Macadam, using the Bunsen photometer:

Common tallow or dip	1.25
Composite (stearine) candle	1.1
Paraffin candle	1.49
Sperm oil in flat-wick lamp	1.306
Rape (colza) oil in flat-wick lamp	1.05
Whale oil in flat-wick lamp	0.9
Sperm oil in Argand lamp	13.0
Rape oil in Argand lamp	11.33
Whale oil in Argand lamp	9.8
Paraffin (kerosene) oil in cottage lamp	6.0
Paraffin oil in parlor lamp	9.0
Paraffin oil in dining-room lamp	12.0
Paraffin oil in ¾-inch Argand lamp	9.0
Paraffin oil in Doty's Argand lamp	11.33
Coal gas, No. 5 jet	28.0
Coal gas, No. 4 jet	20.8
Coal gas, No. 3 jet	13.8
Coal gas, No. 2 jet	7.8
Coal gas, No. 1 jet	3.0
Coal gas, No. 0.5 jet	1.0

Dr. Macadam also calculated the relative cost of the various means of lighting, based on the amount of light provided by one pennyworth of the fuel. Allowing for the differences in light production by large and small lamps and gas jets, and for the variations in price of the fuels in different parts of Scotland, he concluded that coal gas and paraffin oil averaged out to about the same cost. In North America, where petroleum for the manufacture of kerosene did not have to be imported, the oil lamp user may have enjoyed a somewhat greater economy. However, when one considers the convenience of gas lighting, with no wicks to trim, chimneys to clean, or fonts to fill, it is easy to see why this was the popular means of illumination in cities until the introduction of the electric light. In small communities and on farms, where gas distribution was not practical and electricity was slow in coming, kerosene lighting remained predominant until well into the twentieth century.

NOTES

1. Fire-making by Europeans prior to the nineteenth century was usually by means of "flint and steel," i.e., the striking of sparks from a small bar of steel by blows with some hard mineral such as flint or chert. The steel and the flint were contained in a tole canister, along with tinder made of scorched linen and wooden sticks tipped with sulphur. The operation was tedious and in damp weather almost impossible. Chemical methods of fire-making were introduced early in the nineteenth century. The first of these used wooden matches tipped with a mixture of potassium chlorate and sugar. These were ignited by dipping into sulphuric acid. The first friction match appeared in 1826.

2. *Transactions of the Royal Scottish Society of Arts,* Vol. 8, pp. 325–343, 1873. For assistance in locating this important paper I am indebted to Dr. W. E. Swinton, Centennial Professor of the History of Science, University of Toronto.

BIBLIOGRAPHY

d'Allemagne, Henry-René. *Histoire du luminaire depuis l'epoque Romaine jusqu'au XIXe siècle*. Paris: Alphonse Picard, 1891. 702 pp. 80 figs., 80 pls.

Clegg, Samuel, jun. *A practical treatise on the manufacture and distribution of coalgas, its introduction and progressive improvement*. Illustrated by engravings from working drawings, with general estimates. 2nd ed. London: John Weale, 1853. 298 pp., 42 textfigs., 19 pls.

Hough, Walter. Collection of heating and lighting utensils in the United States National Museum. U.S. Natl. Mus., Bull. no. 141, 113 pp., 99 pls., 1928.

O'Dea, W. T. *The social history of lighting*. London: Routledge and Kegan Paul, 1958. 254 pp., 59 text-figs., 32 pls.

Perry, D. H. *Out of darkness. A history of lighting*. Rochester, N.Y.: Rochester Mus. & Sci. Center, 1969. 22 pp., 36 figs.

Rapp, H. W. *Early lighting. A pictorial guide*. Boston: The Rushlight Club, 1972. 129 pp., 447 figs.

Russell, L. S. *A heritage of light. Lamps and lighting in the early Canadian home*. Toronto: Univ. Toronto Press, 1968. 344 pp., 220 figs.

Steward, E. G., (Science Museum) *Town Gas: Its Manufacture and Distribution*. London: Her Majesty's Stationery Office, 1958. 55 pp.

Thwing, L. L. *Flickering flames: a history of domestic lighting through the ages*. Rutland, Vt.: Charles E. Tuttle Co., 1958. 138 pp., 54 text-figs., 97 pls.

Watkins, C. M. *Artificial lighting in America: 1830–1860* [?]. Ann. Rept. Smithsonian Institution. 1951, pp. 385–407, 8 pls., 1952.

CHAPTER 12

Building the Capitol

MARIO E. CAMPIOLI

MR. CAMPIOLI is Assistant Architect of the Capitol, Washington, a Fellow of the American Institute of Architects, and a former Director of Architecture at Colonial Williamsburg.

The United States Capitol is the architectural symbol of the world's greatest experiment in democracy. For nearly 200 years it has dominated the Federal City. Its creation involved the intense devotion of many gifted people. Washington, Jefferson, Lincoln and the others—even Jefferson Davis—were deeply and personally concerned. Besides the usual political turmoil, the Capitol was to witness a long succession of professional triumphs and disappointments. While its architects were outstandingly able men, none ever departed happy.

Physically this vast, complicated structure well illustrates the progress (sometimes faltering) of the nation's building technology over 7 decades. Before the "Second Completion" in 1865—when this account ends—it generally reflected the best of most periods. Since then, while admitting from time to time such conveniences as elevators, electric lights, telephones and air conditioning, it has, for sentimental reasons, held its original architectural character. Later concepts may be seen in office dependencies that now surround the Capitol.

WILLIAM THORNTON AND HIS ASSOCIATES

Some nine years before the Federal government moved from Philadelphia to its new seat on the Potomac a general outline for the city had been laid down, approved and published. Its author, Major Pierre Charles L'Enfant, had in October of 1791 been requested "to prepare a draft of the public buildings"[1] and two months later he actually started grading the hilltop site set aside for the Capitol.[2] "L'Enfant had no plans prepared for the Capitol" wrote Thomas Jefferson at the time, but "he said he had them in his head."[3] Shortly afterward the impetuous major fell into controversy with his Washington associates and withdrew entirely. At Jefferson's urging President Washington directed the Commissioners to hold an open competition for an architectural design.[4]

In an advertisement prepared by Jefferson and dated March 14, 1792, the Commissioners called for plans. Ground plans, elevations and sections were requested as well as an estimate for the total cubage of brickwork—all to be submitted by July 15. A first award of $500

202

and a lot in the city was promised; second prize, $250 and a lot.

Some sixteen designers responded; the submissions were highly mixed as to quality (Figs. 12.1 and 12.2).[5]

Not until a year later—and after some very informal proceedings—the award was made to Dr. William Thornton, a Philadelphia physician, who had heard about the competition while in the West Indies.[6] President Washington commended Thornton's plan for "grandeur, simplicity and convenience." Secretary of State Jefferson declared it "simple, noble, beautiful, excellently arranged and moderate in size."[7]

Although the Commissioners found the outward appearance of Thornton's design "striking and pleasing" George Washington ordered Hallet to combine the most desirable parts of several plans which allowed all rooms to be lighted and ventilated through windows in the outer walls.[8] On April 5, 1793, Thornton's design won the award but criticism continued.

The professional architects seemed to be united against the amateur Thornton. Washington was much disturbed by their objections and requested Stephen Hallet and James Hoban (architect of the President's House) to hasten to Philadelphia for a conference.

"It is unlucky that this investigation of Doctor Thornton's plan, and estimate of the cost

Fig. 12.1   James Diamond of Somerset County, Maryland, submitted a design that was fortunately not selected. *(Maryland Historical Society) Photographs from the Architect of the Capital unless otherwise noted.*

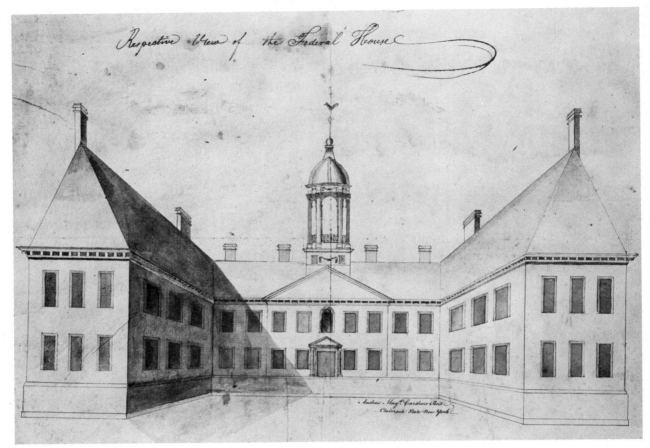

Respective View of the Federal House

Fig. 12.2 Andrew Mayd Carshore of Claverack, New York, another unsuccessful competitor, tried his hand at perspective drawing. *(Maryland Historical Society)*

had not preceeded the adoption of it . . . wrote Washington, "Nothing can be done to the foundations until a final decision is had . . . The case is important. A Plan must be adopted; and good, or bad, it must be entered upon."[9]

Jefferson conducted the meeting and reported that Dr. Thornton had with him Thomas Carstairs, a Philadelphia builder, and a Colonel Williams as advisors. Hallet's modifications were generally approved "so that it is considered as Dr. Thornton's plan reduced into practical form . . . it is on the whole a work of great merit." It was also agreed that the "reformed plan" would reduce the cost by one half.[10] There were still lingering doubts about the lighting of the interiors which made skylights necessary and created a serious practical problem for many years to come. The profes-

sionals were right but both Hallet and Hadfield finally lost out. Thornton was not a man to be lightly opposed.

To Washington and Jefferson a visual confirmation of progress on the ground was urgently needed. Even the move to the Potomac continued to be disputed by powerful interests and the survival of the new national government was at times a matter of anxiety.

Thornton, appointed first Architect of the Capitol by President Washington, served less than a year and became a commissioner (September 2, 1794) along with Gustavus Scott and Alexander White. The logistical problems they faced in trying to build great public buildings — not to mention a whole new city — on the rural reaches of the Potomac were formidable. Surviving correspondence shows that inquiries for mechanics were made as far away as Germany, France, Italy, Ireland and the Netherlands. Stone cutters would be needed in numbers. Jefferson learned that while as many as a thousand house carpenters might be available from wooden Connecticut for two-thirds of a dollar per day plus board, the outstanding candidate from that state had never cut a column.[11] Jefferson tried to interest James Traquair "a Capetal stone cutter" of Philadelphia who imported his workmen from Edinburgh.[12] He also showed Washington a model of a machine for sawing and polishing stone by a Mr. Milliken and compared it with a French invention. This inspired Leonard Harbaugh of Baltimore to design one worked by horses or oxen.[13] The founding fathers were intensely interested in all new building technology—even such as the earlier forms of concrete.

The cornerstone of the Capitol's North Wing was laid with impressive ceremonies on September 18, 1793, but the next two years were troubled. While the design committee had been able to arrive at a compromise on the plans, Hallet, even though engaged to make the detailed drawings, continued to resist Thornton's concepts and finally had to be dismissed. The relative advantages of work paid for by the piece versus work by time were debated. By the spring of 1795, though the foundations had been laid, it had not yet been decided if there would be a basement.[14] At the end of June it was agreed that a foot of new wall would have to be taken down to get a better bonding in the masonry.[15] The Commissioners wrote "Those not acquainted with the Motley Set we found here . . . can form no adequate idea of the irksome Scenes we are too frequently compeled to engage in." Then, toward the end of September, it appeared that the money might run out.[16]

In November President Washington, then in Philadelphia, had to lecture architect George Hadfield for getting away from Thornton's plans. It was already being said that "the present plan is nobody's but a compound of everybody's." Yet the project struggled on. At the end of the building season of

1796 the finished stonework was up to the second floor,[17] and by November of 1797 the roof was going up.[18] By the end of 1798 the roof had been boarded, shingled and painted and the gutters leaded. Inside, the floors were being framed, the doors and windows were under way and "clear northern pine" was on hand to make flooring and trim.[19]

When George Washington died at the end of 1799, just a year before Congress met in the Capitol, the exterior masonry was all up, including the chimneys and their lightning rods. Inside, the plastering was nearing completion and the carpenters and joiners were expected to finish by the first of February.

But in April a conference had to be called to investigate the damages to the plastering caused by the carpenters[20] and it was reported by July the leadwork of the roof was leaking.[21] Earlier, in May, the platforms around the lanterns on the roof were papered, tarred and sanded.[22] At last by November 22, Congress could meet in its new home.[23] "Although there is cause to apprehend that accommodations are not now so complete as might be wished," observed President John Adams hopefully, "yet there is great reason to believe that this inconvenience will cease with the present session."[24]

The urgency to complete the House or south wing was evident. Early in 1801 the Commissioners requested space for the Supreme Court to sit and it was allowed to use a committee room on the first floor.[25] By summer President Jefferson was considering alternative plans prepared by Hoban for completing "an apartment" for the House of Representatives in the south wing,[26] which at that time had hardly risen above ground level. But Benjamin Henry Latrobe—only 5 years from the drafting rooms of London—was available.

BENJAMIN HENRY LATROBE: BEFORE AND AFTER THE FIRE

Latrobe's first architectural commission for the Federal government (November, 1802)

East Front of the Capitol

Fig. 12.3   William Thornton. What remains today of the original drawings is problematical. This elevation, engraved by C. Schwarz for Robert King's map of Washington published in 1818, is thought to be a faithful copy of the Thornton elevation.

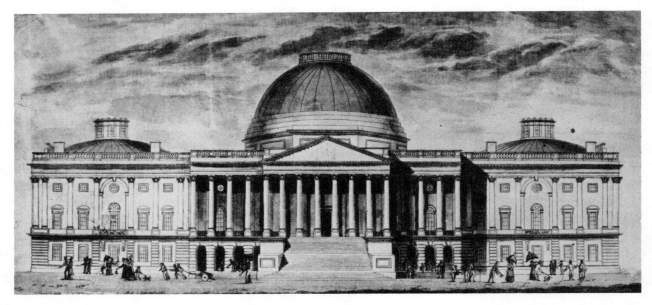

Fig. 12.4   Charles Bulfinch. At the end of the year 1822 the wooden dome of the great rotunda was completed, joining the earlier wings. Aquatint by H. & J. Stone from the architect's drawing. *(Stokes Collection, New York Public Library)*

Fig. 12.5   Thomas U. Walter. The growth of the United States and its government called for a much enlarged structure. This watercolor approved by President Fillmore in 1851 proposed extension to over 750 feet in length.

Fig. 12.6   The "Second Completion." Under the direction of Thomas U. Walter and Captain Montgomery C. Meigs, the magnificent iron dome was designed and built 1855–63. View ca. 1900, photographer unknown.

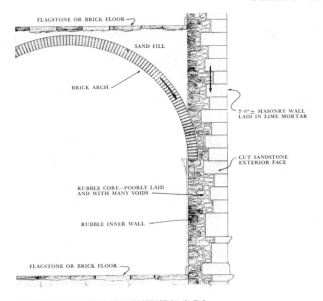

FLAGSTONE OR BRICK FLOOR

SAND FILL

BRICK ARCH

3'-0"± MASONRY WALL
LAID IN LIME MORTAR

CUT SANDSTONE
EXTERIOR FACE

RUBBLE CORE—POORLY LAID
AND WITH MANY VOIDS

RUBBLE INNER WALL

FLAGSTONE OR BRICK FLOOR

EXTERIOR WALL & FLOOR ARCH CONSTRUCTION IN THE
ORIGINAL PART OF THE UNITED STATES CAPITOL.

Fig. 12.7 Cross section through typical wall and floor construction of the original wings. Such fireproof construction, topped by sheet iron roofing in 1803, resisted the efforts of the British to burn the building in 1814.
(Courtesy Architect of the Capitol)

was to plan for President Jefferson a huge covered drydock to mothball the capital ships of the Navy. Although never built, the design so pleased Jefferson that he offered (March 6, 1803) Latrobe the position of Surveyor of Public Buildings of the United States, which was accepted readily. Because of other projects, mainly the Delaware and Chesapeake Canal, Latrobe did not move to Washington immediately but he employed John Lenthall as his representative at the Capitol. Hallet, Hadfield and Hoban had left the project but William Thornton remained in Washington; he and the new architect soon became implacable enemies.[27]

Congress had appropriated $50,000 to start work on the south wing to accommodate the House of Representatives, which met for one session in temporary accommodations built inside the exterior walls, known unfavorably as "The Oven." It was necessary to make an early show of progress. The exterior of the new—or south—wing had to match the old—or north—wing. Symmetry was obviously required. Unfortunately, Latrobe's first report

of April 4 was a catalog of Thornton's shortcomings as an architect. The foundations were poorly built, he said. The drawings did not agree with each other. There were interior spaces hard to light and not enough committee rooms.[28] But Latrobe was advised by the President to stay as close to the original plan as possible. In the meantime Thornton was attacking in the public press and Congress was restive, though it appropriated another $50,000 and then $110,000 more.

Among the practical problems were the old ones of getting enough freestone and finding competent workmen. "Stone cutters have been collected from distant parts of the Union . . . for almost every material we use, we are dependent upon distant places;—for our lime on the new England States,—for our lumber on the Delaware & the Eastern shore of the Chesapeake,—for our iron on Pennsylvania; for many other articles for which there is no demand here, excepting for the supply of the public buildings, on Baltimore & Philadelphia—and that even for our freestone, the most important article in the work, the wants of the public buildings are not sufficient to encourage the employment of much capital and labour in the quarries . . ."[29] In addition there was the need to find proper heating stoves and window glass.[30]

While economy was always publicly advocated, Jefferson had ambitious ideas as to the quality of stone carving. That problem was finally solved by bringing from Italy the two young sculptors Giuseppe Franzoni and Giovanni Andrei. These men proved to be talented and they earned very favorable public notice. While much of their work was to be destroyed in the War of 1812, Franzoni's "Corn cob" capitals are still shown to admiring visitors.[31] When the matter of flooring came up, both inexpensive marble tile from Italy[32] and hexagon paving tile from Bordeaux were considered.[33]

With the employment of George Bridport of Philadelphia in 1811 as decorative painter for the domed ceiling of the Hall of Representatives, the wing was finally completed. Talbot Hamlin called it "undoubtedly the most beau-

Fig. 12.8   The north wing was the first unit completed; it was occupied in 1800. Stone cutters are working in the foreground. Watercolor by William Birch, 1800. *(Library of Congress)*

tiful legislative chamber in the Western world."

Along the way, though, Latrobe had difficulty with that perennial problem at the Capitol: leaky skylights. Jefferson had been intrigued by the lights in the dome of the Halle aux Blés in Paris and wanted to imitate the effect, but Latrobe worried that they would drip water part of the time and blind eyes on sunny days. Jefferson got his way (with some compromise) but the unfortunate results soon proved that he was wrong. Then there was the serious matter of a large cost over-run in the appropriation of 1807, which drew public charges of extravagance.[34]

The collapse of the vaulting in the North wing (due to premature removal of the supports) caused the death of the ever loyal John Lenthall and cast doubts on Latrobe's vault design. While he won a libel suit against the vitriolic Dr. Thornton, these emotional experiences greatly affected Latrobe.[35] The handsome appearance of the Hall of Representatives was much admired but the acoustics remained difficult. There was much controversy and only at the last minute did Congress grant the architect his pay up to July 1, 1811. He then had little more to console himself than the unflagging support of Jefferson, now living in retirement at Monticello, and Washington friends outside of the opposition in Congress.

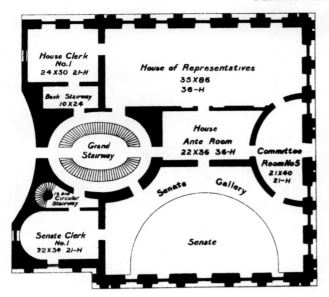

Fig. 12.9 North wing principal floor plans, ca. 1800. The original wings of the Capitol were so deep that all the spaces would not be lighted through windows. Skylights suffered serious leaks and expensive repairs over many decades. Plans reconstructed by Glenn Brown FAIA, 1902.

Construction was stopped by the war with Britain, which brought the invader to our shores. Late in the summer of 1814 came sudden disaster. Veteran British troops sailed up Chesapeake Bay and the Patuxent River and made a surprise attack on the Capital city. The American militia, routed at the Battle of Bladensburg, left Washington open to the mercies of Major General Robert Ross and his naval colleague Rear Admiral George Cockburn. Their mission being to "chastise" the Americans, it was decided to burn the public buildings and the Capitol was high on the list.

The troops of the 3rd Brigade did the job. After firing a volley through the windows of the East front to warn any sharpshooters, they broke in and searched the building. Lieutenant James Scott was much impressed by Latrobe's great Hall of Representatives though he thought it revealed an "unseemly bias for monarchial splendor."[36] Lieutenant John Gleig, another of the last persons to visit the doomed structure, left us a description. "It was built entirely of free-stone, tastefully worked and highly polished; and, besides its numerous windows, was lighted from the top by a large and handsome cupola. Perhaps it could not be said to belong to any decided style of architecture. . . ."[37]

The following spring Latrobe, after inspecting the ruins, wrote a vivid report to Jefferson including a pen sketch. He heard that an enemy team of three men had gone through the south wing, chopping the woodwork, throwing chemicals on it and lighting fires without doing much damage in the smaller rooms. Rockets had then been launched through the roof without setting it on fire. Latrobe's sheet iron roofing, rolled at his own mill in Philadelphia, held fast. It must have given the architect much satisfaction (Fig. 12.10).

Finally a fire had been started in a great pile of furniture in the House of Representatives with rocket fuel added. "The devastation has been dreadful" wrote Latrobe, ". . . so intense was the flame, that the glass of the lights was melted, and I have now lumps, weighing many pounds of glass, run into mass." In the Senate chamber the marble columns "were burnt to lime . . . a most magnificent ruin."[38]

Congress, meeting again on October 20, instructed the Superintendent of Buildings to inquire into the cost of rebuilding. Architects and master builders were then consulted and it was their opinion that the burnt-out walls could be used again.[39] On deliberation it appeared that it would be more economical to restore the old structure rather than to abandon it and the local banks offered to loan half a million dollars. On February 8, 1815, Congress voted to rebuild[40] and five days later it borrowed from the banks at a rate not to exceed 6 per cent.[41]

Latrobe, who had by then failed in the steamboat business out at Pittsburgh, wrote to President Madison offering his services for the restoration of the public buildings and was accepted.[42] By April the architect was back in Washington for a few days spent in studying the burnt walls, in ordering stone and in organizing the force of nearly two hundred men that had already been engaged.[43]

A great deal had to be done, starting with the demolition of unsafe masonry. Latrobe

Fig. 12.10   On August 24, 1814, British troops captured Washington and fired the Capitol. Fireproof construction enabled the old walls, though damaged, to be reused. Drawing by Chittenden in *Documentary History of the United States Capitol.*

had no draftsman and had to make all of his own drawings. There was a shortage of freestone at the public quarries. To save time and money it was decided to have some of the marble Corinthian column caps carved in Italy and in addition three more carvers were brought over.[44]

At this point the architect was made responsible to a single commissioner, Col. Samuel Lane, with whom he was later to quarrel. The common laborers struck over working hours and the Senate decided to enlarge its quarters, requiring the demolition of interior walls which had survived the fire.[45] Under extreme pressure at the Capitol—and in attending to private commissions in other cities—Latrobe's troubles grew. His clerk of the works Shadrach Davis was abruptly discharged by Lane. Newspaper criticism started up again and a year and a half later President Monroe succeeded Madison.

In the interest of efficiency the newly elected Monroe brought in two Army officers, General J. G. Swift and Colonel G. Bomford to consult and expedite matters. Misunderstandings increased and on November 20, after a scene in the President's office, Latrobe resigned, though he spent the remainder of the year completing the drawings.[46]

CHARLES BULFINCH AND THE FIRST COMPLETION

On January 8, 1818, President Monroe appointed Charles Bulfinch of Boston to 'be Architect of the Capitol.[47] A period of administrative peace and of construction progress was to follow.

In his first annual message the President had recommended the completion of the Capitol "on a scale adequate to national purposes."[48] Eight weeks later he was able to approve an appropriation by Congress for "repairing the public buildings" in the amount of $200,000[49] and things now moved fast. Only nine days after that the new architect from Boston made his first report. He had studied Latrobe's numerous drawings (which he intended to follow in large part), conceding "that the public rooms of Congress will be very splendid, and exhibit favorable specimens of correct taste." Already on hand was a large quantity of free stone, fifty Italian marble chimneypieces, most of the doors, window frames, sashes and glass, as well as enough copper for covering the roof; all that plus a quantity of lumber.[50]

Hesitant at first, it was not long before Bulfinch could find ways, he thought, to improve on Latrobe's plans. He certainly improved on his public relations. One of the first moves was to employ Solomon Willard, a carver and model maker from Boston who happened to be working in Baltimore. A 4-foot scale model, showing various proposals, was built right in Bulfinch's office and it made a hit. In Willard's words, the exhibit seemed "to give satisfaction, and Congress has shown uncommon liberality in the appropriations for carrying the work into effect."[51] Notably, there was a hundred thousand dollars to begin construction on the "Centre building" which was to connect the wings for the first time.[52]

The building season of 1818 was altogether a very active one, with well over three hundred mechanics and laborers working on the job. There were two misadventures: a strike by the workmen for higher wages which held up progress for better than a month; and the threatened failure of a great arch intended to support a minor dome and no less than eighteen chimney flues over the north wing, which cost another anxious month. On removing the centering, the great arch sagged ominously and the workmen "left it in alarm." The structure of the dome was then redesigned by Bulfinch in the form of a shallow brick cone. The

Fig. 12.11 House of Representatives, hall in the south wing, designed by Latrobe and completed in 1819 by Bulfinch, featured Corinthian columns of a richly variegated local marble. It has been called "the handsomest legislative chamber in the New World." Today it survives as "Statuary Hall." Brown, *History of the United States Capitol.*

two Army engineers, Swift and Bomford, again brought in as consultants, approved it so the work could proceed, though adding considerably to the cost. In spite of all that the Commissioners of Public Buildings thought it had been the best year ever.[53]

There is today a scarcity of drawings by Bulfinch and the extent of his work is not completely understood. He generally stayed out of controversy and within the general lines laid down earlier. Though Bulfinch was not content with the "intricate and dark" passages inherited from Thornton's plan,[54] the wings were finished by 1820.

For the central and final section of the building the first appropriation was made on April 20, 1818. The new plans reduced the exterior dimensions of Latrobe but increased the number of rooms.[55] By the end of the building season of 1822 the west front was complete (including an elegant new library) and the big wooden dome—which was to dominate the skyline for over 30 years—was sheathed and ready for coppering. Its design was something of an accident; Bulfinch preferred a lower profile but certain members of the cabinet a taller one. The architect lost out and had to carry out a design he didn't like.[56]

By the end of October, $4,000 had been spent on excavations and foundations for the new center structure and its great "Rotundo." Plans for winter work included the cutting of freestone elements and the making of doors and windows by the carpenters.[57]

In this period the decorative stonework—inside and out—was continued in the shops of the resident mechanics. Freestone for the exterior still came up from Aquia Creek and the decorative marble from the quarries opened by Latrobe on the Upper Potomac near Point of Rocks. Certain interior marble work was cut in Philadelphia and New York. What little remains today of the stonecutters' art on the exterior is competently executed.

The first of the many paintings commissioned by Congress were the four great historical scenes for the Rotunda by Colonel John Trumbull (Fig. 12.12). These were painted on strong canvas and fitted to the curved walls 1819–1824. But problems of vandalism and dampness were encountered almost from the first. In only 4 years mildew appeared and the paintings had to be taken down and treated with hot beeswax. The niches behind them were then plastered with hydraulic cement.[58] To reduce excessive humidity two "powerful and improved" warm air stoves were ordered from Safford and Low of Boston and installed under the floor. This effete improvement was made over the objection of a homespun congressman from Kentucky who thought the Rotunda would then become only a "more comfortable resort for loungers and idlers."[59] There are a number of stove designs still among the Bulfinch Papers. We reproduce here a monumental one that was used in the new Library of Congress (Fig. 12.13). On a more sophisticated level, Lieutenant Humphreys called attention to the warming and ventilating of Sir Charles Barry's new Reform Club in Pall Mall, London, where a small steam engine circulated air by a special kind of fan.[60]

One of Bulfinch's last innovations was the installation in 1828 of a visitors' gallery in the Senate Chamber supported on light cast iron columns (Fig. 12.14). It had been proposed in the architect's annual report for 1827 "to prevent the necessity of admitting strangers on the floor."[61] This early use of structural iron was a prelude to its extensive use in the Capitol during the 1850s.[62]

At the end of the Bulfinch period a certain amount of landscape work, especially tree planting, was carried out. It even impressed Mrs. Trollop, that difficult English visitor. The general plan that has survived shows that Bulfinch was carrying out a scheme of Latrobe dated 1815. The only new feature in it is the "Botanic garden" at the foot of the western slope where the great Mall of the McMillan plan now takes off for the Lincoln Memorial on the bank of the Potomac.[63]

Bulfinch, rather sooner than he expected or wished, was retired by President Andrew Jackson in 1829.[64] But the Capitol had been "com-

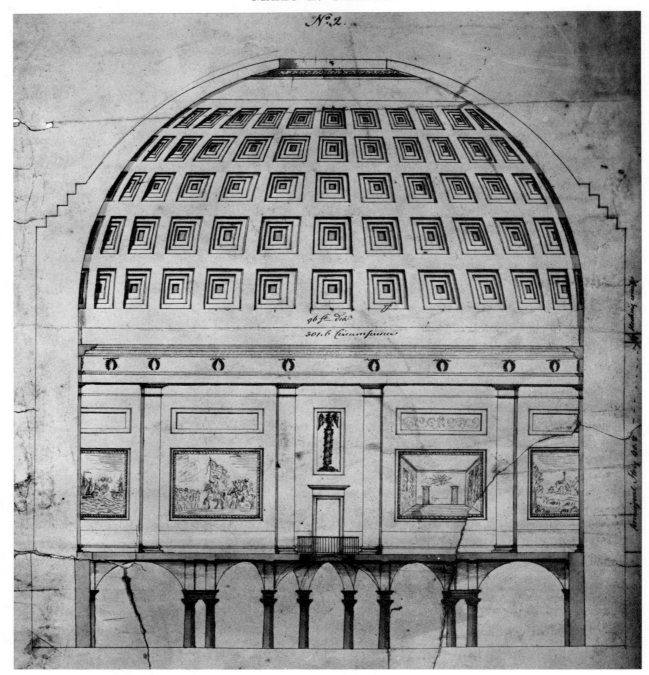

Fig. 12.12 The great Rotunda, 96 feet in diameter, built under Bulfinch in 1818–1822, incorporates the four great historical paintings by John Trumbull. The design for the high wooden dome overhead (Fig. 12.4) was selected over the architect's objections. Drawing by Bulfinch. *(Library of Congress)*

pleted" for the first time. As Glenn Brown summed it up:

That a structure as interesting and harmonious as the Capitol proved to be should have been produced when we consider the various hands through which it has passed, is remarkable. This was due both to the skill of the architects employed and the determination of the many Presidents who took a

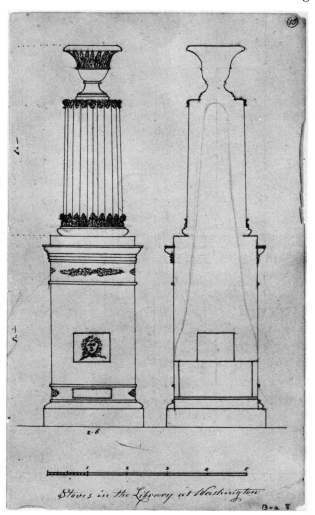

Fig. 12.13 Design for a stove used in the Library of Congress. The heating devices used before 1860 ranged all the way from open fireplaces to air heated by anthracite coal fires and circulated by power fans. Bulfinch drawing. *(Library of Congress)*

domes and lanterns. A great convenience was the introduction of gas lighting. A comprehensive report by Architect Mills on April 5, 1840, summarized the arguments for the use of "carburetted hydrogen gas," although it was 7 more years before gas was admitted.

We may assume it as a general fact, now well established: 1st, that the expense of lighting with coal gas is reduced to one half, at least, of that with oil, under the same intensity of light; 2d, that it is more safe in combustion than any other material producing light; 3d, that it is more easily managed and attended; 4th, that, from its economy it can be more easily extended in lighting up the public walks, &c., and thus render them more secure as promenades at night; 5th, upon the same principle of economy, fancy and other stores would introduce this light, and thus add to the cheerfulness of the public ways; 6th, these, with other facts, will prove the expediency of lighting up the public buildings and grounds with gas.

"It contributes greatly to warm as well as to illuminate the apartments" added the report[66] — and that turned out later to be only too true. But even the fall of the great chandelier in the Hall of Representatives under the weight of its oil (in 1840) did not hurry the change.

THE SECOND COMPLETION: THOMAS U. WALTER[67]

When Bulfinch finished his work at the Capitol in the summer of 1829, the two original wings had finally been joined under a central dome with monumental stairways fore and aft. It was probably the largest structure in the United States; it dominated old Jenkins' Hill and the whole seat of government as envisioned by Major L'Enfant the planner.

But the United States continued to grow. New representatives came from the West and there was jostling for space. At one point it had even been proposed to subdivide the great Rotunda into committee rooms. Archi-

personal interest in the work to see it executed, if not in strict conformity at least in harmony, with the original design.

For 7 years after Bulfinch there was no architect in the Federal establishment. Then, on July 6, 1836 Robert Mills of South Carolina was appointed by President Jackson as "Architect of the Public Buildings."

The following years at the Capitol witnessed various ineffectual attempts to correct bad acoustics.[65] A water supply was piped in while unwanted water continued to come down through the leaky roofs with their assorted

Fig. 12.14   The Senate Chamber visitors balcony of cast iron was installed in 1828 to replace another. The painting by G.A.P. Healy depicting a speech of Daniel Webster in 1830 hangs in Faneuil Hall, Boston.

tecturally, the Capitol had not been designed with a view to extension. It almost seemed that it was easier for the nation to grow West to the Pacific Ocean than to add to the home of Congress. Then California entered the Union with its flood of virgin gold and something had to give. The addition of the huge "Extensions" and the spectacular iron dome under the direction of Architect Thomas U. Walter and Army engineer Montgomery C. Meigs was a technical achievement worthy of the great Railroad Era in which they were built.

In the year 1850 Architect Mills had submitted designs for the addition of large wings to the north and south. These were approved by the Senate Committee on Public Buildings but not the *whole* Senate.[68] After much discussion, Congress appropriated $100,000 and left the choice of a plan up to the President. The bill was approved on September 30, 1850, and the same day the Committee advertised for architectural plans, offering a premium of five hundred dollars for a design and estimates.

The extension might consist either of new wings to the north and south or of a separate building to the east. The Committee did not wish to inhibit "the free exercise of architectural taste and judgment" but reserved the right to combine parts of several plans in a final decision. Little is known of the drawings which resulted and few of them have been preserved. Early the next year Senator Jefferson Davis of the Committee acknowledged that it was "a problem of extreme complexity" and Mills was then asked to prepare new plans combining features of all four competition sets. At the same time it was recommended that the Bulfinch dome be replaced with a taller one, more in proportion with the wing extensions.

But President Fillmore considered the action taken by the Senate to be *not* within the law. In accordance with an act of Congress which had empowered him to appoint the architect, Thomas U. Walter of Philadelphia, one of the competitors, was sworn in on June

11, 1851 as the Architect of the Extension of the Capitol.[70] After that, things moved quickly.

To make brief a complicated story, Walter's plans were adopted by President Fillmore on June 10.[71] A sum of $100,000 was available for construction by the end of the month and there was a rush to stake out the South Wing. On July 4 the cornerstone was laid with a two-hour oration by Daniel Webster, Secretary of State.[72]

Walter soon completed his staff. In June the German August Gottlieb Schoenborn, who produced the magnificent set of rendered drawings so admired today, was employed. He was to make six thousand drawings in that office over the next 47 years.[73] Z. W. Denham "clerk of the office" was appointed on July 21 and the following day Samuel Strong "superintendent of the work."

Then there was the matter of organizing the construction program. Walter decided that the work should be handled by contract rather than by day labor or "the days-work system" as he called it.[74] Plans were hurried and bids for stone, lime, sand and cement taken.

A great deal of competition developed among suppliers of building materials. The selection of freestone for the exterior finish was a matter of intense public interest. Early on, Walter had recommended that marble "as nearly white as can be obtained" be used above a basement of the lightest colored granite. The old Capitol with its walls of Aquia Creek sandstone was to be painted to match the new work.

Probably anticipating trouble from disappointed suppliers and their friends in Congress, extraordinary pains were taken to test the various kinds of marble offered.[75] Walter started his own experiments but soon decided to convene a blue-ribbon committee of experts which included General Joseph G. Totten, Chief Engineer of the War Department, Andrew Jackson Downing, the celebrated landscape architect and author, Professor Joseph Henry of the Smithsonian Institution, and Thomas Ewbank, Commissioner of Patents.[76]

Walter's report at the end of the year (Dec. 23, 1851) announced the near completion of the foundations, which were immense. Eight feet, nine inches thick at the base they went down as deep as 40 feet through the made ground in the western slope of the hill. The rough building stone was bought under contract and laid in the walls by "days-work."

The old Capitol plus the two new wings was now reaching its present length (751 feet, 4 inches, from north to south). The Hall of Representatives was planned to seat 400 members at desks and 1200 spectators in galleries. The whole to be lighted by fifty ornamental glass ceiling panels. In the north wing the Senate Chamber was planned to seat 100 Senators and the gallery 1200. The whole was designed to be lit with twenty-six windows supplemented by a glass ceiling as in the South Wing. All other rooms were to be vaulted with brick and therefore fireproof, except the corridor skylights framed with iron rafters.

Walter recommended warming by hot water in heat exchangers in the cellar, the air to be driven by fans attached to small steam engines. In addition, fireplaces were to be introduced throughout the building "as many of the members are used to open fires, and would not be willing to dispense with them." Both halls were designed according to "acoustical principles"; details still to be worked out would assure complete success. Barring any increase in architectural embellishment—the architect believed that the Extensions, as they were called, could be done for two and a half million dollars, taking 5 years to execute.[77]

By March 27, 1852, a set of drawings including a perspective and a model were ready to be exhibited. The hundred thousand dollars already appropriated had been spent and a new set of contracts was ready for awarding. Granite for finishing the basement walls was to come from Richmond, Virginia.[78] The marble selected for the upper stories—from Lee, Berkshire County, Massachusetts—had "exceeding beauty," a low price and a warm recommendation from the sculptor Horatio Greenough.[79] Alexander R. Boteler of Shepherdstown, West Virginia, won the award for hydraulic cement,[80] and Christopher Adams of Albany, New York, one for six million well-

Classical Stonecutting Under Four Architects

Exterior Corinthian pilaster caps show interesting variations on a familiar theme. The examples in Figs. 12.15 through 12.17 are of Aquia Creek sandstone, Fig. 12.18 of Berkshires marble.

Fig. 12.15   Thornton, 1794–1803.

Fig. 12.16   Latrobe, 1803–1817.

Fig. 12.17   Bulfinch, 1818–1829

Fig. 12.18   Walter, 1851–1865.

burned bricks of tempered clay.[81] The roofs were to be framed with wrought iron and covered with copper.[82]

On April 14, 1852, an appropriation of five hundred thousand dollars was approved to continue the work. Measurers of materials were stationed at both the north and south wings; the delivery of marble by schooner began in July and by the end of the year over

80 per cent of the new appropriation had already been spent.

A sudden diversion occurred at the old building—on the day before Christmas, 1851. The Library of Congress just to the west of the Rotunda caught fire from a defective flue and burnt out with a tragic loss of books and manuscripts. The hand-pumped engines worked from the floor of the rotunda which was covered by several inches of water. A month later Walter submitted plans for reconstruction using, as he said, "no combustible materials whatever." On March 19 $72,000 was appropriated for the purpose and work proceeded quickly. The remarkable room is now gone but Walter's drawings have survived, together with photographs, and show that it was fitted with two galleries elaborately ornamented—all of iron, including the bookshelves—and a hung ceiling of decorative glass and highly enriched iron paneling (Fig. 12.19). On June 21 the contract for the iron fittings was let to Janes, Beebe & Co. of New York, lowest of eight bidders and the work was started as soon as patterns could be prepared. Nearing completion at the end of the year, the room was warmly praised by Secretary of the Interior Stuart.[83]

But praise from the executive establishment was not enough. Walter's adversaries were busy and his competence in general was being questioned. The masonry of the new foundations was already under attack. In the Senate Mr. Borland reported that with the toe of his boot he had kicked out mortar after the work had been standing for months.[84] Three weeks later Senator Walker claimed to have shoved

Fig. 12.19 The Iron Library, built 1852. Designed by Thomas U. Walter to replace the completely burned out Bulfinch library. The galleries and ceiling elements were modelled and cast in New York in a rich Italianate style.

the full length of his walking stick into the masonry and that he could scoop up the mortar like mush with a common teaspoon.[85] Four Army engineers were detailed to examine the problem. It was found that the lower courses of stonework had been laid in hydraulic lime and were quite solid. Higher up more fat lime had been used and some of it had been damaged by frost before it could set.[86] The engineers cleared the work and the architect was upheld by President Fillmore.

Enter Captain Meigs

But the administration changed at that point and the new President, Franklin Pierce — in deference to the Senators — decided to put an Army engineer in overall charge. On March 29, 1853, Secretary of War Jefferson Davis (West Point, 1838) detailed Captain Montgomery C. Meigs (West Point, 1836) — already in the city working on the great Washington Aqueduct — as Superintendent at the Capitol over Walter's head.[87] That, unwittingly, initiated a contest of professional prerogatives that was to climax dramatically nearly 7 years later.

Captain Meigs[88] was particularly instructed not only to examine the strength of the foundations but to look into "the arrangements for warming, ventilation, speaking and hearing" — problems which had annoyed Congress for half a century. He was especially enjoined to see that "no vitiated air shall injure the health of the legislators" and he took the opportunity of examining some of the principal public halls in the larger cities of the East.[89]

Meigs came back to recommend radical changes in the Capitol's floor layout which were ultimately approved by the President. These resulted in placing the Senate Chamber and the Hall of Representatives in the *center* of their respective wings. This eliminated windows to the out-of-doors and made the great halls entirely dependent for light from the ceilings and on artificial ventilation. It was a fateful decision, as we shall see later.

The architectural taste of the mid-fifties was a rich one, owing a great deal to high style Ital-

ian work. The prevailing prosperity encouraged costly and elaborate ornament in all media. Secretary of War Davis invited Congress to contemplate the embellishment of the interiors. "The original plan and estimate" he wrote on December 3, 1855 "were for a finish similar to that of the main building, but this style would not be a fair sample of the present state of architectural skill . . . It has therefore been thought proper to have prepared for inspection specimens of encaustic tiling, instead of brick and sandstone, for the floor; of painting, instead of white-washing, for the walls and ceilings; these and other contemplated improvements may be introduced."[90] Captain Meigs' report (October 14, 1855) showed that the staff did, indeed, have some splendid ideas. Specimens of Tennessee and Vermont marble were on hand. The use of fresco painting was under consideration. Rich patterns of Minton floor tiles imported from England were being laid to brighten the corridors.[91]

By the end of October it had been decided to use iron trusses from which would hang ceilings of cast iron plates. The span of 96 feet in the Hall of Representatives called for a bold design. The rooms would be lighted only from above, and at night by gas jets through panels of ground or stained glass. Meigs was confident that these halls would be "the best debating rooms of such size ever built." Among other notable features of the new circulation plan were marble staircases, according to Meigs "the most Stately in the country" and a corridor especially assigned to telegraph operators of the national press.

Already a sample fresco was being painted by Constantino Brumidi in a committee room. "It is the most appropriate and beautiful mode of finishing the building and it will afford a field for the talents of artists never before offered in this country . . . the designs and cartoons are being made . . ." The designs of sculptors Randolph Rogers and Thomas Crawford were being executed in bronze and marble. Even the boiler fronts in the cellar were decorated. Those were heady times at the Capitol; Congress had become a real patron of the arts and seemed to relish the role.

On the exterior the eastern porticos of both wings were modified by introducing central pediments in order to break the long horizontal silhouette of the roof balustrade[92] and these were to be filled with sculpture.

The House of Representatives moved into its new hall on December 16, 1857, and the Senate on January 4, 1859. But in the meantime relations between Walter and Meigs had deteriorated to the point where they separated their offices and Walter was refusing to share the drawings. Meigs was also feuding with the powerful Secretary of the Interior, John B. Floyd, in whose favor President Buchanan made the final decision. "I am removed because I am not supple enough for the purposes of the War Department" noted Meigs as he received orders of dismissal on November 2, 1859.[93] Walter continued at the Capitol under the easy-going direction of Captain William B. Franklin of the Topographical Engineers.

The Dome

Once the wings were well under way, it had become more apparent than ever that the old wooden dome of Bulfinch was no longer adequate to dominate the great mass. Furthermore, architectural styles were changing. Robert Mills had been the first to propose a tall dome on the Italian model, abandoning the shallow saucer models of Thornton and Latrobe. Walter, in Mills' footsteps, was to create the masterpiece that we know today. He had before him the famous precedents of St. Peter's in Rome and St. Paul's in London. Just as important, he also had engravings of the new St. Isaac's Cathedral in St. Petersburg.

The structural problem in Washington dictated the final material: iron. A decade or two later it might well have been steel, but in the 1850's American technology was not yet so advanced. It would have been impossible to support a *stone* masonry dome on the old walls of the Rotunda below. But iron had

proved itself practical in the great dome of St. Isaac's (built 1840–42 by the French architect Richard de Monferrand), where the problem was aggravated by the swampy ground on which the structure rested. The Russian example, which had three concentric layers of iron, was one of the spectacular successes of its period. Walter, adapting the suggestions of his assistant Schoenborn—and always under the review of Captain Meigs (who was in turn assisted by Adolf Faber and O. Sonnemann "iron constructors and computers")—produced the design we know today.

The basic "skeleton" of the new dome was essentially a sort of convex cone of cast iron ribs made up of openwork, triangular-webbed iron voussoirs bolted together and working much like masonry ribs. These sprang from a ring of new brickwork on top of the old Rotunda walls. They soared upward to support the tholus, itself a circular cupola which carries the great bronze statue of Freedom, topping the whole monumental composition.

From the Rotunda below only inner domes meet the eye. First, there is a light, coffered dome which opens to reveal above it a great, smooth saucer bearing the great painting called "The Apotheosis." The exterior dome is the outside shell or "skin" which encloses and protects the whole.[94] The whole elaborate ensemble is skillfully joined and ornamented—a magnificent composition of hardware, the tour de force—both structural and aesthetic—climaxing the American Age of Cast Iron.

The erection of this colossus required great ingenuity at the site. The old wooden dome was removed and a temporary roof built over the opening during the summer of 1855. In July Captain Meigs invited bids from fifteen foundries scattered from Richmond to New York City. Prices were first asked for 36 fluted Corinthian column shafts cast in 27-foot lengths. The capitals (with foliage to be attached by screws or rivets) and the bases were to be made separately. All but two of the firms bid. On a second round Betts, Pusey & Co. of Wilmington lost out to Poole & Hunt of Balti-

The Great Iron Dome

To design a dome tall enough to dominate the huge mass of the Capitol as extended — yet light enough to ride the old masonry walls safely — was a major challenge, both structural and aesthetic. A tall, Italianate silhouette had been proposed by Robert Mills as early as 1850. After a study of St. Peters (Rome), St. Paul's (London) and St. Isaac's (St. Petersburg) the Washington design was arrived at by a team of three: Architect Walter, Engineer Meigs and Draftsman Schoenborn. Construction began by demolishing the old wooden dome in the summer of 1855 and was capped off by the installation of the Crawford statue of Freedom on December 2, 1863. The foundries of Poole & Hunt of Baltimore and Janes, Beebe & Co. of New York prepared most of the patterns and castings.

Fig. 12.20 The section shows successively the massive footings, the "Crypt," the Bullfinch Rotunda, the two light inner domes, the "skeleton" and the highly architectural outer dome or "skin." In contrast, on either side may be noted the small masonry domes of earlier construction: small, expensive, structurally suspect and generally given to leaking. Drawing signed by Walter December 7, 1859.

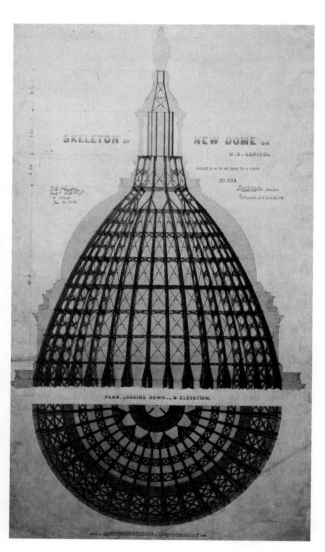

Fig. 12.21 The "skeleton," or basic frame of the dome, was composed of open-work cast iron units bolted together to form ribs. It rode a ring of new brickwork and was tied at the bottom with a wrought iron tension band. Drawing signed by Walter December 12, 1859.

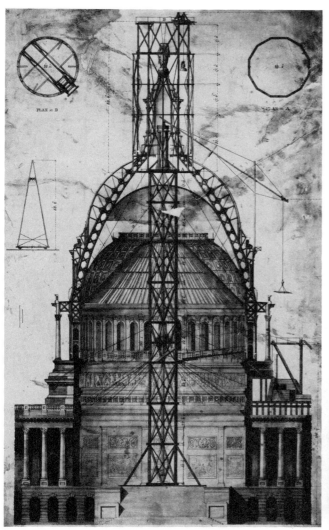

Fig. 12.22 The various scaffolds, cranes, derricks and hoisting apparatus were the special concern of the engineers who were having parallel problems on the U. S. Post Office Extension and the Cabin John Bridge. The iron founders who supplied the units at 6¢ per pound were paid an additional 3¢ for hoisting, fitting and securing them in place. Drawing signed by Walter May 8, 1863. (Note the head of Freedom being raised — an accomplishment actually performed 7 months later.)

Fig. 12.23 Elevation showing the completed composition. Both iron and sandstone were painted a marble color to match the wings. Height from grade to the top of the 7-ton bronze statue stated to be 287′5½″. Drawing signed by W. B. Franklin (Meigs' successor) December 9, 1859.

Fig. 12.24 The completed interior of the Capitol dome.

more. The latter were awarded the contract on September 13 at $.034 per pound, the customary way of pricing such work.[95]

The long campaign of procurement began a year later. The cast work was parcelled out in units to various foundries, numbered drawings — or photographs of them — being sent out from Washington. Janes, Beebe & Co. of New York (who had previously done the castings for the iron Library room) got the cornices at 9¢ per pound "including all expenses of patterns, freight and drayage (the government to furnish all the hoisting apparatus and hoist them) plus the necessary staging."[96]

Wrought iron naturally had its place in the design, too. For the great tension band around the base of the dome a contract was made with Murray & Hazlehurst of the Vulcan Works, Baltimore, to use "Abbott's puddled plate, from charcoal pig," carefully tested.[97] Besides the main structural and ornamental elements water pipes, "hot-pressed" nuts, floor plates, iron sash and a host of other elements had to be assembled. On February 15, 1860, after considerable negotiating, Janes, Fowler, Kirkland & Co. (the older firm with a new name) got the contract for completing the dome.[98]

The outbreak of the Civil War did not stop the work for long; the completion of the dome had become a symbol of faith in the survival of the Union.[99] On November 1, 1862, Walter could report that the ironwork had reached a height of 215 feet, with 71 feet, 3 inches yet to go, including the crowning statue. The contractors had over a million pounds of castings on the ground when they were advised that they could proceed only at their own risk "trusting in the future to the justice of the government. This was alike creditable to their perseverance and to their patriotism. "They", wrote Walter, "have prevented the sound of the hammer from being stopped on the national Capitol a single moment during all our civil troubles."[100]

Progress was slowed by the general preoccupation with war and by the want of such workmen "as could be trusted to manage heavy masses of iron at such fearful heights." By the end of 1863, however, Constantino Brumidi was drawing the cartoons for the paintings in "the eye of the inner dome" and arrangements had been made to light it all up by means of Gardiner's electromagnetic gaslighting apparatus.[101]

Noon of December 2, 1863 — just after the Union victory at Chattanooga (at which battle Meigs, now a major general, was present) — saw the dramatic capping of the dome. Orders issued the day before called for a national salute of thirty-five guns fired by a field battery on Capitol Hill; the last gun to be answered by others in the long Washington defense line counting Forts Stanton, Davis, Mahan, Lincoln, Bunker Hill, Totten, De-Russy, Reno, Cameron, Corcoran, Albany, and Scott.

The architect, pleased, reported:

> Precisely at 12m. . . the crowning feature of the statue was started from the ground in front of the Capitol, by means of the steam hoisting apparatus which has been successfully used for the construction of the entire dome, and in twenty minutes it reached the height of three hundred feet, when it was moved to its place, and firmly attached to the remaining portion of the figure; as soon as it was properly adjusted, the American flag was unfurled over its head, and the national salute was fired . . . The effect was thrilling, and grateful to every loyal heart.[102]

Exit Walter

On April 14, 1865, President Lincoln was assassinated; his body lay in state under the great dome for a day. The arrangements were made by General Meigs, who had spent the night at the President's bedside and marched in the funeral procession.[103]

Walter was to depart from the Capitol only six weeks later. When the new Secretary of the Interior cancelled the contract with the iron fabricators, Walter resigned in protest after 14 years on the job.[104] "In taking this step", he wrote stiffly, "I am moved by considerations of self-respect." Schoenborn years later revealed that "He went away in a rage taking all the

Artificial Light and Artificial Air

On April 4, 1853, after construction of the wings was well under way, Captain Meigs was directed by the Secretary of War to restudy the arrangement for warming and ventilating the House and Senate chambers—as well as their acoustics. Six weeks later, after a visit to leading theaters and other notable places of assembly in the East, a drastic change in the floor plan of both wings was recommended. The two great halls were thereupon moved away from exterior walls with windows that could open, to become interior spaces lit only through glass ceilings and ventilated through air ducts. Four and a half years later Meigs, speaking of the new installations, reported: "This apparatus is one of the most extensive and complete in the world. Its arrangement and details have required a vast amount of study, of scientific and mechanical knowledge, and experience, in which I have been ably assisted by the manufacturers, Messrs. Nason & Dodge, and their agents."

The halls were soon in use and in the summer of 1858, Speaker of the House James L. Orr complimented Meigs on the acoustics:

The ventilation is equally successful. The densest crowd in the galleries, during the most protracted sittings, breathed a fresh atmosphere, free from all heaviness or impurity.

The heating apparatus is so perfect that the engineer had only to be notified what temperature was desired, when in a few minutes it was supplied.

The arrangement for lighting the hall is admirable.

Not a burner is seen, and yet such a flood of softened light is poured down through the stained glass ceiling of the hall that it was difficult to distinguish when the day ended and the night commenced.

The hall and its fixtures are a splendid triumph of your professional skill, and will ever remain a proud monument to your genius.

But Congress was not all that pleased and as time went by, the euphoria diminished. As early as March, 1860, there was an abortive move in the Senate to move their chamber back to the outer walls with windows.

Four years later a great deal was heard in the House along the same lines. One congressman declared that "this is an unfit place to do business in, especially when the furnaces are going . . . it is impossible to stay in this House any length of time without going out to seek the fresh air . . . Members must breathe, they must have fresh air. Here the air is artificial and the light is artificial." Mr. Trumbull later (July 23, 1866) thought that the Hall should be moved back to the exterior of the building. "What a relief it would have been during the late hot and oppressive weather . . . to have the benefit of windows upon the exterior . . . to look out upon the world and not to be shut up here like prisoners in a jail or school-boys in a school-room."

The problem was studied and restudied for years afterward as new experts and new types of equipment were considered. These issues which arose here so early are almost universal today and very much alive.

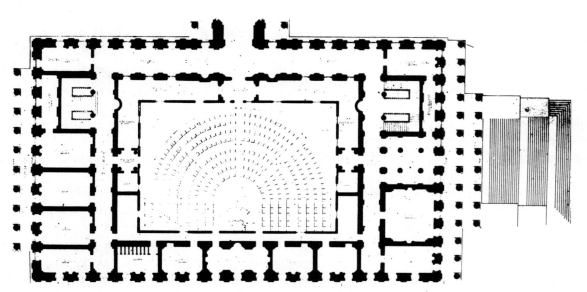

Fig. 12.25 "Plan of Principal Story, South Wing." The new floor layout approved by the President on June 27, 1853, moved the Hall of Representatives to the middle of the wing.

Fig. 12.26   The House of Representatives meeting in its new hall December 3, 1860.
Harper's Weekly for December 15, 1860.

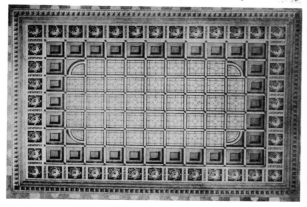

Fig. 12.27   Plan of ceiling dated March 28, 1856. The visible ceiling was of cast iron elements cast in New York and stained glass from Philadelphia. Drawing dated March 28, 1856.

Fig. 12.28   Photograph dated Feb. 2, 1858. Between the glass ceiling and the skylight was a blazing inferno. Suspended gas pipes were arranged in curves with "the jets near enough together to light each other from a small perpetual burner."

Fig. 12.29   Ornamental design for ventilating register.

leading drawings with him, saying that it was customary that architects never parted with their drawings." Walter really expected to be called back from Philadelphia but his former assistant, Edward Clark, maneuvered to get the job "and so barred him out."[105]

NOTES

1. Glenn Brown, *History of the United States Capitol,* Washington, 1900, I:4.
2. The procurement of workable stone for the new Federal buildings was from the very first recognized as a major problem. L'Enfant, as early as December 19, 1791 had requested the quick development of the public sandstone quarries on the Potomac with barracks built for twenty workmen. Timber was to come from a designated property in the forest nearby, a new sawmill was proposed and wheelbarrows were even ordered for the excavations. H. Paul Caemmerer, *The Life of Pierre Charles L'Enfant, Planner of the City Beautiful,* Washington, 1950, pp. 187–192.
3. Jefferson to Thomas Johnson, March 8, 1792. Saul K. Padover, ed., *Thomas Jefferson and the National Capitol,* Washington, 1946, p. 111.
4. As early as 1790 Jefferson had considered a competition. Padover, *Op. cit.,* p. 33.
5. They called for a brick building to include "a room for Representatives" (300 persons plus lobby), a room for the senate (1,200 square feet plus anteroom and lobby), a conference room (300 persons), and twelve rooms for committees and clerks (600 square feet each). Brown, *op. cit.,* I:5. The drawings are in the collections of the Maryland Historical Society, Baltimore. Some have never been published.
6. Dr. Thornton as an architectural designer was something of a dark horse outside of Philadelphia where in 1789, he had won the competition for the design of Library Hall. Charles E. Peterson, "Library Hall: Home of the Library Company of Philadelphia, 1790–1880," *Transactions of the American Philosophical Society,* New Ser., Vol. 43, Part I (March, 1953), pp. 132–133.
7. Jefferson to Daniel Carroll, Philadelphia, Feb. 1, 1793. Padover, *op. cit.,* p. 171.
8. Jefferson to Washington, March 26, 1793. Padover, *op. cit.,* pp. 180–181.
9. Washington to Jefferson, Mount Vernon, June 30, 1793. *Ibid.,* pp. 182–183.
10. Jefferson to Washington, July 17, 1793. Padover, *op cit.,* pp. 184–186.
11. U.S. Congress, House of Representatives, 58th Congress, 2nd Session, Report No. 346, *Documentary History of the Construction and Development of the United States Capital Buildings and Grounds,* Washington, 1904, pp. 20–21. (Hereafter *DH*).
12. Padover, *op. cit.,* pp. 163–164.
13. *Ibid.,* pp. 162–165.
14. *DH,* p. 34
15. *Ibid.,* p. 35
16. *Ibid.,* p. 36
17. *Ibid.,* p. 76
18. *Ibid.,* p. 78
19. *Ibid.,* p. 85
20. *Ibid.,* p. 89
21. *Ibid.,* p. 91
22. *Ibid.,* p. 99
23. *Ibid.,* pp. 93–94
24. *Ibid.,* p. 91
25. *Ibid.,* p. 94
26. *Ibid.,* pp. 96–97
27. In a letter to Jefferson on February 27, 1804, Latrobe describes a visit to Thornton in which the latter declared that there were no difficulties in his plans "but such as were made by those too ignorant to remove them" and refused to discuss the subject. *Latrobe Papers.*
28. A year later Latrobe, in sending the President a roll of new drawings, was still attacking Thornton's plans. In a long letter he criticized the layout of the south wing ". . . the space becomes a dark cellar . . . the doors backing into it are useless if not absurd . . . none of the rooms can be furnished with fireplaces . . . a room, which though measuring 50 feet each way, seems to be of no possible utility & which can be lighted by only one Window in a Corner . . . innumerable bad consequences of a design radically defective . . . It is greater labor to correct errors in Art than avoid them by beginning a new design" and so on. Latrobe to Jefferson, Newcastle, Delaware, March 29, 1804. *Latrobe Letterbooks* (MS) Maryland Historical Society, courtesy the Papers of Benjamin Henry Latrobe, Edward C. Carter II, Editor.
29. Latrobe to Jefferson, Nov. 25, 1806. Padover, *op. cit.,* pp. 373–374.
30. Latrobe wrote to Jefferson on January 26, 1805, that window glass would probably have to be imported. "Very excellent window glass has been made at Boston, I believe by remelting the broken glass collected in our cities. I am uncertain whether the furnace is now at work." *Latrobe Letters* (MS) Library of Congress.

On August 31 the discussion continued. "German Glass is however deservedly getting into very bad reputation. Almost all the 10 by 12 & 8 by 10 Glass imported last Year into Philadelphia from Hamburg is so thin & Knotty as to be scarcely saleable.—I am told that this is owing to the sudden demand for Geman Glass for the American Market, which has had an effect on the speculative views of the Glass Manufactorers [sic] in Bohemia,—for, in the french Market nothing but good Glass is Saleable.—English Glass is better but very dear." *Latrobe Letterbooks.*

31. At one point Latrobe proposed the use of cast iron Greek capitals which "may easily be cast in one piece in Iron with the upper row of plain leaves. The lower may be cast separately & fixed with Copper rivets." Latrobe to Jefferson, March 29, 1804. *Latrobe Letterbooks.*
32. Latrobe to Jefferson, January 26, 1805 (LC).
33. Authorization June 16, 1805 cited in Brown, *op. cit.,* I:40.
34. Latrobe vented his frustration in a letter to his fellow countryman Lenthall from Washington on January 7, 1805. "You & I are both blockheads. Presidents & Vice presidents are the only Architects & poets, & prophets for ought I know, in the United States. Therefore let us fall down & worship them (as for the Ladies behind the piers, and the respondent at the long table, they shall ogle the vice president, & the vice president may ogle them as much as they please—but the latter shall pay for your brains. If you ever get into the Eastindies [sic] in consequence of the transmigration of Yourself into an Hindoo Rajah,—provided the British have any of them alive,—you will see little Hamilton & little Burr standing in the temple of Lingam (the Hindoo priapus) like the Columns of Jachim & Boas in the temple of Solomon in eternal & basaltic erection with frution,—for their Sins . . ."
35. A sampling of the letter written to Thornton from Newcastle, Delaware, on April 28, 1804: "Sir, Open hostility is safer than insidious friendship. I cannot therefore regret the declaration of War contained in your letter of the 23d April. For a considerable time I have been convinced than an

open rupture with you would be more honorable to me than even that show of good understanding which has prevailed between us . . . It seems however necessary to remind you of your conduct to me the day after you consented to the alterations proposed in the South Wing of the Capitol in April 1803, and the insult you then offered me, of the coolness which since subsisted between us: of the frequent attempts which I notwithstanding made to come to an amicable discussion of the Subject, and especially of the insulting audience with which you honored me in the public passage at the door of the Patent office . . . I now stand on the Ground from which you drove Hallet, and Hatfield to ruin. You may prove victorious against me also, but the contest will not be without spectators. The public shall attend and judge I shall not court public discussion. It is in my *power* however, more than in my inclination to show you in a more ridiculous light, even than were I, as is the fashion after such a correspondence, to call you to the field. But you have other accounts of that sort to settle before it can come to my turn . . ."

36. Walter Lord, *The Dawn's Early Light*, New York, 1972, pp. 162–165.

37. ". . . its *tout-ensemble* was light, airy and elegant. After traversing a wide and spacious entrance-hall, you arrived at the foot of a handsome, spiral hanging staircase; on the left of which were two magnificent apartments; one above the other, which were occupied as sitting chambers, by the two houses of representatives. From these branched off several smaller rooms, fitted up as offices, and probably used as such by the various officers of state. On the right of the staircase, again, were two other apartments, equal in size to those on the left, with a like number of smaller rooms, branching off from them. These were furnished as a public library, the two larger being well stocked with valuable books . . ." [John Gleig], *A Narrative of the Campaigns of the British Army*, London (2nd Ed.) 1826, p. 134.

38. Padover, (undated) pp. 473–5.

39. Thomas Munroe, Superintendent of Buildings, summing it all up on October 29 reported that nearly eight hundred thousand dollars had been spent on the two wings of the Capitol. *DH*, pp. 173–174.

40. *DH*, p. 184.

41. *Ibid.*, p. 185.

42. Hamlin, *op. cit.*, p. 434

43. *Ibid.*, p. 438.

44. *Ibid.*, pp. 440–441.

45. *Ibid.*, p. 443.

46. *Ibid.*, p. 452

47. The Bulfinch salary was to be $2,500 per year with the assistance of a draftsman at $500 and moving expenses for his family and furniture. Monroe had visited Boston in the summer of 1817 and had been impressed by both the architectural accomplishments and the modest personality of Bulfinch. A feeler had been extended as early as November but Bulfinch considered it unethical to accept the office before Latrobe's resignation. Harold Kirker, *The Architecture of Charles Bulfinch*, Cambridge, 1969, pp. 321–323.

48. Dec. 2, 1817 (*DH*, p. 199).

49. Jan. 28, 1818 (*Ibid.* p. 200).

50. Feb. 5, 1818 (*Ibid.*, p. 201).

51. William W. Wheildon, *Memoir of Solomon Willard, Architect and Superintendent of the Bunker Hill Monument*, [Boston] 1865. pp. 38–41. Willard was invited to assist with the stone carving in the Capitol but declined.

52. Further funds were soon provided. In an act approved April 20, $80,000 more was appropriated to continue on the burned wings. Kirker, *Op. cit.* p. 323. This included the development of the marble quarries where the housing, feeding and even the clothing for workmen had to be provided. The handsome marble columns eventually cost more than three times the original estimate. *DH*, pp. 219–220.

53. *DH*, p. 206

54. Kirker, p. 323.

55. *Ibid.*, p. 326.

56. It was even suggested that a Gothic design be used to achieve greater height. "Architects," wrote Bulfinch, "expect criticism and must learn to bear it patiently." Kirker, *op. cit.*, pp. 327–329.

57. *DH*, pp. 207–210

58. Charles E. Fairman, *Art and Artists of the Capitol of the United States of America*, Washington, 1927, pp. 32–37, 91–92. Hydraulic cement had only recently been discovered on the Erie Canal and was very new as used in buildings.

59. Wheildon, *op. cit.*, pp. 48–50. The author claimed that his friend Willard had invented the house-warming furnace, though he could find no patent. (See also *DH*, pp. 291, 295.) At first a 10-foot hole in the center of the Rotunda floor opened down into the Crypt. There were no outside doors at the Crypt entrance, only a grilled gate. The cold and damp whirled up through the opening, causing temperature and humidity changes that affected the Trumbull paintings. When the floor was closed and doors added to the Crypt, the problem was greatly alleviated.

60. *DH*, p. 416.

61. *DH*, p. 295. For "iron gallery to Senate Chamber and stairs leading there to . . . $2,045.50." *DH*, p. 297.

62. Cast iron columns had been used in a Liverpool, England, church as early as 1772. Turpin Bannister, "The First Iron-Framed Buildings," *The Architectural Review* (London), April, 1950, pp. 245–246.

William Strickland used them in Philadelphia in his Chestnut Street theatre built 1820–22. Agnes Addison Gilchrist, *William Strickland*, Philadelphia, 1950, p. 61.

63. Kirker, *op. cit.*, pp. 330–331. The "extensions" of the 1850s wiped out the immediate setting shown in the Latrobe-Bulfinch drawings. Bulfinch's gatehouses nicely done in Aquia stone to harmonize with Thornton's design of the main building survive.

64. Brown, *op. cit.*, I:61, 62. Jackson gave his reason as "economy." As early as 1827 the curmudgeon from Kentucky had complained about the "four or five gentlemen who are drawing an annual salary from the public treasure . . . they will never finish it as long as you will furnish them the money to waste upon it." *DH* p. 290.

Bulfinch left Washington on June 3, 1830, after completing some private work.

65. The acoustics of the House of Representatives ranged from difficult to impossible. There were complaints from the very beginning and two of the leading American architects—Robert Mills and William Strickland of Philadelphia—made extensive studies of the problem without solving it. Engineers of the United States Army added their opinions, too. Low, flat ceilings of glass, cloth and plaster were among the solutions proposed. A cloth ceiling was tried and removed as unsatisfactory. Kirker, *op. cit.*, p. 325.

66. The first gas at the Capitol was manufactured by James Crutchett in 1847. After a year, a contract was entered into with the newly formed Washington Gaslight Company for gas made from oil or rosin at $6 per 1,000 cubic feet, later reduced to coal gas at $4. (Report of C. W. C. Dunnington, Superintendent to J. B. Blake, Commissioner of Public Buildings, October 11, 1855).

Before gas gave way to electricity the *Evening Times* for October 28, 1895, reported that 'the heat generated above the magnificent glass panels in the Senate Chamber was so great as to ruin every year the glass and metal skylights in the roof. . .'

67. For Thomas U. Walter (1804–1887) there is no adequate study because

his papers have not been available. Professor Robert B. Ennis of Philadelphia is now making a comprehensive investigation up to the year 1851, the year Walter went to Washington.

For General Montgomery C. Meigs, USA (1816–1892) Professor Russell F. Weigley of Temple University has done an excellent military life, *Quartermaster General of the Union Army* (1956). It is very helpful on the politics in Washington during Meigs' duties at the Aqueduct and the Capitol. But a major study in the tremendous archives will be necessary to evaluate his work at the Capitol, untangling the rival claims of the Walter faction.

68. Brown, *op. cit.*, II:115–16.

69. *Ibid.*, I:116–18. *DH*, pp. 430–448.

70. Brown, *op. cit.*, II:119. The day before a plan by Walter had been approved by the President. On June 14, even before Walter had a chance to get off to the capital city, he was offered a specimen of artificial stone by one R. McCabe. *Walter Letterbook* (Bound MS), Architect of the Capitol. June 14, 1851.

71. Brown, *op. cit.*, II:119.

72. *DH*, p. 448.

73. Bannister, *op. cit.*, pp. 17, 23.

74. This was actually done as far as expedient. But it is interesting to note that Walter (himself once apprenticed to his father as a bricklayer) recommended that the brick at Washington be laid by day labor. In building the Girard College structures at Philadelphia day-labor was used for the first 8 years but afterwards changed to the contract procedure to save money. *DH*, pp. 449, 450.

75. Report of July 29, 1851. *DH*, p. 450.

76. The members were invited November 5, 1851. *Walter Letterbook* (MS), Office of the Architect of the Capitol. Cubes of marble were cut from specimens submitted by twelve bidders and crushed in a testing apparatus at the Navy Yard. The official report showed that the strongest came from East Chester, New York; Lee, Massachusetts; Hastings, New York; and Baltimore; they resisted crushing in that order.

The apparatus was invented by Major Wade, an ordnance expert, and tests made with the assistance of Lieutenant Dahlgren of the Navy. The absorption of water as related to erosion by freezing and thawing was also considered. *DH*, p. 558.

On March 11, 1852 a long and ultrascientific report on the foundation stone was turned in by Walter R. Johnson of the Franklin Institute, famous for his investigation of the strength of iron and of steamboat explosions *DH*, pp. 559–566.

77. Report by Walter, Dec. 23, 1851. *DH*, pp. 464–468.

78. Exall and Emery, whose quarry was located about 3 miles southwest of Richmond on the James River Canal, won the award with a low bid of $22,400.

79. The records relating to the selection of marble are extensive. See especially the report of the special commission of December 22, 1851, and Walter letter to Secretary Stuart of December 27.

80. The price for 1000 barrels was $1.12-½ per barrel of 300 pounds. A break in the C&O Canal the following spring interrupted delivery. Walter to Stuart, May 6, 1852. *Walter Letterbook.*

81. Promoters of "steam-pressed" brick urged their product but the architect (himself a former bricklayer) thought that a lighter brick easily cut on the job was more suitable.

82. *DH*, pp. 516–517

83. Report dated Dec. 4, 1852. *DH*, p. 347.

When completed, it will present the first specimen of a room constructed entirely of iron; and I think I may add, that for convenience and beauty of arrangement it will be without a rival. The workmanship is of the most admirable quality, and when we look at it as it now stands, and reflect that it consists of more than ten thousand separate pieces, of an aggregate weight of four hundred tons, and that it was planned in Washington, and executed in New York, more than two hundred miles from the hall in which it was to be placed; and when we see that every part of it fits together with the precision of cabinet work, we are at a loss whether to ascribe most honor and praise to the genius and taste of the architect who conceived and marked out the design, or to the skill and fidelity of the contractors who performed the work. Among other splendors on view was a ceiling covered with gold leaf. It was a good example of the rich decoration favored in that period.

84. *DH*, p. 499 (March 15, 1852).

85. *Ibid.*, p. 528.

86. *Ibid.*, p. 578. But Walter had his defenders, too. In the House Mr. Stanton of Kentucky, who had had some practical experience (and presumably was coached by Walter) gave an extended lecture on the history of cement, quoting from the *Transactions* of the Royal Institute of British Architects, and alluded to the heavy concrete footings under the new Houses of Parliament at Westminster on the marshy bank of the Thames. He "blushed" for the intelligence of his fellow-members and declared the workmanship at the Capital better than anything in the city. *DH*, pp. 473–477.

87. Brown, *op. cit.* II:126.

88. Meigs entered on duty April 4, 1853.

89. *DH*, p. 587. Professor A. D. Bache of the Coast Survey and Joseph Henry of the Smithsonian, two of the most eminent scientists of their time, were enlisted as advisors and as a party they promptly visited a number of well known buildings. The acoustics in each were tested and conclusions worked into the plans for the Capitol.

In Philadelphia they saw, among other places Musical Fund Hall, the Circular Church in Sansom Street and several theatres. In New York and Brooklyn they viewed the Metropolitan and Niblo's Halls, in Boston the State House chambers, Faneuil Hall and other rooms. *DH*, p. 587–590.

90. Report dated December 3, 1855. *DH*, p. 627.

91. No history of the tile works of Minton, Hollins & Co. of Stoke-upon-Trent has been located. The Capitol tile was procured through Miller & Coates of New York. In a letter of March 26, 1855, Sir Charles Barry endorsed them warmly:

I hereby certify that I have made extensive use of Messrs. Mintons tiles in the New Palace at Westminster, and that I have reason to entertain the highest opinion of their quality and durability, as well as of their beauty in color, and the fidelity with which the original designs for them have been carried out. The first employment of those tiles was in the floor of the Lobby of the House of Lords in the year 1846 and in the year 1851 they were employed in the Central Hall and main thoroughfares of the Building; and although perhaps not less than 1,500,000 persons have passed over the central portions of those floors, I cannot perceive the slightest abrasion of the original surface of the tiles.
(Copy in A. & R. Library. Office of the Architect, US Capitol)

The designs were made at the potteries in England and sent to Washington for approval. As the work neared completion (October 29, 1859), Meigs reported that 95,806 square feet had been laid in the two wings at a cost of $131,037.19. *DH*, p. 739.

92. Meigs report October 27, 1859 (*DH*, p. 736).

93. For an account of these proceedings see Russell F. Weigley, *Quartermaster General of the Union Army, a Biography of M. C. Meigs*, New York, 1959, pp. 911–99. Meigs retired to his other

Washington responsibilities, was later ordered to duty in the Dry Tortugas and finally served a brilliant tour as Quartermaster General during the Civil War. Meigs' abilities as an architect may be adjudged from buildings he later designed on his own.

94. Three of the most published original drawings are (No. 1901) "Section through Dome," (No. 1938) Skeleton of Dome"—both dated just after Meigs' departure and signed by Walter and Franklin—and (No. 2079) "Section Showing Scaffolding" carrying Schoenborn's name and signed by Walter May 9, 1863.

95. In Meigs' office M. Sonnemann carefully computed the weight of each piece ordered. Meigs to Poole & Hunt, February 12, 1857. The prices naturally varied with the complexity of the patterns and the difficulty of fitting and erecting the elements. The deliveries had to be carefully scheduled, too.

96. As a part of the agreement, Meigs stipulated 6¢ per pound for iron casting delivered at the job and 3¢ additional for putting it up. Meigs to Janes, Beebe & Co., October 8, 1856. The foundry sometimes got detailed advice from Washington. "Before casting the leaves and eggs and other ornaments", insisted Meigs, "Send me one from the pattern . . . I wish

to approve each leaf before it is cast . . . There are certain leaves in the cornice of the Senate which I would not have accepted without alteration, had I seen them in time. They show too light on the edges, and to the eye look like stamped-work, instead of architectural enrichment."

97. Meigs to H. Abbott & Son, October 21, 1857.
98. Endorsement by J. B. Floyd, Secretary of War.
99. With the outbreak of war, work at the Capitol was suspended (May 15, 1861) by order of the Secretary of War whose personnel had plenty to do elsewhere. On April 16, 1862 control passed to the Interior Department. Instructions were to carry on such construction as needed to protect the work already in place. Quarrying had stopped in Massachusetts and had to be reorganized. There was a scarcity of vessels to carry the marble down from the railroad terminal at Bridgeport, Connecticut. Some staining had resulted at the site by laying the white marble blocks in asphaltum.

Rogers' bronze doors were still at the foundry in Munich, awaiting instructions. Twenty-nine of the great monolithic columns quarried in Baltimore County had been delivered. Altogether, 46,941 barrels of hy-

draulic cement had been used and 177 man-days of photography accomplished in the proceeding year. Thomas Crawford's great statue of Freedom—19 feet, 6 inches, in height and weighing 7 tons—had been cast in five sections at Clark Mills' foundry. The bronze was composed of tin from Phelps Dodge & Co. ($570) and copper from the Revere Company ($3,328.70). It was set up in the Capitol grounds for public inspection (Walter report, 37th Congress, 3rd Session, House Ex. Doc., Vol. 2, pp. 604–611.)

100. An act for the relief of Janes, Fowler Kirland and Company was finally passed in 1871 in the amount of $66,000. *DH*, p. 1030.
101. Walter report, November 1, 1863. *DH*, p. 1025.
102. *DH*, p. 1026.
103. Weigley, p. 323.
104. On May 26, 1865. Brown, *op. cit.*, II: 139–141.
105. Sad to relate. Walter never collected the fees he felt due him in the amount of $113,000 "for extra work" on other government projects. He died with honors but poor. For years his daughters pressed the Walter claim, going down to Washington annually to lobby (Bannister, *op. cit.*, pp. 28, 29). The architect's papers remain with the family.

PART TWO

Building Preservation

CHAPTER
13

Preservation of Historic Monuments in the United Kingdom

PATRICK A. FAULKNER

MR. FAULKNER, Superintending Architect in the Directorate of Ancient Monuments and Historic Buildings, Department of the Environment, London, is a Fellow of the Society of Antiquaries and author of several works on medieval buildings.

A co-ordinated system of preservation requires a formal identification with legal status of those buildings or monuments whose merit justifies special care. In the United Kingdom the necessary legislation[1] has been embodied in successive Ancient Monuments Acts and Planning Acts administered[2] by the Directorate of Ancient Monuments and Historic Buildings; Department of the Environment[3].

At present, some 18,000 *scheduled monuments* are listed by reason of their architectural, traditional, artistic or archaeological interest, and about 200,000 occupied houses and churches are listed as *historic buildings*.[4] The latter are categorized by merit as Grade I, II or II* and the lists include (a) all buildings prior to 1700 which survive in anything like their original condition, (b) most buildings dated 1700–1840 with some degree of selection, (c) buildings of definite quality built be-

tween 1840 and 1914 and (d) a rigorous selection of buildings after 1914.[5]

Publication of both lists gives them statutory force. Owners of monuments are notified individually, but owners of listed buildings are expected to review the list at Local Authority Offices.

Another category, *Conservation Areas*, identifies areas of special architectural or historic interest[6] whose character or appearance it is desirable to preserve or enhance. Usually Local Authorities designate these areas, although the Secretary of State may do so if warranted.

A major responsibility of the Directorate is for *Royal Palaces and Parks*, *public buildings*, *national galleries and museums*. Ownership and control of these is complex but the Directorate is responsible for their historic fabric.

The UK recognizes two types of preservation: negative and positive. Negative preservation is merely protection from willful or ignorant damage; hence owners of scheduled monuments must give three months' notice of intention to repair, alter, demolish or do any work affecting it. This gives the Department time for appraisal and persuasion; only in exceptional circumstances would the Secretary of State use his powers for taking the monument into his care compulsorily, the ownership remaining unaffected and compensation being payable.

Fig. 13.1   Hadrian's Wall, Northumberland, is now preserved as a Scheduled Ancient Monument. *(All photos in this chapter courtesy of the Ministry of Works, Crown Copyright reserved, unless noted otherwise.)*

Negative protection for listed buildings is the responsibility of Local Authorities, who must notify the Department of the Environment before permitting changes to Grade I or II buildings. Again, as a last resort, the Local Authority may purchase a disputed listed building compulsorily, or repair it and recover the cost from the owner. Conservation areas are accorded the same negative protection.

As ecclesiastical buildings are exempt from this legislation, Diocesan Advisory Committees have been set up to advise Parish Church Councils about work on historic churches. The Department is not directly involved.

Positive protection involves actual repairs and preservation. Scheduled monuments may be taken into state care by purchase or gift, or under guardianship in perpetuity. Most monuments in State care are open to the public, preserved, maintained and managed by the Department with funds[7] voted annually by Parliament. There are some 700 monuments in State care in England, Wales and Scotland, ranging from prehistoric to nineteenth-century industrial examples of great technological achievements. Preservation work on monuments is usually done by the Department's own labour force, which averages 800 persons, directed by a team of architects and technical officers with the archaeological guidance of the Inspectorate of Ancient Monuments; a Presentation Section manages public admissions and publicity.

For monuments, preservation rather than restoration is the policy. Missing floors, stairs and unbridged moats are not rebuilt, although alternative access is provided for visitors if possible without impairing the visual or historic quality. Interpretation varies from providing a simple ticket office to complex sales and control points complete with car parks, toilets and refreshment facilities.

Fig. 13.2 Hopton Castle, Shropshire, was built in the fourteenth century. A Scheduled Ancient Monument, it is shown here in a 1964 photo after preservation work was completed.

Another function of the Department is to give advice and grants toward preservation of scheduled monuments. The grants are

Fig. 13.4 Bramshill House, Hampshire, 1605–12, a police training center, is a Listed Historic Building. The main entrance is shown in this 1953 photo.

Fig. 13.3 Stokesay Castle, Shropshire, built in the late thirteenth century, is a Listed Historic Building. The 1964 photo shows its condition after preservation.

Fig. 13.5 Carlisle Castle, Cumberland, built in the twelfth and fourteenth centuries, is
an example of an Ancient Monument in State care.

often in the form of direct labour and allow the Department's expertise to assist many minor projects.

Positive protection to listed buildings is always by grant-aid to the owners, who receive the benefit of advice from the Department's experts. Grants average 50 per cent of the cost of historic repair, but exclude improvements such as bathrooms, kitchens and utilities. The grants were originally designed to protect the great country houses, but now extend to all listed buildings.

Historic towns are covered by Town Schemes[8] which recognize that historic value may depend on buildings that themselves are not necessarily outstanding. Within a specific

Fig. 13.6 Castle Acre Priory, Norfolk, dates from the late twelfth century. It is now an Ancient Monument in State care.

Fig. 13.7 A grant aided the owners of Barn Hill House, Stamford, Lincolnshire, to preserve this late seventeenth century structure, a Listed Historic Building.

Fig. 13.8   Market Hall is part of the Designated Conservation Area in Thaxted, Essex. Such classification protects the entire area from modern intrusions, even though it contains buildings which do not conform to the standards.

area, designated buildings may benefit by grant-aid recommended by the Local Authority. The great merit in such grants is that small sums may be distributed quickly and somewhat informally. Local Authorities also receive grant-aid for positive protection that will improve or maintain the historic character of Conservation Areas: these grants may vary from burying overhead wires to reinstating paved surfaces of streets, as well as financing work on buildings.

CONSERVATION IN ACTION

Because of the extremely wide scope of the Department's work, the examples included here cannot be considered "typical" or comprehensive. They do, however, illustrate the various types of preservation outlined above.

THE PRESERVATION OF A MONUMENT IN STATE CARE:
Bolsover Castle, Derbyshire

Bolsover Castle is now a castle in name only. The great mansion whose remains now stand on the site was built by the Cavendish family on the site of a medieval castle. The buildings were started in 1608 and continued with many changes, to the successive designs of Robert, John and Huntingdon Smithson until 1640. A long Terrace Range to the west of the great court, now roofless, was the mansion proper. South of the court the surviving Riding School, forge and stable are entirely the work of William Cavendish, Duke of Newcastle. To the north is the Little Castle, a romantic folly with vaulted chambers, frescoed walls and battlemented turrets, which still contains

239

much of its important early seventeenth-century interior decoration. The castle was abandoned as a residence in the early eighteenth century and finally presented by the seventh Duke of Portland as a gift to the Ministry of Works (now Department of the Environment) in 1945.

The buildings stand on a rocky bluff overlooking the valley with the tower-like Little Castle at the outer end. The site has a long history of instability due to faults in the rock formation and to subsidence following deep mining in the area, one of the richest coal fields in the country and long a source of wealth. It is known that the present castle's medieval predecessor suffered collapses and, in the late nineteenth century, the nearby parish church was severely damaged by movements along a fault line which crosses the Riding School range and passes beneath the Little Castle.

When the castle came into the Department's care in 1945, immediate precautions were taken to deal with the two principal points of danger: the walls of the Riding School were shored to contain a serious outward movement (Fig. 13.9); reinforced concrete beams were introduced to tie together the foundations of the Little Castle across the worst fractures on its outward or cliff face. Thereafter, work was concentrated on clearing the site of undergrowth, laying grass and making the area fit for public access.

As the work was to be carried out by the Department's direct labour force and there was clearly a long-term programme in hand, craftsmen were recruited locally and trained to use the preservation methods developed within the Department. The number varied between ten and fifteen men. From 1950–51, work was concentrated on the Riding School range to utilize the limited funds and labour to the best advantage. Two principal problems were to be faced: the distortion and decay which endangered the fine early seventeenth-century timber roof of the Riding School and the condition of the walls which, though massive, were badly fractured. The roof was treated in sections, each in turn being covered

Fig. 13.9 The Riding School at Bolsover Castle, shown with shoring during preservation work in 1950. Begun in 1608, Bolsover was abandoned as a residence in the early eighteenth century and is now a Monument in State care.

with a temporary scaffold roof while the great trusses were dismantled, repaired by piecing in new sections where the old were decayed, and reassembled without the iron ties and straps added at various times. At the same time the gabled stone dormers, which rested on the trusses, were carefully taken down and a reinforced concrete ring beam cast at wall-top level with cantilevered extensions to carry the weight of the reerected dormers free of the timber roof (Fig. 13.10). Individual fractures, mostly at window heads and sills, were stitched with reinforced bonders set behind the face. All timber was treated with a pesticide before replacement and truss ends isolated from the surrounding masonry and set on lead bedding. The roof was treated in five sections over some 12 years.

At the Little Castle, the lead domes on the

Fig. 13.10   By 1964, the Bolsover Castle Riding School had been reinforced and the fractures consolidated. The seventeenth century timber roof was dismantled, repaired and returned; a reinforced concrete-ring beam now supports the stone dormers.

turrets and central lantern and the main lead flats of the roof were lifted, the understructure repaired and the lead recast and reformed. This work and the repairs of leaded-light windows were necessary to protect the decorations within, pending the more radical treatment that would eventually be necessary. Small fan heaters were installed at strategic positions to reduce potential damage to the wall paintings through condensation. In the ruined Terrace range the dangerously loose masonry was secured (the site is exposed to high winds), and rubble that filled the lower levels to a depth of some 15 feet was excavated. Little of interest was recovered from this operation, as the rubble fill proved to be surprisingly recent.

The next major operation was to consolidate the Little Castle, the movement of which

had been kept under close observation over a period of nearly 15 years. The emergency reinforcement first placed in the foundations had not arrested this movement, nor had the more local bonding later introduced across the worst fractures. Much of the interior wall surfaces was concealed by panelling; such of this as was removable was taken out for conservation treatment and preliminary steps were taken to fix the wall paintings and the plaster ceilings. A vault, constructed of remarkably thin grey and white marble, was suspended from a steel grid above prior to the next stage. In 1968–69 a more ambitious proposal for total reinforcement was devised using the Fondedile method. Prestressed sheradized steel rods 30mm diameter were grouted into long drillings 4 ½-inch diameter on each external wall and through cross walls at

---

PATRICK A. FAULKNER

three levels. Sixty-five of these rods stabilized the building so that it could move, as it undoubtedly will, as a whole. In addition the short stitching rods were laced diagonally across the principal fractures. The system of reinforcing is now being extended to the high curtain walls while archaeological excavations will determine the line of their medieval predecessors. Meanwhile the work of cleaning and conserving the painted panelling and other decorations continues both on site and in studio.

Beyond this capital work, constant maintenance continues on the buildings and grounds. The original path layout has been recovered, lawns laid and the yew hedges of the charming fountain garden restored to their clipped form. Bolsover being adjacent to a town, extensive visitor facilities are not nec-

essary except for a car park to one side of the entrance avenue which has been reformed to its original width in keeping with the scale of the buildings, and a small ticket and sales office. A handbook to the castle has been published and is on sale at the entrance.

MONUMENT INTERPRETATION:
Hailes Abbey Museum, Gloucestershire

In 1948 the Department accepted in guardianship the ruins of the thirteenth-century Cistercian Abbey of Hailes and inherited with this a small private museum, primarily built to accommodate lapidary and other items found in the course of excavation and clearing on the site. At the Dissolution, part of the monastic buildings had been converted into a mansion which had, by the early eighteenth century,

Fig. 13.11   The Hailes Abbey museum displays medieval bosses recovered from the site as well as glass and other decorative objects. A future museum will be devoted to architectural displays.

242

deteriorated into two farmhouses. By 1794, these too had been demolished.

In the small museum many architectural fragments from the site were stored rather than displayed. This has recently been reorganized to achieve the double object of interpreting the site to the visitor and displaying some of the mass of high quality material associated with the medieval building. Included are some particularly fine carved bosses from the Chapter House vault and a good series of encaustic tiles (Fig. 13.11).

The display of lapidary fragments presents special problems. Their original architectural context must be made clear and at the same time they must be presented as the works of art they are in their own right. The technique used here is to present them against a contrasting coloured background at such an angle that they are seen as their designer intended. As always, a site museum relies for its original material on chance survival and this is rarely balanced, either typologically or in quality. The choice here has been to illustrate quality carving, mostly of the thirteenth century, and encaustic tiles. Panel displays, using reproductions of original documents, illustrate the history of the site. There remains a large number of purely architectural fragments destined for a future extension to the museum which will illustrate the Abbey's architectonic rather than decorative qualities.

AN HISTORIC BUILDING IN STATE CARE:
The Banqueting House, Whitehall, London

The Banqueting House, standing on the east side of modern Whitehall, is the last complete surviving building of the vast and sprawling Palace of Whitehall that succeeded the Palace of Westminster as the metropolitan residence of the Sovereign and his court (Fig. 13.12). It is also the sole reminder of Stuart ambitions to grace the capital with a great classic palace worthy of the dignity of both city and crown. The building was begun in 1619 to the designs of Inigo Jones and introduced into England at one stroke the full purity of Palladian classicism. Externally of two storys with a basement, the interior is a single double-cube chamber, 110 feet in length and 55 feet in height and width, with a low vaulted undercroft below. The interior is severe with galleries carried on console brackets along the sides. This severity is relieved by the ceiling, richly coffered by nine great panels (the largest 40 feet by 25 feet) filled by Rubens paintings depicting the benefits of James' I rule and commissioned by Charles I in 1634 (Fig. 13.13). From the restoration of Charles II until 1698, the building served a ceremonial purpose; after the destruction in that year of the greater part of the Palace by fire, it served as a Chapel Royal until, in 1890, Queen Victoria permitted its use as a museum for the Royal United Services Institute.

In 1962 the decision was taken to restore its original ceremonial purpose. Battle trophies, showcases and divisions were removed from the hall itself, the basement cleared out, toilets and cloakrooms installed at that level and the interior cleaned and redecorated. This left the building suitable for very little and still bearing many blemishes disturbing Inigo Jones' original design, leaving the way clear for the more definitive work of 1971.

Objectives of the general restoration work undertaken in 1971–73 were to restore the interior to its seventeenth century appearance, improve its acoustic qualities and reposition the incorrectly assembled Rubens paintings. The work was carried out under the direction of the Historic Buildings' Architects and Conservation Studio, both part of the Directorate, aided on the research side by the Inspectorate of Ancient Monuments. The building work was placed out to contract, but work on the paintings was by Conservation Studio staff.

Because its large windows overlooked Whitehall, traffic noise made the building acoustically unusable, and the large areas of hard plaster aggravated the condition. It was necessary to provide sound insulation by double glazing and by reducing internal reverberation. In order to be aesthetically acceptable, the double glazing was designed to be invisible. Single sheets of glass were fixed to frames

Fig. 13.12   The Banqueting House, Whitehall, London, after its cleaning in 1965. Final restoration was not completed until 1974.

sunk into the deep reveals of the windows, which were cut back for the purpose and then relined with acoustic board. Acoustic board was also introduced into the soffits of the galleries, which had to be extensively reconstructed because of decay in the timber bearers.

The causes of both acoustic improvement and architectural restoration were served by the treatment of the south wall of the hall. On conversion to a museum and in order to link it to an adjoining building, an additional gallery had been inserted at this end, new doors made and the original window altered. By removing the additions and replacing the original concept of a draped Royal Dais as the focal point of the interior, based on a design by Sir Christopher Wren, the dual objective was achieved.

The interior colour scheme had long been considered unsatisfactory. Portions of the interior had been given additional coats of plaster: upon the removal of these, some of the original paint surfaces were revealed. This evidence was confirmed by a report made by Sir Christopher Wren as Surveyor of the Royal Work in about 1700, who concluded that Inigo Jones intended that the main wall surfaces should be painted to represent creamy Bath stone and that the architectural elements of pilasters, cornice, etc., should represent white Portland stone. Lighting also received attention. The existing chandeliers were lowered to improve the scale of the interior and give better sight lines to the ceiling, which was brightened by concealed floods from the galleries.

244

Fig. 13.13   Interior of the Banqueting House after restoration. The Rubens paintings have been repositioned on the ceiling according to the artist's original plan and the Throne Canopy has been recreated to serve as a focal point to the great room. Dual pane windows and acoustical board were installed to cut down traffic noise from the street.

Exhaust ducts were also installed to compensate for lack of ventilation resulting from sealing the windows.

The work was completed in time for the building to be reopened with an exhibition marking the tricentenary of Inigo Jones. Many of his original drawings for the Banqueting Hall and his designs for the court masques held here were displayed.

Major repair and restoration to the Rubens paintings had been undertaken about every 50 years, starting in 1686 when Wren was redecorating the interior. They were taken down in 1729, cleaned and relined and considerably restored by Cipriani in 1776–77. In 1830–31 the roof was repaired and further work done on the paintings by Sir John Soane. By 1906 it was found that their great weight had caused the canvas to sag, so they were again taken down, restored and mounted on laminated board. The paintings were again removed, this time for their safety, during the 1939–45 war. Before replacement they received their first careful and scientific treatment. Old, darkened varnish was removed, as well as later overpainting including Cipriani's work, revealing much of Rubens' work intact beneath; this was secured and retouched with more care and respect than the original work had before received. After restoration carried out by the Directorate's Conservation Studio, the paintings were replaced in the ceiling in 1950.

In May 1970, in the light of recently discovered original drawings by Rubens for the ceiling, the theory was put forward[9] that at some stage in their removal, the position of some of the panels had been altered. Rubens' drawings show that his plan was to treat the ceiling as if viewed from different viewpoints as a domed effect, with the bottom edge running around the ceiling rather than at one end. In April 1972, while the work of restoration and redecoration were in progress, the paintings were repositioned by the Conservation Studio to complete the operation of bringing the interior of the Banqueting Hall back to its condition of Stuart times.

GRANT-AID TO A SCHEDULED MONUMENT:
Great Yarmouth town walls, Norfolk

Yarmouth, a town of consequence in the Middle Ages, is still a thriving port. The medieval town which stretched along one bank of the River Bure was defended on its landward side by walls punctuated with thirteen towers and ten gates, about 1⅓ miles long and originally 25 feet high. None of the gates survive but there are remains of eleven towers. The wall was built between 1320 and 1390 with no provision then for cannon. It is built of brick and flint rubble with a facing, typical of East Anglia, of squared cut flints interspersed on some towers with a pattern of chequered brickwork. In later years it was engulfed by a miscellany of buildings, gaps were torn through and stretches collapsed through sheer neglect. By the present century only fragments were visible and much lay buried within and beneath later structures.

By 1945 more had become visible as a result of war damage and further lengths were revealed by subsequent development. Piecemeal, the surviving stretches were acquired by the Local Authority and after 1953 the visible portions and new lengths, as they were exposed, were consolidated (Fig. 13.14). This is being funded annually by the Local Authority with a matching grant by the Department of the Environment. Work on the wall is carried out by the Department's direct labour force while, in later stages, the associated landscaping has been the responsibility of the Local Authority.

The nature of the work precluded a progressive programme from one end of the wall to the other. The order of work was set by the urgent need to stablize the more insecure areas and the random manner in which fresh portions became accessible. Primary repairs consolidated loose facework that had become detached from the inner core and secured friable wall tops. Where the face was entirely free, it was photographed and carefully removed, the rubble interior grouted, and the face flints replaced in their original position.

Fig. 13.14   Great Yarmouth, Norfolk, was once ringed by a medieval wall, but later construction enveloped it and passages were forced through some of the towers. A long-term restoration program has saved the wall piecemeal and reinforced the pierced towers.

Blemishes where, for instance, later buildings had cut into the wall were not restored, but the medieval work was secured in position as found. A trial excavation in 1961 revealed that the outer face of the wall extended some 16 feet below present ground level to the floor of a moat. It was not wholly practicable to clear this moat.

A major task has been treating a section of the wall, approximately 350 yards long and some 35 feet high, which had been heightened in the sixteenth century. It has the original arrow slits in the lower part, many much mutilated, and later Elizabethan gun ports above. It was treated in sections by two gangs of three men leap-frogging one another as each section was completed. Again the work required was mainly straightforward consolidation by grout and tamping, refixing loose stonework and picking up overhanging areas by corbelling out beneath them in corework.

During 1972 the condition of one of the towers, Blackfriars Tower, caused concern. In the past an opening had been forced from front to back some 7 feet wide and equally high. This was supported by an oak lintol carried on heavy oak posts, the bottoms of which had rotted allowing the lintol to drop, causing severe fractures in the already loose and friable wall above. Internal and external shoring was used to secure the walling in position whilst a reinforced concrete lining was cast. With the loose work still held in position, the interior face was carefully taken down and three reinforced concrete-ring beams were inserted in the core of the wall spreading 4 feet on either side of the fractures. The whole area was then grouted, the facework replaced and the exterior and interior faces pointed up.

Gradually these town walls, which prior to 1953 were little appreciated, are emerging as a considerable asset to the townscape; their true value as an important surviving example of medieval town fortifications is revealed. This example of the use of grant-aid under the Ancient Monuments Act illustrates how much can be achieved by the Local Authority and the Department working together over a period of time within a very limited budget.

GRANT-AID TO AN HISTORIC BUILDING:
Great Cressingham Priory, Norfolk

Great Cressingham Priory, listed Grade II, is a priory only in name: its history, as is often the case, only became apparent during repair operations. The present house is only a fragment of a house known, by a monogram in the rich decoration on the building, to have been built by one John Jenny ca. 1545. We now know that this house never reached completion although excavation revealed that footings had been prepared for a considerable courtyard house. The fragment that remains is but a half of one side of the quadrilateral

and the arch of the entrance gate. The outer face of the building is enriched with elaborately moulded terra cotta and is a notable survival of the early use of this material (Fig. 13.15). During the course of the work it was found that the inner or courtyard elevations were decorated in patterned coloured plaster, a decorative technique of which few other examples survive (Fig. 13.16). The value of the building lies in this decorative use of terra cotta and plaster, coupled with the intrinsic interest as an example, albeit incomplete, of a mid-sixteenth century house still unaffected by the new renaissance forms.

In 1955 the building was partly derelict, with only a later added farmhouse wing occupied. Its condition was brought to the attention of the Historic Buildings Council for England who—through the then Ministry of Housing and Local Government—made efforts, including an offer of a grant of £500, to persuade the owner to find a suitable tenant who would repair and occupy the entire building. Such was the concern for the possible loss of the terra-cotta work at this stage that proposals were prepared for its removal to Norwich Museum.

A change of family circumstances in 1966 finally led the owner to decide to repair and occupy the building herself. She approached the Ministry for a grant toward work for which she had obtained an estimate from a local builder. The Historic Buildings Council called for an examination and estimate from the Ministry Architect and, on the basis of his report, a grant of £4,300, 50 per cent of the cost of historic repairs, was offered to the owner with a recommendation that an architect be appointed.

In July 1966, architects Purcell, Miller & Tritton carried out their preliminary investigations, during which new significant historical facts emerged. Parallel with these investigations, the University of East Anglia conducted a research excavation on the site. Following these investigations the architects submitted new plans and a schedule of work for approval by the Ministry with estimates of £9,000 for the historic repairs and £10,000 for improve-

Fig. 13.15   Great Cressingham Priory, Norfolk, is a rare example of a prerenaissance sixteenth century building with terra cotta decoration. State aid enabled the owner to repair
and restore it. *(Hallam Ashley photo)*

ments not eligible for grant-aid. These were accepted on the advice of the Historic Buildings Council and an increased offer of £6,750, 75 per cent of the cost of historic repairs, made.

Work was started on repairs and improvements in March 1967 and continued until April 1968. As in the great majority of buildings grant-aided under the Historic Buildings Act, there were no major structural repairs to be done and no restoration or reconstruction—only a large number of minor repair works which called for careful workmanship and supervision and which made the difference between dereliction and preservation.

Although in the course of the work, a number of terra-cotta panels and some carved timber were identified as unused at the abandonment of the grand design in the sixteenth century, no attempt was made to incorporate these into a hypothetical reconstruction.

The repairs necessary were set out in a detailed schedule by the architect, leaving a minimum amount of work dependent on *ad hoc* decisions as the work progressed and assuring the closest possible control of expenditure. Grant-aided works included repair of roof coverings on the farmhouse wing by stripping and retiling; repair of roof and floor timbers by cutting out fungal rot and replacing de-

Fig. 13.16 A present interior at Great Cressingham Priory (formerly a courtyard wall) is decorated with colored plaster work which has now been preserved.
(*Hallam Ashley photo*)

cayed members, and general treatment of the remainder with fungicide; repairing external rendering and repointing the terra-cotta panels (no new panels were inserted nor damaged panels restored provided repointing alone resulted in a weathertight wall). Original blocked doorways, windows and a fireplace were opened up and exposed, but later inserted sash windows were repaired as found. Leaded lights were repaired and a damp-course inserted beneath the ground floor. Walls and floors damaged in the course of investigation were made good and some historic features redecorated.

Improvements carried out by the owner, which were not grant-aided, consisted of clearing away unwanted modern outbuildings, new partitions to form bathrooms and kitchens, the necessary services for these and their equipment, redecoration and landscaping.

GRANT-AID IN A CONSERVATION AREA[10]:
Lower Bridge Street, Gamul House and Gamul Place, Chester

The greater part of the walled city of Chester is a conservation area. In Lower Bridge Street, one of Chester's worst examples of area decline, lies Gamul House and to the rear

of that, Gamul Place. Gamul House, a grade II listed building, has a facade ca. 1700 but enshrines some good Jacobean work including the former Great Hall (Fig. 13.17). In 1967 this survived partly empty and derelict and partly as an upholsterer's store. Gamul Place is a courtyard of eleven nineteenth-century terrace houses of no particular architectural merit, not listed but nevertheless characteristic (Fig. 13.18). These houses too were derelict and intended for redevelopment.

The combined property has since been purchased by the City of Chester out of its Conservation Fund, a special fund set aside by the city for conservation purposes, the product of a penny rate. Repairs to the house were initiated under the Chester Town Scheme, funded by contributions of 25 per cent from Chester Conservation Fund, 25 per cent grant from Department and 50 per cent from the Local Authority as owners. Subsequently, to assist with a large number of buildings which the city had to purchase and support as "owner," Departmental aid was increased in some cases to 50 per cent.

Works to the house included stripping, re-slating and structural repair of the roof, underpinning rear extensions which had no foundations, and provision of services. In the hall, the handsome Jacobean fireplace was cleaned and an added gallery removed, restoring the hall to its proper proportions and appearance. The interesting historical discovery was made that the "hall" had in fact once been an open courtyard surrounded by earlier buildings.

During 1975 — European Architectural Heritage Year — the City intended to use the building as an Exhibition and Conference Centre, starting with the annual conference of the Society for the Protection of Ancient Buildings (which has contributed William Morris curtain material). After 1975 tenants will be sought to make the enterprise self-sufficient, perhaps as an arts workshop and sales centre.

Gamul Place has not been redeveloped as originally suggested but the houses have been

Fig. 13.17 Gamul House, Chester, is shown during repair of its Jacobean Great Hall, which had originally been an open courtyard. This Grade II Listed building is now used as an exhibition and conference center. *(Gamul House photos courtesy of Donald Insall Associates.)*

Fig. 13.18 Gamul Place, which adjoins Gamul House, Chester, has been preserved not for the intrinsic value of the houses, but because it is important to the ambiance of the Conservation Area. This photo was taken before improvements were made.

modernized to form a pleasant, if undistinguished, close; the same inhabitants were returned to them in the majority of instances.

PRESERVATION OF A PARISH CHURCH:
St. Lawrence, Blackmore, Essex

To omit an example of preservation work carried out on a Parish Church would ignore a very large percentage of the total architectural preservation work carried out in the country. The work is not the direct concern of the Department of the Environment but falls under the Inspection of Churches Measure 1955, an enactment by the General Synod of the Church of England. Every church must be inspected at least once every 5 years by an appropriately qualified architect, ensuring a

continuing conservation programme within foreseeable budgetary limits.

Blackmore Church,[11] one of the most interesting in Essex, contains work dating from the twelfth to the sixteenth centuries. Its most magnificent features are the late fifteenth-century timber bell-tower and spire, shown in Figures 13.19 and 13.20. These posed especially difficult problems, as the massive timbers had been seriously weakened by beetle attack and wet rot.

The defects were particularly severe at the base, where the main oak plate supporting the entire structure had rotted on its underside through rising damp; beetles had eaten away large internal cavities, especially at vital joints. Past attempts had been made to prevent decay by covering the outside of the base story in

Fig. 13.19 St. Lawrence Church, Blackmore, Essex, with its late fifteenth century timber tower. While churches are not within the scope of the Department's work, their owners may call on Departmental expertise for advice and assistance. *(Courtesy Donald Insall & Associates)*

Fig. 13.20 Interior timber work at St. Lawrence Church, Blackmore, after repair.
(Courtesy Donald Insall & Associates)

hard cement rendering, but these had been both unsuccessful and unsightly, tending to hold in dampness rather than exclude it. Repair by conventional means would have been difficult and expensive owing to the large size of the oaks, the necessity of shoring up parts of the tower whilst timbers were renewed, and the need to retain the appearance of the original timber engineering unimpaired.

An effective and economical answer was to inject compounds of synthetic resins and powdered slate into the decayed cavities through small drillings. This compound hardens into a material of great strength which bonds securely to the oak and is dampproof. By this means strength was inconspicuously restored to the weakened joints and timbers, at the same time protecting them against damp.

Related Work Carried Out by Directorate of Ancient Monuments and Historic Buildings

In addition to the work of preserving the historic structures in which many of the National Museums and Galleries are housed, the Directorate carries out associated works on these buildings. Two examples are noted here to indicate the scope of such work.

The Natural History Museum, South Kensington, built in 1880 to the design of Alfred Waterhouse, required cleaning and restoration of the Main Hall interior, cleaning of the exterior and construction of a new East Wing to accommodate the Museum's Paleontology Department. In designing the new work, no attempt has been made to copy the existing style of architecture, but a sympathetic relationship between the old and new has been achieved by the use of compatible materials. A new lecture theatre seats over two hundred and includes facilities for the disabled, a projection room equipped with cine and multiple slide projectors and closed circuit television to enable the audience to view small-scale demonstrations at the lecturer's bench.

The National Gallery fronting Trafalgar Square, first opened in 1838, is one of the landmarks of London's West End. There has been steady expansion to provide for the Gallery's growing collections. The latest extension contains over 23,000 square feet of new gallery space, scientific laboratories, offices, a conservation studio, lecture room, seminar room, extension to the library and, in the basement, the main air-conditioning plant room and boiler house to serve both new and existing buildings. Part of the new extension, which is contemporary in design, involved completion of the original facade. In the new galleries natural daylight is used as far as possible; this required special glass with ultraviolet filters and automatic sunblinds to control excess sunlight.

A conservation studio, situated in London, is a part of the Directorate's organization. Work is carried on in the repair and conservation of paintings on canvas or board that decorate buildings in State care or are part of the collections included in these buildings. There are facilities for full cleaning and restoration techniques, including photographic x-ray and vacuum table equipment. In addition, the staff carries out repair, cleaning and restoration *in situ* of wall and ceiling paintings and large canvases that cannot easily be transported.

The staff also repairs and regilds picture frames and decorative furniture, and there are separate workshops for repairing decorative wrought iron and stone carving.

The Inspectorate of Ancient Monuments is responsible generally for archaeological excavations sponsored by the Department. These excavations are usually part of a massive programme of "Rescue Excavations" designed to investigate and record sites threatened by new road works or other development. They are funded or grant-aided by the Department and supervised by the Inspectorate or by supervisors under contract to the Department. Publication of reports arising from these excavations is generally through the transactions of Local or National Archaeological Societies assisted by grants from the Department or, in the case of major excavation, reports produced by the Department are published by H. M. Stationery Office.

Excavations on sites in the Department's

care are usually carried out and supervised by the staff of the Directorate and may arise either in the course of monument consolidation or become necessary as a means of elucidating the historical development of the site.

A laboratory provides a scientific service to the Directorate, mainly in connection with excavations undertaken on behalf of the Department on sites of ancient monuments in advance of destruction or in the course of conservation and consolidation. Geophysical surveys may reveal features previously unknown and examination of buried deposits and soil analysis provide environmental evidence and may yield comparative information on early technology. Conservation of buried objects found in the course of excavation includes their scientific examination, cleaning, stabilization recording and, if required, reconstruction.

The Directorate acts as a focal point for, and initiates, research into materials and techniques used in the conservation of structures. Pure research is carried out in collaboration with the Building Research Establishment (not in itself a part of the Directorate of Ancient Monuments and Historic Buildings), and an advisory service is provided.

Stone preservation is a primary objective with research into the effectiveness of preservatives available on the market, as well as further investigation into application techniques and their effect on stone durability. To arrest surface decay in historic buildings, particular attention is currently being paid to the removal or neutralization of deleterious salts absorbed into masonry. Laboratory research is matched by field work on the many sites in the Department's care which provide an unusual opportunity to study stone behaviour under a wide variety of known environmental conditions.

Stabilization of fractured or shattered stone by epoxy resin injections and reinforcement is being developed as a technique permitting the retention of vital orginal detail that would otherwise be lost. Research is also being done on techniques of cleaning buildings, particularly stone, exposed to polluted atmospheres. In this field the Directorate acts in an advisory role to the Department of the Environment generally. Among other topics being considered at present are timber treatments and reinforcement, the matching of mortars and renderings and the use of fungicides.

The Presentation Section of the Directorate with the Central Office of Information is responsible for the publication (generally through Her Majesty's Stationery Office) and distribution of material for sale on monuments in the Department's care. For most monuments an Official Handbook is published in standard format containing a full historical account of the monument, illustrated by photographs and plans. These form the basic series of guides to the monuments and are supplemented on the larger or more popular sites by short leaflet guides and by souvenir guides intended for the visitor who prefers a pictorial reminder of his visit.

A series of regional guides is also published to form a gazeteer of those monuments in the country open to the public. Each contains an introduction giving the architectural and historical background of the region. Special monographs are published from time to time on special subjects, such as Castles and individual buildings, while colour slides and postcards are produced for sale at most monuments.

NOTES

1. The legislation is contained in the Ancient Monuments Protection Bill 1882; Ancient Monuments Consolidation and Amendment Act 1913; Ancient Monuments Act 1931; Historic Buildings and Ancient Monuments Act 1953; Field Monuments Act 1972.

2. The Ancient Monuments Act 1913 requires the Secretary of State to "prepare and publish a list containing such monuments as are reported by the Ancient Monuments Board as being monuments, the preservation of which is a matter of public interest . . ." There are separate Ancient Monuments Boards for England, Scotland and Wales, and their members represent the Royal Commissions on Historic Monuments in England and Wales; The Society of Antiquaries of London; The Royal Academy of Art; The Royal Institute of British Architects; The Trustees of

the British Museum; The Secretary of State for Education and Science.

3. The Department of the Environment assumed the responsibilities of the former Ministry of Works and Public Building, Nov. 12, 1970, under Statutory Instrument No. 1681 1970.

4. Local Authorities (Historic Buildings) Act 1947. Town and Country Planning Act 1944 as amended 1947, Sections 54–58.

5. Insofar as churches are exempt from listed building control, a distinction is made by grading them a, b, or c, with criteria based on quality rather than date of building: a) First class churches that are still more or less complete; b) churches of special interest (the majority of listed churches are in this grade); c) churches of statutory quality.

6. Town and Country Planning Act 1971, Section 277 reenacted and amended by Town and Country Planning Act 1974, Section 1.

7. 1974–75 expenditures were approximately £2,350,000 on monuments, preservation and management; £950,000 on excavations; £500,000 on grants, compensation and miscellaneous.

8. Each Town Scheme is arranged between the Local Authority and the Department, each working within existing powers. The first Town Scheme was introduced in Bath in 1954.

9. Julius S. Held, article in *Burlington Magazine*, May 1970.

10. Information supplied by Donald Insall and Associates, Conservation Consultants to the City of Chester.

11. Information supplied by Donald Insall and Associates, Architects and the Council for Places of Worship.

CHAPTER 14

Restoring Three Cathedrals: York Minster, Norwich, and St. Paul's

BERNARD M. FEILDEN

DR. FEILDEN is a Fellow of the Royal Institute of British Architects and of the Society of Antiquaries. He serves as restorationist to St. Paul's and Norwich cathedrals and to York Minster.

Conserving and restoring historical monuments in Great Britain, due to their greater age and the British habit of building literally on top of earlier works, may seem to be quite different from restoring historical buildings in North America. But on both sides of the Atlantic we have learned that the principles are the same, no matter what the size, age or condition of the fabric we must conserve. The principles listed here, which emerge from my own practice, are probably universal. They were learned during work on, among other restorations, the three cathedrals discussed in this presentation.

Every historic building is a unique problem which must be assessed in the cultural, historical, archaeological, social, economic and even political situation of the country in which it is situated. Therefore, each case must be treated on its own merits.

The historic building is a cultural document whose integrity must be respected; one should retain the original design and as much original material as possible. Forgeries are not permissible, but copyism is allowed in repetitive work. In repair work new materials should be identifiable on close inspection, but should not disturb the aesthetic intentions of the original designs: e.g., new "painting" with vertical brush strokes on frescoes; slight change of texture in new stones set in old walls; dating new pieces of glass inserted into old windows.

In some cases a new use, together with practical and economic considerations, must modify the scope of *conservation*. The amount of restoration permissible depends on the use and purpose of the building. In certain cases, purely archaeological considerations will predominate; in others, artistic or historical considerations.

Repairs should not disrupt the visual unity of the building; wherever possible they should be reversible, leaving the original material intact. They should respect the intentions of the original designer and his aesthetic concept of the building.

The principle of conservative repair should be followed. Nothing should be replaced unnecessarily, and existing structure and materials should be used as far as possible. New repairs should be "kind" to the old building and not too strong. Too much repair work should not be done at any one time.

In actual practice, restoration work must be based on a thorough visual inspection, followed up by structural investigation in depth. Restoration is a multidisciplinary skill, but it is the architect who must coordinate the team; he has the most experience in making value judgments and his role is to understand the building as a whole. He must maintain the architectural unity of the building, calling on persons who are experts in various aspects of the task but who may not understand the project as a whole.

The work must be organised on a scale comparable with the task and, finally, there must be good communication. Decisions should be fully discussed and no design adopted unless the contractor agrees he can carry it out. Perhaps informal communication over a glass of beer with sandwiches for lunch is one of the best insurance policies. Furthermore, workmen should be told about design problems and their opinions, especially about difficulties encountered, be reported through channels and carefully listened to. On cathedrals, it is imperative to have a genuinely interested Dean, Clergy and committee of laymen.

Lastly, it is impossible to restore a building unless one understands its history fully. Previous restorations present structural hazards, as do earlier buildings partially existing below grade, such as the Roman structures we encountered below York Minster. There is always a fourth dimension in restoration work — that of time.

YORK MINSTER

The work at York Minster was very different from that of Norwich, just as its history and structure is entirely different (Fig. 14.1). In Roman times York was the capital of northern Britain and the military headquarters of the island. A legionary camp was established at Eboracum (York) and in its time was visited by the Emperors Agricola, Septimus Severus, Constantious Chlorus and Constantine the Great, who was proclaimed Roman Emperor from a spot close to the south entrance of York Minster. The Roman buildings were occupied in the Dark Ages and stood for centuries after Rome had left Britain; it is probable that some of the Roman buildings were even reerected by their Saxon successors. The Roman buildings must have seemed vast to these barbarians. The Saxon Chronicle records that the legendary King Arthur celebrated the first Christmas in York. If so, he must have used the Roman buildings, traces of which have been found under York Minster.

I frequently visited and loved the Minster as a boy and was greatly honoured to be appointed Surveyor in 1965. The first task was to get to know the fabric well and this involved a stone by stone inspection which took two years to complete. Halfway through, however, suspicions that all was not well were aroused by the discovery of fresh cracks in the heavily loaded central piers. A building as old as York Minster ought to be in a state of stability, but fresh vertical cracks low down in the main piers indicated that there was some new movement. It should be said of course that there were thousands of cracks in the building which had no significance, but several groups pointed to the possibility of foundation difficulties. There were also cracks in the Choir piers, but these were old, full of dust and dirt, and it was later discovered they were due to the shrinkage of freshly quarried stone which was inserted after the fire of 1829.

As a result of the inspection, consulting engineers were called in and exploratory holes were dug in the crossing area, with the digging camouflaged as an archaeological operation. The holes exposed to view quite large cracks and severe settlements in the foundations. Glass "telltales" were fixed in the first instance and these were followed by setting up strain gauges whose measurements showed that active movements were taking place in the foundations.

There was general despondency over the size of the task and a quandary as to what should be done. However, estimates ranging from £1.67 to £2.67 million were prepared

Fig. 14.1   York Minster during restoration (1968) being carried out on the Central Tower and on the Northwest Tower, while scaffolding has also been erected *(extreme right)* for work on a defective buttress. Also in progress was work on foundations and restoration of glazing. One of the major problems at York was coordinating so many simultaneous activities and organizing reuse of scaffolding as efficiently as possible. *(All York Minster photos couresty of Shepherd Building Group Ltd. except where noted)*

and submitted to the Dean and Chapter responsible for the fabric. It was pointed out that urgent work was necessary: it was not so much the immediate danger of collapse of the central tower or other parts of the Minster, but whether the fabric was strong enough to stand the strain of engineering surgery necessary to save it for posterity. Lord Scarbrough was appointed High Steward and under his leadership a fund raising campaign was started.

No time was lost in beginning a 7-year programme of work. A team of archaeologists, art historians, structural engineers, electrical and mechanical engineers, quantity surveyors and other specialists was brought together under the leadership of the Surveyor of the Fabric. It was essentially a collaborative effort in which time was one of the important elements. A system of glass telltales and measuring was set up and soon some five pieces of glass had broken in the central tower piers and foundations, and movements were being recorded. These movements were carefully monitored throughout the job. In order to design and plan the work, it was found necessary to have fuller information on how the Minster had

Fig. 14.2   Due to the unstable foundations, great cracks had opened up in the stonework during the centuries.

been built. In this sense the archaeologists contributed a great deal of information and enabled intelligent design decisions to be made. Nevertheless it must be recorded that the diagonal design of the foundations that finally emerged, in its harmony with the archaeology, was more due to luck than knowledge of what remains were yet to be found under York Minster.

The historic evolution of the building must be understood. After the Roman legionary fortress there were three Saxon buildings, including an Oratory where King Edwin was baptised in 627A.D. soon followed by the first cathedral built by St. Paulinus, and the second cathedral by Albert in about 790 A.D.. These buildings were used when York was a renowned centre of learning under Alquin, yet no traces have been found; in fact, the knowledge we have acquired is that we know where they are not.

On the other hand, after the Norman Conquest when Albert's building was destroyed by fire, a totally new Norman Minster was built.

Of this we have discovered a great deal, identifying it as a completely new type of English Norman Cathedral with art historical links with Rhineland Churches. It was a great barn 45 feet in span—well over 70 feet in height—with a massive central tower. The Choir was probably narrower than the Nave and surrounded by an ambulatory. Towers were later added at the West End. After a fire, the Choir had to be rebuilt in 1190, and only a few pieces of these early buildings remain visible aboveground. It was this unique Romanesque church which set the pattern for all future buildings in York Minster and determined that it should be such a vast building, with dimensions only a little less than St. Paul's Cathedral.

Work started on rebuilding and enlarging York Minster in about 1220 when Archbishop de Gray and Treasurer John Romanus built the south and the north transepts, respectively, with a campanile or bell tower in the centre which was built on the Norman tower itself. The Chapter House was built in about 1260 but its vault fell and had to be rebuilt about 30 years later. We can still trace the same masons' marks from the Chapter House through the vestibule into the Nave, which was begun about 1300 and finished in 1330. There was some delay in vaulting the Nave but it was probably roofed over and already in use when the vault was finally added. According to Dr. Gee of the Royal Commission on Historic Monuments, this was not an unusual procedure.

At this time York was the military headquarters of operations against the Scots and much visited by the Plantagenet Kings, Edward I, II and III. Edward III's portrait appears above his Royal Coat of Arms at the east end of the Lady Chapel which was begun about 1360. Work proceeded from east to west. About four bays were built in the third Edward's reign, and the work was continued under Richard II until the Choir was rebuilt and the Central Tower was remodelled. But in 1406 there was a disaster—the Central Tower fell. The fall was probably in a southwesterly direction, as there are remains of Romanesque work on the three other quarters. The rebuilt Central Tower, which was substantially finished about 1430, is a magnificent structure some 60 metres high and weighing over 16,000 tons. This weight rested on the original Norman walls and Norman foundations which had never been designed to carry such a mass. At first there were rapid settlements which gradually subsided, then for nearly 500 years the structure remained static, until fresh movements were discovered in 1966. It is difficult to identify a positive reason for the movements, but it is more than likely that the lowering of the water table was a major cause. York Minster sits in a "perched" water table which is extremely subject to local variations and is supported upon clay which is sensitive to any change. The water level had fallen about 4 feet since 1830. However, the most critical point would have been when the level fell below the Norman foundations, probably quite recently.

The first step in the restoration was to organize a full soil mechanic's investigation of the site; the second was to strap up the main piers, start excavation and explore the problem fully. Nevertheless, the danger of possible "uplift" due to removing the surcharge of soil was fully recognized. Due to burials to a great depth, the soil was honeycombed and did not give the old masonry much lateral restraint. Excavation procedure at the lower levels consisted of work first by the archaeologist, followed by the builder. When the old foundations had been measured and inspected, an equivalent weight of the soil was replaced by sandbags, (Fig. 14.3), then heavy planking and strutting were inserted so that the new foundations could be built.

The design of the new foundations presented several unique problems, but the principles of conservative repair were applied in that the existing material was used to maximum benefit and to avoid disturbing the fabric insofar as possible. Various schemes were examined, such as freezing the ground and inserting caissons; underpinning and inserting new concrete foundations; inserting sloping small-diameter piles by diamond core

Fig. 14.3 During excavation, sandbags were used as ballast to stop the concentrated weight of the building from pushing aside the earth around and below the original foundations.

drilling using the Fondedile Pali Radici technique; or simply clamping on large concrete blocks to the existing foundations and improving their weight-bearing capacity with piles. However, the soil mechanics survey showed the possibility of relatively shallow foundations if a loadbearing capacity of 3 tons per square foot was acceptable. In fact, the ultimate loadbearing capacity of the ground was rated at 8 tons per square foot: this load was reached in various parts of the existing foundations around the Central Tower, thus confirming the Surveyor's suspicions that the fabric was in an unstable equilibrium. Messrs. Arups, the consulting engineers, devised the concept of a concrete compression pad, upon which were placed flatjacks and over which conventional point foundations were formed about 17 metres square in effective area. These foundations incorporated all the earlier Norman masonry. They also included a collar at the higher level to prevent the old masonry bursting, as there had been a certain amount of evidence that this was a distinct possibility. As the loads coming down from the Central Tower were eccentric, this was compensated for in the foundation design. The scheme consisted of four large independent footings of new reinforced concrete tied into the old masonry by means of 100 stainless-steel tension rods in each foundation. These rods had to be drilled through the concrete and through the old masonry in order to combine the ancient with the modern (Fig. 14.4). When the concrete had been placed and the stainless steel rods had been tensioned, the hydraulic flat-jacks were to be inflated to pressurise the new foundations to 3 tons per square foot, thus sharing their loads with the old foundations.

Excavation and concreting went well but drilling, despite favourable initial tests on unstressed masonry, presented great difficulties. Four hundred holes, some 50 feet long, had to be drilled; in the first 6 months only eight had been achieved. Even more powerful drilling machinery was tried. The initial difficulties were disheartening; nevertheless, with time matters improved and finally it was found — when the men had acquired all the secret skills that go into such a job — it was possible to drill

a good hole within 2 hours. A "difficult" hole which met obstacles such as the embedded timbers of a Norman raft or particularly loose and friable areas of masonry might take up to 8 hours. However, any hole might have to be redrilled several times — up to eight in the worst cases — so the extreme range of drilling times was from 2 hours to a week.

The men learned to change the drilling bits as necessary, achieve acceptable accuracy and, by organizing a night shift, keep the drilling up to the overall programme. Once the Cathedral Organist was asked what note the drills were playing; without a moment's hesitation, he said "E Flat." At the end of the restoration he played a sonata which he had composed entitled the *Rebirth of a Cathedral in E Flat.*

While work was going on at foundation level the public was allowed access at ground floor level and services continued without interruption. The tea break was organized at 4 o'clock: at 3:55 some 100 workmen trouped out and the choir, followed by the Dean and Chapter, moved in to celebrate Evensong. A long lesson or psalm might overrun the allot-

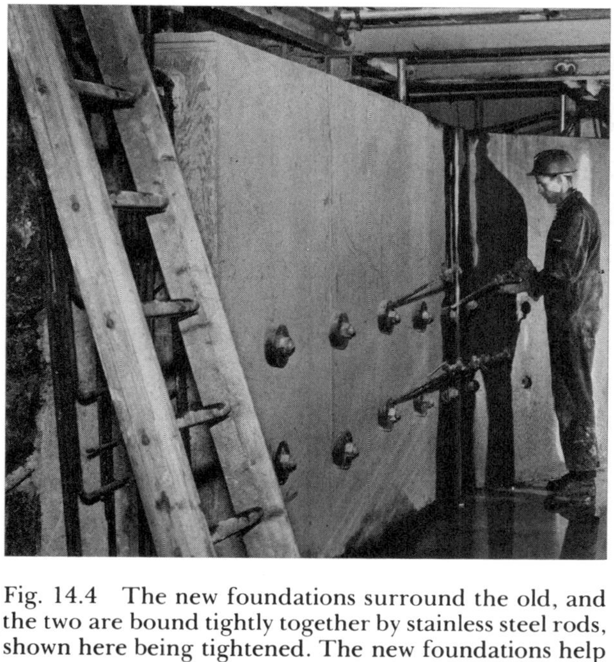

Fig. 14.4 The new foundations surround the old, and the two are bound tightly together by stainless steel rods, shown here being tightened. The new foundations help relieve the old ones of the excessive loads to which they have long been subjected.

Fig. 14.5 The Central Tower was strengthened by inserting stainless steel ties in each side to prevent bulging and further movement of active cracks. First, however, extensive borings were required to determine the true condition of the walls, as shown here.

ted half hour but the Dean was well aware that each minute of lost time would cost the Appeal Fund about £2.50. Work was proceeding at the rate of about £30,000 per month.

At the same time that excavations of the foundation were started, the middle level of the Central Tower was strengthened by insertion of twelve stainless-steel ties in each of its four sides to prevent bulging and further movement of active cracks (Fig. 14.5). At the top of the Tower a *ring beam* was inserted, to-

gether with a new roof covering of lead. The Central Tower had always been given first priority; it was work of a special dimension because of the risk that, should anything go wrong, there was no means of obtaining sufficient bearing area of ground to erect temporary supports, the ground bearing generally being so poor and the weights so large.

Visual inspection showed that in the course of 6 centuries the East End of York Minster had leaned out: when measured, the distance

Fig. 14.6   The East End, which had leaned outward required emergency shoring which rested on hydraulic jacks so that constant pressure could be maintained on the fabric even if the building expanded or contracted.

was found to be some 625 millimetres, clearly rather critical; nevertheless it has always been assumed that it could be satisfactorily shored up should trouble arise. In one week six tell-tales broke and it was deemed expedient to explore the foundations. A 12-foot hole revealed nothing but stones with no mortar. Shores were quickly ordered and installed. These shores had a special feature in that they rested on four hydraulic jacks with a constant pressure mechanism, which meant that if the building moved or if the shores expanded or contracted with heat the same horizontal pressure would always be applied to the masonry, avoiding damage to the fabric (Fig. 14.6). Special precautions had to be taken to protect the Great East Window, which is full of priceless stained glass (Figs. 14.7 and 14.8). Then a plan was devised to insert a new foundation nearly 40 metres long by 6 metres wide and 3 metres deep. This was inserted using conventional underpinning methods in lengths of 1.3 metres. The consulting engineers phased the work very carefully to avoid differential settlements which might have damaged the Great Window, above which were already strutted horizontal flying shores of great strength.

The work went well but there was some

Fig. 14.8 Steel ropes supported the window against pressures which might have caused it to bulge outward.

Fig. 14.7 Inside, special precautions were taken to protect the Great East Window and its priceless stained glass.

rather rapid settlement when the new foundation took up its loading; this prompted a further study of the soil at this level. It was found that the clay had very thin layers of sand in it, which meant that it would compress under load extremely quickly. We rather wished we had included the flatjack idea in this foundation to counteract this effect. Nevertheless, this characteristic of the subsoil probably accounts for several features in York Minster, firstly the collapse of the first Chapter House vault, and secondly, the resolve of the builders of York Minster not to use masonry vaults but to rely on timber construction in spite of the fire risk. For this a heavy price was exacted in 1829 when the Choir was burnt down by a madman called Jonathan Martin, who disapproved of the music being sung in the Minster, and again in 1840 when the Nave was burnt down by a careless workman.

Bearing these lessons in mind, we installed fire detector systems and new ring fire mains so that the additional risks arising from workmen during the restoration would be covered. It was fortunate that this was done, because when all the work was completed in 1972 a careless workman left a powerful electric light close to a curtain, starting a fire which was quickly put out. The lightning conductors also received early attention in the programme of work (it is worth remembering the pioneer work of Benjamin Franklin in this field, which must have saved many a church).

Having got the Central Tower and East End well under way, the West End was considered and it was decided that this needed similar treatment to the Central Tower and that it would be wise to continue the building operation now that a certain momentum had been achieved and specialist skills acquired. An early start was planned so as to give the archaeologists more time for a proper excavation; some attempts were made to estimate the cost of the excavation and to assist the archaeologists by mechanising their work with conveyor belts to get rid of the spoil.

The procedures were essentially the same as for the Central Tower, but it was found that one of the southern pier foundations was badly cracked below the floor level and this meant placing the concrete collar around the crack in advance of building the concrete foundation. However, the experience that the restoration team had acquired on the Central Tower proved of great benefit in the Western Tower scheme, which was conceived to avoid unnecessary work and designed so as to make the drilling task easier. Some large quantities of concrete were inserted in the foundations, one record day being when three men managed to lay some 270 cubic metres of concrete pumped into the Minster from a queue of mixing lorries. Indeed the contractor, by his skill and use of the mechanical plant, made a major contribution to the efficiency of the whole operation. The Western Towers had also been strengthened with girdles at three levels and had been grouted to consolidate the

masonry, some 3,500 gallons being inserted in each tower representing about one fortieth of their volume. All structural movements were carefully monitored and there was some anxiety when the towers showed signs of tilting. The shores which had been taken down from the East End and kept ready for this contingency were reerected: the horizontal pressure which was applied corrected the situation within one month.

When work was finished at the East End, overhaul of the interior was initiated by cleaning and repairing masonry (Fig. 14.9). Unfortunately the Minster had been covered with a century of dirt which concealed many defects, such as cracked bosses, ribs with open joints and loose voussoirs, badly cracked stonework and loose plaster on the vaults. After careful washing, these defects were made good and the Minster cleaned from end to end.

Many interesting discoveries were made during this cleaning and all the bosses in the Minster were painted and gilded in the traditional manner (Fig. 14.10). The stained glass was carefully washed and temporary leaded glass inserted instead of black panels. The nave clerestory was reglazed and the general tone somewhat lightened to admit more illumination for the vaulting and thus reveal its

Fig. 14.9 Craftsman carves a replacement piece for use where original material had completely deteriorated. (*British Tourist Authority photo*)

266

Fig. 14.10 A complete cleaning and refurbishing of the interior included repainting and regilding of bosses, shown here after a century's worth of dirt and grime had been removed. The cleaning also revealed many defects, previously unnoticed, which were repaired.

Fig. 14.11 The old foundations, here consolidated with the new below the Central Tower, are part of the Roman military headquarters which originally occupied the site. This lower area has been set aside as a museum.

beautiful form. The main works were finished in 1972 and Her Majesty the Queen visited the Minster at the time she presented the Maundy money at Easter. The work was finally completed, apart from the exterior at the West End, by St. Peter's Day of 1972, exactly 500 years after the original Gothic Minster had been finished with the building of the two Western Towers.

The reconstitution of the foundations had required working space for the insertion of the stainless-steel rods into the predrilled holes (Fig. 14.11). The question was studied as to whether it was cheaper to backfill with weak concrete or to build retaining walls, thus forming a useable void. The latter course was adopted; the void was used as space for a museum to outline the history and development of the site from Roman times to the present day, including display of the new founda-

tions with the nuts and bolts all exposed. This museum is combined with a Treasury and has proved to be an extremely popular tourist attraction which has helped to raise funds for the preservation of the Minster.

On the contracting side, high efficiency was maintained throughout the years and labour costs only exceeded that of the materials in the last three months when clearing up was the order of the day.

The principal operations in the work consisted of excavating 30,000 tons of soil, placing 7,000 cubic yards of concrete and over 120 tons of high tensile steel reinforcement, injecting 200,000 gallons of cementacious grout to fill voids and fissures, drilling over 12 miles of holes in masonry and inserting 6 ½ miles of stainless steel reinforcement. Seven hundred and fifty tons of steel were used in shoring and strutting, including the 80-foot long heavy duty raking shores. Sixteen miles of scaffold tubing, over 13,000 scaffolding fittings and 29,000 feet of scaffold boards were used: this had been erected and dismantled eight times during the progress of the works. In restoring the fabric, 350,000 square feet of masonry had been cleaned. The Minster masons fixed a record amount of 4,267 cubic feet of stone in 1971, which represents fifteen times the average through-put of previous years.

BERNARD M. FEILDEN

COSTS

Estimating and cost control are difficult but important aspects of restoration work. Estimates should always express doubt by giving a top and bottom figure. Cost control should be on a budgetary basis but must be sympathetic, as so many unknowable factors have to be met and solved without delay: otherwise these costs will escalate. On such large projects which may take years to complete, sufficient allowance must be made against inflation when preparing estimates; this too involves "crystal ball gazing." The original estimates of January 1967 are compared with actual costs of the finished work:

	Gross estimate January, 1967	Actual cost, January, 1974
Central Towers	£450,000–930,000	£713,000
East End and Choir Piers	120,000–140,000	103,000
North West Tower including Shores	300,000–430,000	237,000
South West Tower including Shores	300,000–430,000	237,000
Roof Repairs and Insulation (works not completed)	138,000–155,000	79,000
Engineering Services	50,000–59,000	102,000
Organ Reinstatement	10,500–13,000	24,000
Miscellaneous Building Work	69,000–77,000	83,000
Cleaning and Scaffolding	69,000–80,000	89,000
Undercroft and Treasury	(not included)	
Stoneyard and Glaziers	210,000–245,000	366,000
Archaeology	15,000–18,000	*7,500

*A further sum of £47,809 was provided by grants from the Department of the Environment, the British Academy, the Pilgrim Trust, the Russell Trust and the Leverhulme Trust. The cost of writing up is estimated to be between £30,000 and £40,000 and will take up to three years.

Note: To arrive at Gross Costs, Administrative Costs of £62,000 approximately, Professional Fees of £189,000 approximately and Direct Costs of £36,000 have been allocated on a percentage basis of 11.6% to the Actual Costs as at December, 1973.

Now that the urgent works are complete, there is a 30-year programme for masons to renew defective copings and weatherings, rebuild decayed pinnacles and buttresses and, above all, maintain the pointing of the masonry so that it does not suffer from the ingress of rain and damage by frost. This work of daily care is planned to stave off decay.

NORWICH CATHEDRAL

The building of Norwich Cathedral, which is essentially a Norman fabric, was started in 1090 and finished in 1130. The Cathedral has been less altered than most in Britain and much of its Romanesque character remains. The Central Tower is one of the supreme monuments of late Romanesque architecture (Fig. 14.12). The spire of 325 feet, second in height only to Salisbury, is the third such structure and replaced one that fell in 1460, destroying the Presbytery roof. The Transepts and Nave were heavily restored in 1845 by the Victorian architect Salvin and the level of the Presbytery floor was raised in Victorian times to meet the current ecclesiological fashions. But perhaps the most important modifications were carried out in the late fifteenth and early sixteenth centuries when the windows were enlarged and the vaulting inserted over the Nave and the Transepts.

Despite all these alterations, the character of the Cathedral is essentially Romanesque. The foundations are of flint concrete resting on good firm gravel capable of carrying 4 tons per square foot. The walls are faced with stone and filled with flint rubble. Nearly all the interior stonework is original Caen Stone from Normandy, but almost all the original Norman external facing stonework has been replaced in a continuous series of repairs and renewals. One square metre of masonry may well represent six different types of stone used on different occasions. About ten different types of stone have been used in the Cathedral altogether, all being limestones—as it were, cousins.

A major programme of repairs was initiated in 1952 when a Lay Committee of Friends of Norwich Cathedral under the High Steward, an ancient and honourable office, took over

268

Fig. 14.12   Norwich Cathedral, essentially Norman. Its spire (the third topping this structure) dates from the fifteenth century and presented the most difficult problems during the latest restoration work. It was strengthened, rather than rebuilt, by grouting the existing fabric and using reinforced concrete in stitches and rings. This view, after restoration, is from the southeast. (*All Norwich photos courtesy of Feilden and Mawson Architects*)

the responsibility for maintenance of the fabric. Since then nearly a half million pounds has been spent on various building operations, principally the roofs. First the Presbytery roof, then the Cloister roofs were renewed. Since 1954 I have been responsible for major works, the first of which was the north transept, followed by the ambulatory, followed by work on the spire and the tower, reroofing the south transept, the long nave and, currently, the north and south aisles of the nave. Of course many other incidental works have been carried out during the period, in particular cleaning and relighting the Presbytery vault. The Friends of the Cathedral raise approximately £12,000 a year but are assisted by grants from private trusts and benefactors.

The most difficult work was that carried out on the spire in 1962 (Fig. 14.13). An inspection showed that all eight sides were badly cracked, the outer facing of stonework was bulging and the spire was in extremely delicate condition. A detailed inspection was

Fig. 14.13 The spire, during restoration, required about one quarter of its stone replaced. This in itself was traditional, as nearly all the original Norman external stonework had been replaced in a continuous series of repairs with limestone very similar to the original Caen Stone from Normandy.

made using a telescope, as it was not practical to erect scaffolding. The defects in the spire arose from its unique method of construction. It is approximately 180 feet high: the first 50 feet consist of 350mm brickwork, the last 120 feet are an octagonal cone of 112mm medieval brick, all of which are faced with 50–75mm of stone. This composite construction had been erected from internal staging working "overhand." About five courses of bricks would be laid, then a facing stone, the gap between the internal brick and the external stone being filled with mortar. Until the top 40 feet were reached, no external scaffolding was necessary. In the course of time cracks occurred and dust wedged down, forcing the outer casing away from the stonework as much as 100mm in places. The cracks also allowed the ingress of rain and the decay of mortar and masonry.

My predecessor suggested that the spire would have to be rebuilt. Engineers were consulted and one suggested a steel frame inside the spire but this was rejected as it would have been too strong and immensely difficult to erect. The second engineer suggested a concrete structure, which might have been easier to erect but would have been too heavy and much too rigid in comparison with the tower. In the end it was argued that, as the spire had stood 500 years, it could be strengthened in an *ad hoc* piecemeal way using the existing material to the best advantage. I prepared schemes and these were used as a basis of consultation with Mr. Bertrand Monnet (now Inspector General of Historic Buildings in France) and Mr. Patrick Faulkner (now Chief Architect of the Department of Ancient Monuments and Historic Buildings).

A semipermanent ring of reinforcement tensioned with springs in a central ring was installed in the weakest part of the tower as a precaution against collapse (Fig. 14.14). The grouting holes were only 300mm apart to avoid danger of building up a bursting pressure which could have destroyed the delicate wall. The procedure that was followed was first grouting with cement, lime and sand to consolidate the masonry and fill the voids,

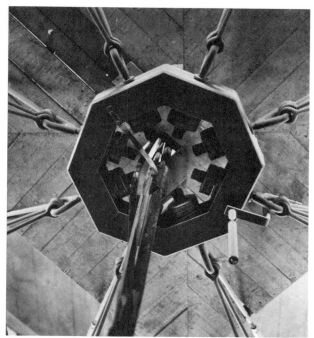

Fig. 14.14 This octagonal device restrained the Spire at its weakest point. The ties, each going to one of the eight external sides, and each equipped with a turnbuckle, pull on springs which in turn pull against the ring. In this way a flexible and strong support was devised. Note the lightning conductor attachment and the plumb bob *(right)*.

then inserting reinforced concrete stitches across the cracks and reinforced concrete rings at intervals, each stitch or ring being about a metre apart. Then it was possible to strengthen the internal brickwork by renewal where necessary and repair the internal staging which had become rotten in many places due to the ingress of rain. The brickwork was inserted in tall narrow strips about 300mm wide, work which had to be done extremely carefully in order not to dislodge the outer stone facing. The internal work proceeded in advance of the external work, which consisted of cutting out and replacing defective stone, and cutting out all joints of stonework for complete repointing.

The idea of Monnet was used to insert approximately 3mm stainless-steel wire of great strength into the horizontal joints which would act as reinforcement. This reinforcement was put in every horizontal course of stone at approximately 35cm centres. The object of this binding wire was to prevent reoccurrence of the original defect, which was the separation of the stone facing from the brick core. About a quarter of the external facing stone had to be renewed, and there were considerable difficulties with regard to scaffolding and hoisting which had to be overcome. However, all the work was carried out in two years at a cost of £30,000 (1962–64 prices) and when Monnet made his second visit he commented that the work had been done fast and at low cost, which was mainly due to the excellent cooperation of local builders and masons.

Following work on the spire, 4 years were spent on consolidating masonry in the central tower by the inserting of grout and reinforced concrete, reinforcing rings and stainless-steel wire rope (Fig. 14.15). Large voids were filled in the masonry of the tower—one grouting probe receiving as much as 120 gallons. This was followed by reroofing the south transept and the nave, using reinforced concrete trusses and frame work with fireproof timber, purlins and rafters covered with the traditional 7-pound cast lead. Additional insulation was laid over the main roof vault.

The last design, which is of interest, is for the nave aisle roofs which have had to be reconstructed because the oak timbers had be-

Fig. 14.15 Grouting the west face of the Tower required as much as 120 gallons for each probe. The original Caen Stone is badly decayed but still structurally sound, although new stone has been inserted where necessary.

come rotten beyond repair. It was possible to restore this roof to its original design, with one bay of original material (Fig. 14.16). Samples of timber joints have been preserved with full record drawings.

The new roof construction looks like the old, but all timbers sit upon copper damp-proof courses and the ends are ventilated to avoid dampness and timber decay followed by deathwatch beetle attack, which were the primary cause of decay. The secondary cause of decay was condensation on the underside of the lead roof, which was to a large extent induced by the central heating installed in the nineteenth century. This problem was overcome by inserting a vapour barrier, ventilated air space and wood wool insulation. By using contemporary building technology and traditional skills it is hoped that this restoration will be truly durable, save energy and increase the comfort of those who use the Cathedral.

Still to come are repairs to the masonry of the West Front, reglazing the West Window, replacing pinnacles which have been missing since World War I, repairs to the masonry in the Cloisters and a rolling programme of

Fig. 14.16 The North Aisle roof, as well as the South, had to be rebuilt to the original design as it had deteriorated beyond repair. All new timbers are damp-proofed, and the roof insulated and ventilated to prevent future decay of the oak. This contemporary building technology also saves energy and increases the comfort of those who use the Cathedral, a side effect which is appreciated.

masonry repairs working to a planned programme around the exterior of the fabric.

In the interior, improved sound reproduction is necessary and in the future a considerable amount of money will have to be spent on the heating installation. The electric lighting needs improvement—which has been started—and, lastly, it is desirable to reseat the nave. Confidence is high among the Friends of the Cathedral because they know what the maintenence plan is, what their responsibilities are, and that they have a continuous programme of work in front of them for the foreseeable future. The care of cathedrals is a never-ending task, with each generation doing its part.

ST. PAUL'S CATHEDRAL

Although St. Paul's Cathedral and York Minster have similar overall dimensions, the two buildings are entirely different. St. Paul's is dominated by a dome with a lantern some 365 feet high (Fig. 14.17). It is an extremely heavy, inflexible building and has had foundation problems which were mainly solved in the Great Restoration of 1925 to 1931. This restoration work strengthened the main piers, which were settling and cracking, and inserted a stainless-steel chain at the base of the drum to the Dome.

Since completion, a full and continuous programme of measurement shows that the dome has settled a further 8 or 9mm and that the western towers have moved apart. There is an envelope of thermal movement within which no permanent movement occurs. Perhaps the most striking movement is a 18mm bulge in the south portico wall which is closest to the heavy traffic flow. Underground water levels are recorded and a rapid rise was recorded in 1953, probably due to a burst water main, but a general fall of some 375–400mm has occurred. The 1–2m of potters clay is well compressed and the 6m of ballast is well graded and permeable, lying over the hardened clay which has shrunk due to water extraction and in which may lie future problems for those maintaining the fabric. The Cathedral

Fig. 14.17   St. Paul's Cathedral, crowded by modern London, suffers multiple problems of settling, thermal movement, deterioration from traffic vibrations, and decay from atmospheric pollution. Its entire salvation still lies in the future. *(British Tourist Authority photo)*

suffered severely in World War II from two direct bomb hits and from one near miss and must have been badly shaken, but due to the efforts of the St. Paul's Watch it survived the hail of fire bombs which destroyed a large part of the City of London.

It was washed in the 1960s and it was realized that further restoration works would be necessary, as many defects were uncovered. Again a meticulous and detailed report was made as a basis for the work and estimates were prepared. Some engineering work was found to be necessary to the Western Towers and to the south portico which had been dam-

273

aged, probably by traffic vibration, but a great deal of the trouble was due to atmospheric pollution.

The problem of traffic vibration is a difficult one, as it is almost impossible to prove that a certain amount of vibration input causes a specific amount of damage to a historic building. On the other hand, it is equally impossible to prove that vibration does not cause damage. The difficulty is that the damage caused by vibration is identical with that due to the natural aging process of the building and the fact is that vibration accelerates the aging of historic buildings according to its intensity. There has been some interesting and relevant research in Czechoslovakia on newly built houses: according to those figures, the intensity of traffic around St. Paul's Cathedral is likely to halve the life of the fabric. It is certainly true that compared with other buildings of similar construction, St. Paul's Cathedral is in poor overall structural condition. There are a great many local cracks and a great many local settlements, particularly in the lower parts of the walls. The south transept portico is of particular interest, as a crack in its wall has opened up some 18mm in the last 40 years. The bulging of the wall has caused serious cracking and

Fig. 14.18 Both Western Towers are shown here scaffolded for restoration work, which is primarily a program of necessary maintenance and preservation of architectural detail before it is lost to pollution. The City of London Corporation plans to prohibit traffic along the south side (this view is from the southeast) as part of a series of civic improvements initiated since World War II (*All subsequent St. Paul's photos courtesy of Feilden and Mawson Architects*)

spalling, and it is no good attempting to repair this wall until the fundamental defects are rectified. It is hoped that in the near future traffic will be routed away from the south side of the Cathedral, as this is the Dean and Chapter's policy.

The principal problem of St. Paul's, however, is the very rapid erosion and weathering of exposed masonry (Fig. 14.19). Most of the walling is well protected by projecting cornices, but these cornices have suffered and it has been the policy to restore them to their original dimensions in the southwest tower, the northwest tower and the central pediment.

The technical and engineering problems of renewing the face of the principal cornice are extremely difficult to solve. Several schemes have been worked out and tested and these have been found unsatisfactory. It is now hoped to find a solution using glass-reinforced

Fig. 14.20   Statues on the west front are particularly subject to damage from motor vehicle exhaust, and even the very durable Portland Stone is decaying. So far, conservation of statues is an extremely difficult problem.

Fig. 14.19   While much of the walling is protected by cornices, there is rapid erosion and weathering of exposed masonry at St. Paul's.

cement or possibly to reconstruct the stone, using permeable resins and stone dust as done on the Renwick Building in Washington, D.C.

Studies have been made of the rate of decay and it would seem that on some exposed urns at low level, the rate has been 27 mm in the last 45 years. This may have been accelerated by CO_2 and other traffic fumes but the principal agent is SO_2. At the upper levels the present rate would seem to be about half as much as the lower levels over 50 years. This is much greater than my average experience on other cathedrals of about 3–4mm in 100 years. The decayed stone can be brushed off in one's hand, and can be seen in the gutters after rainfall or observed dissolved and recrystallised where it has passed through cracks in the

Fig. 14.21 Some stonework is being replaced at St. Paul's. Here Ken Gardner sculpts an urn base.

fabric. Exposure to these intensities of pollution and traffic fumes mean that special protective measures must be taken to preserve the building and this is the current policy.

Others works carried out at the Cathedral include subdividing the dome into eight fireproof compartments and providing a fireman's lift and other fire precautions to protect the public. The Cathedral organ is being reconstructed and improved, the interior lighting is being redesigned and the mosaics in the choir are being cleaned.

To sum up, St. Paul's Cathedral is a building in a hostile environment but surrounded by friendly people. It is a constant source of amazement and delight what affection people have for these architectural monuments which are symbols of our continuity and our social consciousness. A city without its monuments is like a man who has lost his memory.

The Restoration of Independence Hall

CHARLES E. PETERSON

PREFACE

The great landmark built as the State House of Pennsylvania has stood for nearly two and a half centuries. As an architectural background for great events in provincial, national, city and county politics it has often been described and illustrated. But the story of its design and construction and remodellings has had scant attention from writers. And its narrow escapes from demolition—after the national and state governments moved away—are nearly forgotten. Thanks to the City of Philadelphia, Independence Hall is still with us. Vigilant civic leaders managed to keep it standing though often battered by what are now called "adaptive uses."

As the nation readies for its Bicentennial, the story of the fabric should be of public interest. It is not a simple one. Over the last 150 years there have been many attempts at restoration. These failed when modern architects were engaged and each, in his own time, sought to put the stamp of his personal talent upon its appearance. Some actually did, though later generations, in turn, attempted to remove those "restorations" as mistakes.

Some of the long struggle to maintain—or to regain—authentic architectural character is related in the two following chapters. They attest to the growing public interest in the preservation of historic landmarks, the evolution of architectural taste and understanding

and the laggard support by public funds. The compromises that were made along the way are in themselves instructive. After its removal to Center Square during the 1890s, the municipal government somehow kept the old buildings open, gradually converting the rooms to museum spaces as historical appreciation grew, as valuable relics were collected and donated, and as the American public came to see.

The Hall was first designated a National Landmark during World War II. Then, in 1953, the Mayor signed over Independence Hall, Congress Hall and old City Hall together with the Square itself, into the care of the Federal government to be part of the new Independence National Historical Park. After a flurry of repairs to the exterior, including repointing of the brickwork and the removal of the coal-fired heating plant from the Square, the structure was allowed to rest while the architectural staff concentrated on identifying and rescuing other historic buildings still standing in the newly acquired Park area to the east.

The comprehensive re-restoration of buildings on the Square began with Congress Hall, which was having a physical crisis due to its failing roof structure. The intensive architectural study of Independence Hall followed, beginning with the preparation of a large set of measured drawings. This eventually led to the dismantling of the old Haviland "restorations" on the ground floor and to the structural and decorative reworking of the whole second floor. The Assembly Room where the Declaration of Independence was signed was restored and reopened in 1965; the Supreme Court Room in 1969; the second floor in 1972.

The vast accumulation of documentary information utilized was due not only to the initiative of the Park Service staff but to the cooperation and generosity of many Philadelphians and their institutions. It would be impossible today to list all of their names, but two examples were particularly outstanding. In 1951 the late Lester Hoadley Sellers, AIA, came forward and donated the great collection of notes and transcripts gathered by his father, Horace Wells Sellers, FAIA, who evidently had intended to write a history of his own. That never came about, as is the case with most architects who intend to write. The Sellers windfall has now been dispersed so that from that notable hoard (the originals are still mostly in the great collections of the Historical Society of Pennsylvania) the various items are nowadays simply cited as "HSP."

Over the period of study perhaps the greatest single excitement was the revelation of the Joseph Sansom pencil sketch of the original west wall of the Assembly Room. The exact conformation of that room as of 1776 had been somewhat mysterious—even to the patriot artist John Trumbull who painted a view of the Signing of the Declaration for the ro-tunda of the United States Capitol in Washington, where it still hangs. Colonel Trumbull even went to Paris to discuss it wth Thomas Jefferson. But their joint recollections were not enough, for Trumbull's painting did not collate with the structural evidence remaining in the brick walls. It was not until George Vaux, Esq., of Philadelphia produced from his family papers the remarkable Sansom sketch that the documents could finally be reconciled to the physical facts. Good luck, one hopes, will always supplement hard work!

The essays by Mr. Nelson and Mrs. Batcheler which follow—originally given as informal lectures—reveal the interest of the great Park Service project. It was my privilege to bring both of them to Philadelphia years ago. Hopefully, the whole story of the restoration—involving as it did the work of other architects and engineers, constructors, and mechanics, historians, museum curators and archaeologists—will be published *in extenso* so that its lessons can be shared by all Americans entrusted with the care of architectural landmarks.

NOTE

There have been only two published items on Independence Hall by National Park Service architects. They are Charles E. Peterson, "Early Architects of Independence Hall," *Journal of the Society of Architectural Historians*, Vol. IX. No. 3 (October, 1952) pp. 23–26; and Lee H. Nelson, "Restoration in Independence Hall," *Antiques* (July, 1966).

CHAPTER 15

Independence Hall: Its Fabric Reinforced

LEE H. NELSON

MR. NELSON *was in charge of architectural work at Independence Hall from 1960 until 1972. He is now Technical Editor of the* Preservation Handbook for Federal Agencies, *American Editor of the Association for Preservation Technology* Bulletin, *and a member of the American Institute of Architects.*

Building Early America is of special interest to me and I am pleased to have the opportunity to discuss Independence Hall as an example of early building technology, to call attention to its structural systems, and to outline its recent rehabilitation by the National Park Service.

Without going into any details about the architect, the building committee, or the builders themselves, it is evident that the initial construction of the Pennsylvania State House represented a major achievement. To the best of our knowledge it was one of the largest buildings in the colonies at the time of construction, which commenced about 1732 and continued into the 1740s, with the tower being added in the 1750s.

On the first floor, it contained two gener-ous rooms for the judicial and legislative branches of the province (Fig. 15.1). These were the Supreme Court Room and the Assembly Room. These rooms were 40 by 40 feet square with a ceiling 20 feet high, without any interior columns or visible supports.

Although the builders, Edmund Woolley and Ebenezer Tomlinson, were leading master carpenters, it is doubtful that their construction experience had encompassed work of this magnitude. Perhaps it was the structural difficulties and the scale of the building that prompted the carpenters to request an adjustment to their contract claiming that ". . . the Work expected from them was heavy, and to be carried on in an extraordinary manner."[1]

Independence Hall is structurally interesting as well as being an historically significant building. Basically, the building is a simple loadbearing masonry shell measuring about 100 feet long and 44 feet wide. The foundation is of rubble stone laid in lime mortar on a relatively stable sandy clay soil. Above the foundation, the exterior walls are of a good hard-burned brick laid in Flemish bond, with a thickness of 22–23 inches.

There are two interior cross walls, defining the Assembly Room (on the East) with its single doorway and the Supreme Court Room (on the West) with its three arched openings.[2] Both cross walls (17–18 inches thick) are con-

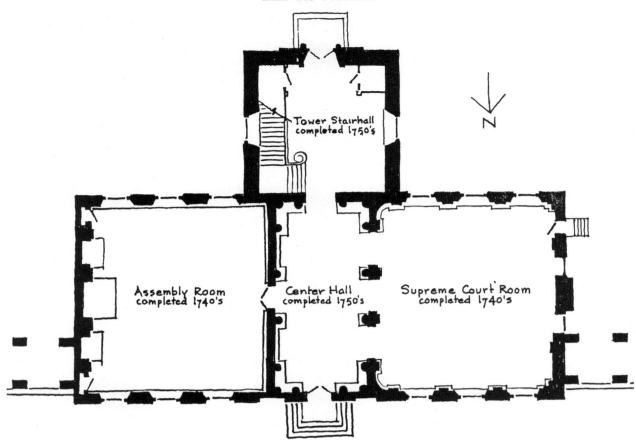

Fig. 15.1   Sketch plan of the first floor of Independence Hall. The exterior walls and the first floor interior walls, constructed in the 1730s, are of hard-burned brick laid in lime mortar on a rubble stone foundation: much of this fabric has survived in good condition. The Assembly Room *(left)* and the Supreme Court Room were large interior spaces for the time. *(Drawing by Penelope Hartshorne Batcheler)*

structed of brick (Fig. 15.2). They relate to the internal floor plan but they *do not* coincide with the exterior door and window arrangement. Each of the cross walls happens to intersect the outside walls at window openings necessitating blind windows on each side of the main entrance door. Because of this awkward juncture, these interior cross walls meet the outside walls with half arches. In fact, arches are an important part of the structural masonry. Concealed arches are used for relieving loads at all openings including doors, windows,[3] and large interior openings into the Supreme Court Room and the Assembly Room. They are also used in the attic to support the enormous end chimney stacks which are among the more conspicuous features of the building. The arches are variously seg-

mental, semicircular, three-centered and flat, apparently reflecting the preferences of the workmen.

In addition, the brick walls were supplemented with numerous wooden grillage beams and wooden bearing pads to help spread the loads above masonry openings and to simplify the installation and seating of wooden girders, beams and trusses.

The various wooden framing systems should be considered as an integrated system (Fig. 15.3) consisting of roof trusses in the attic, concealed wall trusses in the second floor, and what we have come to call the "hand-in-hand" framing system used in the ceilings of the large rooms—the Supreme Court Room and the Assembly Room.

We gave the "hand-in-hand" system that

Fig. 15.2   Brick masonry interior wall dividing the Assembly Room and the Central Hall. The doorway arch seen here has retained the original wooden form or "falsework" used to lay the arch. The half arches (at *right*) were required because the cross walls intersected the front and back walls at window openings. Two of these "blind" windows have survived, flanking the main Chestnut Street doorway. *(Photograph by George Eisenman, J. L. Dillon and Co., Inc., 1964)*

name because the girders intersect each other in an arrangement that looks like the familiar interlocking hands that form the so-called fireman's seat (Fig. 15.4). This interesting system divides the 40-by-40 foot area with 4 timber girders arranged so that no single girder spans more than two thirds the width of the room, with each girder interlocked at third points with other girders at right angles, thus forming the "hand-in-hand" design.

Original girders still in place measure 11 by 11 inches, supplemented with a secondary system of wooden beams. Resulting "bays" were filled with joists (about 2 by 11 inches), all

of which run in a north–south direction. All the various members are connected with mortises and tenons, rather typical of the period. But of special interest is the fact that the main girder intersections were reinforced with iron plates let into each side of a girder and secured with iron pins and cotters, probably to compensate for the reduced cross section resulting from extra-large mortise holes. Normally, the joists have double tenons, the details of which seem to vary according to the carpenter who cut them. There are only single tenons where they would compete with the let-in iron plates (Fig. 15.5).

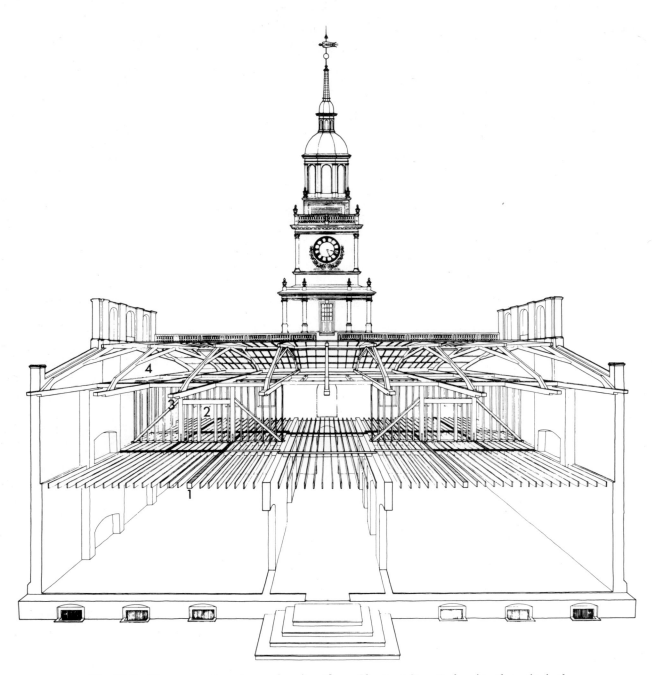

Fig. 15.3 Diagrammatic cutaway drawing (from Chestnut Street) showing the principal structural systems: 1). The "hand-in-hand" first floor ceiling framing; 2). The trussed partitions in the second floor which partially supported the ceiling framing. Both partitions were removed early in the history of the building, therefore the structural details of the trussed partitions illustrated here are conjectural. The evidence for the design shown here consists of the broken-off queen-post tie straps (Fig. 15.6) and the dapps for the diagonal members at the ends of the lower chord; 3) The hand-in-hand framing for the second floor ceiling which is identical to and aligned with the framing at the first floor; 4) The attic roof trusses, which resemble braced arches more than trusses. *(Drawing by Gary Dysert and Robert Mack)*

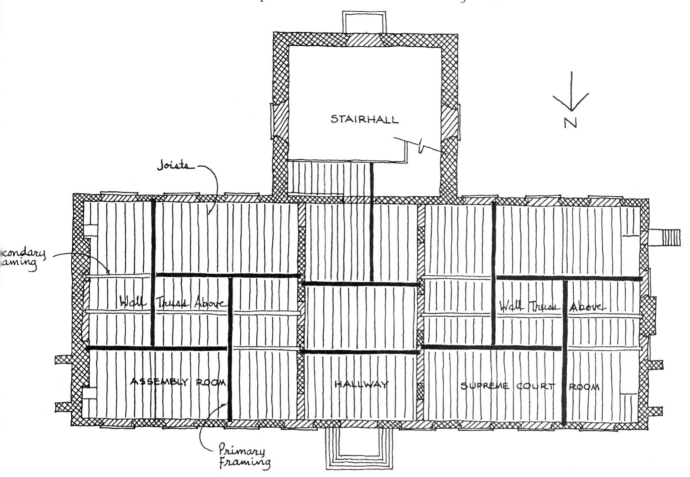

Fig. 15.4   Sketch plan of first floor walls and ceiling framing, including the hand-in-hand framing over the two principal rooms. Supplementary girders were used to reduce the span of the joists in this framing system. The central hallway is simply spanned with two girders and filled in with joists. *(Sketch plan by Penelope Hartshorne Batcheler)*

The origins of this framing system are not known, but a somewhat similar system was illustrated in Batty Langley's book *Ancient Masonry,* published in London in 1734–35 (a trifle late for direct influence in this instance). It was called "a floor by S. Serlio," the Italian architect of the sixteenth century. It is conceivable that the State House ceiling structure may have been the precedent for framing a ceiling in Benjamin Franklin's own house erected in the 1760s. A sketch among the Franklin papers in the American Philosophical Society shows the framing that might have been used or considered for his house, which was about 34 feet square. Perhaps there is another explanation for this penciled sketch. I have mentioned it here because I know of

no other reference to this type of framing in the eighteenth century.[4]

Considering the 40-foot span, even full length timbers would have been subject to an excessive deflection, and the "hand-in-hand" system is more critical in deflection because it is dependent upon mortise and tenon joints to create the effect of continuous girders. The original builders solved this problem by suspending the ceiling framing from wall trusses which were carefully concealed in the longitudinal partition walls of the second floor. Physical evidence suggests the original use of queen-post wall trusses, probably similar to those shown in eighteenth century builders' treatises and supplemented at the third points with wrought-iron tension bars to hold up the

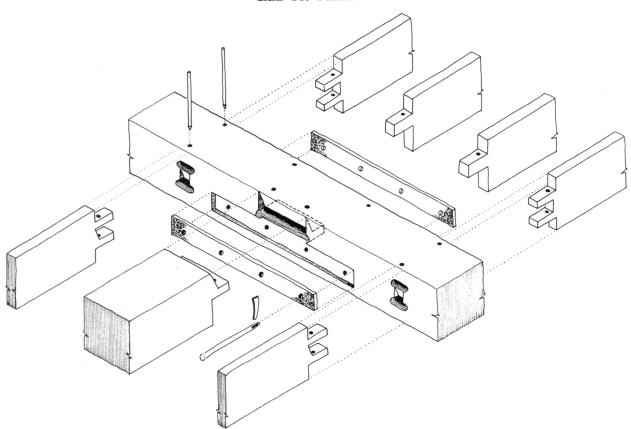

Fig. 15.5 Exploded diagram illustrating typical girder-to-girder connection. Let-in iron plates were used to stiffen the girder and compensate for weakening caused by the large mortises. The plates were secured by iron pins with cotter wedges. Typical floor joists have double tenons. The variety in styles of tenons seems to reflect the personal preference of the carpenters, some of whom provided haunches between the tenons to strengthen the joist against vertical shear. *(Drawing by Author)*

ceiling framing below.[5] Fragments of these iron bars have survived. They served the same function as stirrups except that instead of being U shaped, they are paired bars with large heads which change in cross section from round to rectangular, to secure the ceiling girders to the queen posts. These iron straps were cut off when original wall trusses were removed in the nineteenth century (Fig. 15.6).

The last of the major systems to be mentioned here are the roof trusses themselves and the attic floor framing. While it is hard to discern from the ground, the upper roof is relatively flat with balustrades at the outer edges; the lower roof slopes are much steeper. The roof trusses were contrived to accommodate these different slopes. In fact, they are more like *braced arches* than trusses (Fig. 15.7).

Several survive practically intact but most have been modified or strengthened. Note that they utilize a bent tree shape reminiscent of the medieval cruck frames.[6] These arched members have a cross section that varies, but they average about 8 to 11 inches although the braces are smaller. Structurally, one of the more disturbing aspects is that the bottom chords of the roof trusses are discontinuous and actually form part of a "hand-in-hand" system like that on the first floor ceiling; in fact, they are exactly aligned with each other. It is my opinion that the second floor trussed partitions provided support for the attic floor framing at their midspans. If so, the discontinuity of the bottom chords would have been of little concern originally, but the later removal of the trussed partitions below certainly

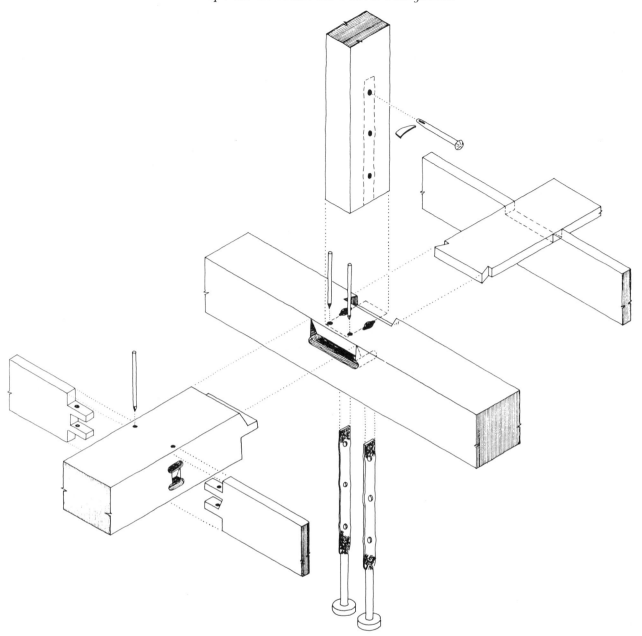

Fig. 15.6   Exploded diagram of probable connection at juncture of hand-in-hand framing and the queen post of the trussed partition on the second floor. The queen post (at *top*) was secured to the girder with large wrought iron bar straps. The heads were let-in (flush with the bottom of the girder) and the bar strap changed in section from a round bar to a rectangular strap at the queen post. When the trussed partitions were removed, the straps were broken off, leaving the round bar portions in the girders. *(Drawing by Author)*

was a matter of concern and necessitated extensive strengthening of the roof trusses.

The attic floor framing had additional girders at the center to support the original cupola or bell turret which was octagonal on the flat above. The location of the eight supporting posts is established by the mortise holes in the attic floor girders.[7] The only representation of that cupola is on an early drawing of unknown attribution (Fig. 15.8). In the 1750s the turret

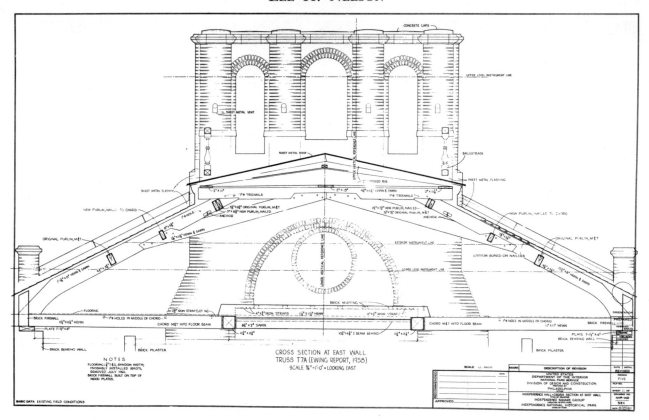

Fig. 15.7   Measured cross section through the attic showing an original wooden truss (actually a braced arch) which supports the double roof slope. The wooden truss members terminate in natural "cruck" forms at the base. The braced arch forms have survived intact only at the end walls; the intermediate braced arches have been heavily supplemented with additional wooden truss members, iron tension rods and other remedial measures. Also visible is the east brick arched pier which supports the enormous multiple chimney stacks. *(Drawing by David Baker)*

was removed, to be superceded by a tall steeple built on a masonry tower appended to the south side of the building.

So much for the original structural system.

We have been able to trace the most significant changes to these systems. The truss over the Supreme Court Room was completely removed in 1830 under the direction of the English-trained architect John Haviland when that portion of the second floor was made into a larger room. To replace the supportive function of the wall truss, Haviland used four cast-iron columns within the Supreme Court Room, the first such intrusion into the first floor.

The wall truss partition over the Assembly Room was removed in 1854 when the entire second floor was remodeled to accomodate

the enlarged Philadelphia City Council when all the boroughs of the County were assimilated. The function of the wall truss was replaced with four iron tension rods so that the first floor ceiling was now suspended from the attic trusses. This installation was possible because the attic girders aligned with the second floor girders. However, this installation overloaded the old attic girders, which had been previously supported from below by the wall truss. It appears that this structural problem was resolved by adding new members within the old arches to form a real truss for the iron tension rods (Fig. 15.9). No known documentation establishes the date of these trusses within the arch, but it seems most likely that it occurred, as suggested, in 1854.

After the City Council moved to Center

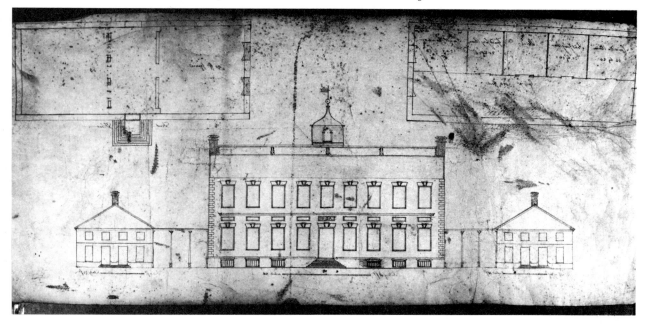

Fig. 15.8 Drawing of the Old State House (ca. 1730s) possibly made during construction, author unknown. It is the only known view which shows the original cupola—as built, an octagonal structure. The cupola was removed during the second stage of construction in the 1750s and was replaced by the tower and steeple in the rear. Architectural evidence for it survives in the attic floor framing in the form of mortises for the eight posts. *(Courtesy Historical Society of Pennsylvania)*

Square—and during the 1896–98 renovation by the Daughters of the American Revolution—the approximate original second-floor plan was reestablished. This made it possible to install new wooden wall trusses in the same location as the originals and to remove the four tension rods together with the four cast-iron columns so visible in the Supreme Court Room (Fig. 15.10). These 1898 wooden wall trusses later sagged and were in turn removed by the National Park Service in 1962 and replaced with two steel plate girders.

THE ORIGINAL TOWER AND STEEPLE

At this point I will backtrack to discuss the steeple and the tower which were appended to the existing building during the 1750s, some 20 years after the initial construction. The tower walls are of rubble stone veneered with brick, creating a masonry structure about 35 feet square and 68 feet high above the ground. The top 29 feet is of solid brick masonry.

The wooden steeple which surmounted the tower was well toward completion by March of 1753 (Fig. 15.11). It too was constructed under the direction of Edmund Woolley.[8] A voucher dated 17 April 1753 covers the sundry items for raising the bell frame and installing the bell. This interesting document covers the expenses of the celebration of that event and includes such items as a peck of potatoes, 14 pounds of beef, mustard, salt, pepper, butter, cheese, 300 limes, 3 gallons of rum, 36 loaves of bread and a barrel of beer (totaling £5.13s.10p).

Despite Woolley's experience as a builder, the much admired and published State House Steeple enjoyed only a short life—in fact only about 28 years. By 1774 the Steeple was considered to be in such "ruinous condition" that it could not be repaired. The problem was not resolved until after the wartime British occupation of the city. It was finally removed by Thomas Nevell, a well-known Philadelphia carpenter, in 1781, and the masonry tower was capped with a pyramidal roof and spire

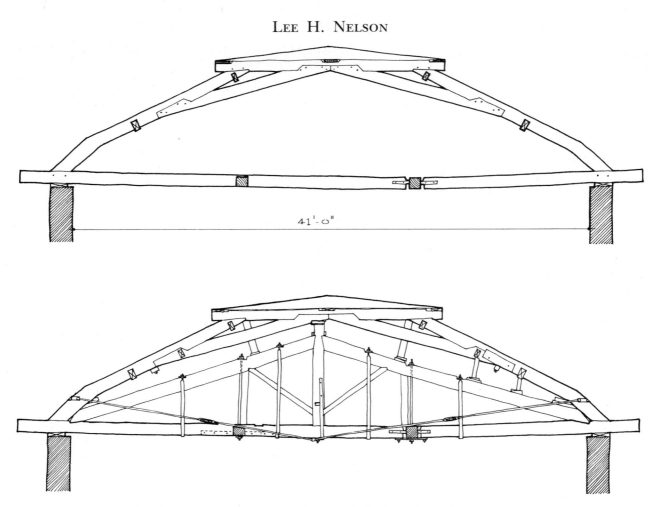

Fig. 15.9   Diagrammatic drawing showing a typical original braced-arch roof truss of the 1730s *(top)*, and as modified with supplementary wooden members and iron bars which were installed at later times *(bottom)*. *(Drawing by Kenneth Anderson)*

with weathervane (Fig. 15.12). It remained that way for the next 47 years, until 1828, when the present steeple was constructed from the designs of William Strickland; that project fairly constitutes one of the earliest American attempts at historical reconstruction. The original steeple, of course, did not have a clock but it was considered a symbolic if not accurate reconstruction — some 147 years ago.

The original steeple structure is not known. Although four of the queen-post trusses survive within the lower tower, their relation to the whole is not established (Fig. 15.13). Basic framing of the present (1828) steeple consists of four great timber posts which measure about 14 inches square at the base, about 12 inches square at the top and are about 48 feet

long. These corner posts are canted inward and thus form a truncated pyramid, supported on specially made cast-iron brackets embedded in the masonry walls. The entire system depends upon the weight of the structure to maintain its stability and resistence from overturning by wind load.

The framing employed the use of "ships knees" for stiffening the structure at the juncture of horizontal braces with the corner posts, all of which was done without mortise and tenon joints, and thus with little removal of wood in the critical members.

Another weak section of the steeple occurred where the structure changed from a truncated pyramid to an octagonal prism. At this point the steeple is extended in height with eight vertical posts which form the octag-

Fig. 15.10   Photograph of the second floor east room (ca. 1896) showing the remnants of the Victorian period Select Council Chamber (1854) prior to the first "restoration" of the second floor. To accommodate this council chamber, the trussed partition had been removed and replaced by iron tension rods from the attic roof trusses. *(Independence Park Collection)*

onal cupola and also carry the lantern and spire. These posts were simply lapped onto horizontal braces and secured with a single thru-bolt.

The original steeple held a 2000-pound bell; the 1828 steeple was built for a 4600-pound bell, which was replaced for the Centennial with a 13,000-pound bell. This, on special occasions such as the Fourth of July and New Year's, was rung equal to the number of years since the Declaration of Independence. It was this sort of use (or abuse), together with natural deterioration, that contributed to the need for a structural intervention in the 1960s. The problems in the steeple, though, were nothing compared to those of the main building.

PLANNING THE STRUCTURAL WORK

In 1951 the custody and operation of the Independence Hall group of buildings was transferred by the City of Philadelphia to the Park Service as a part of the new National Park. By the late 1950s, the combined efforts of historians and architects had accumulated a great deal of documentary and physical information about the building. The Park Service contracted for an extensive engineering survey of structural conditions, or at least of those conditions that were accessible for inspection. While the survey was thorough the first recommendations were drastic and beyond consideration in terms of preservation: they included removal of the original wooden

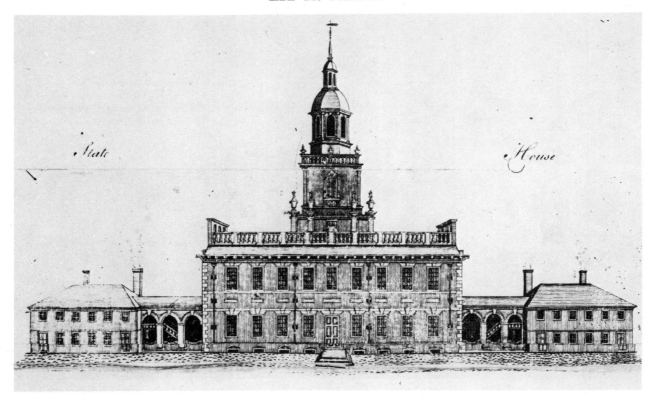

Fig. 15.11 1774 engraving of the State House, one of numerous engravings made after construction of the steeple. It is possibly the most accurate representation of the steeple prior to its removal in 1781. From "Map of the City and Liberties of Philadelphia," John Reed, 1774. *(Independence National Historical Park Collection)*

structural systems, with replacement by concrete and steel. With the rejection of this approach, the Park Service started its own in-house study of preservation problems, beginning with careful, measured drawings of the roof trusses.[9] With the removal of the attic flooring it was possible to really inspect the structural details with all their changes, deterioration, their failures due to defects, cutting through for utilities, etc.

While this recording was taking place in the summer of 1961, a section of the plaster ceiling of the Long Room fell to the floor. It was 12:30 p.m., July 13, an event that is indelibly etched on my mind.

The sounds of cracking plaster, especially as they started directly under one of the attic roof trusses, only reinforced my apprehensions about the structural integrity of the trusses we were measuring. Instead of a real

structural failure, it turned out to be only a very inadequate nailing of the metal plaster lath installed in 1922. The remainder of the second-floor plaster ceiling was cut out, a small section at a time, and removed in the following days. This event had some good and bad consequences, but among the good things that happened was bringing in the late Sheldon A. Keast as consulting engineer. Keast and his associate, Nicholas L. Gianopulos, had just completed the structural analysis and design for the rehabilitation of Congress Hall. It was a happy mix of experience and positive sensitivity to the preservation values of that most important historic structure.

In the ensuing months, we continued to open the fabric. We removed plaster, took up the modern flooring (Fig. 15.14), made more measured drawings, and learned enough about the structural problems so that Mr.

Fig. 15.12   1799 view entitled "Back of the State House, Philadelphia," drawn, engraved and published by William Birch & Son. This engraving shows the State House as it looked for some years after the steeple was removed in 1781 because of its "ruinous condition." It is not known whether the short duration of the steeple was due to faulty construction, or to poor maintenance or both. The present steeple is a reconstruction built in 1828, in the name of historic authenticity, from the design of architect William Strickland.
(Independence Park Collection)

Keast was able to present his proposal to a September 1961 meeting of the Independence Architectural Advisory Committee, which included Sydney E. Martin, FAIA, Joseph T. Fraser, Jr., Dr. Roy F. Nichols, John Lane Evans, AIA, G. Edwin Brumbaugh, FAIA, John F. Harbeson, FAIA, and Grant M. Simon, FAIA, two of them members of the Carpenters' Company. This committee met on several such occasions and gave their valuable time and experience.

We all agreed that portions of the building could remain open to the public during the entire process of structural rehabilitation and

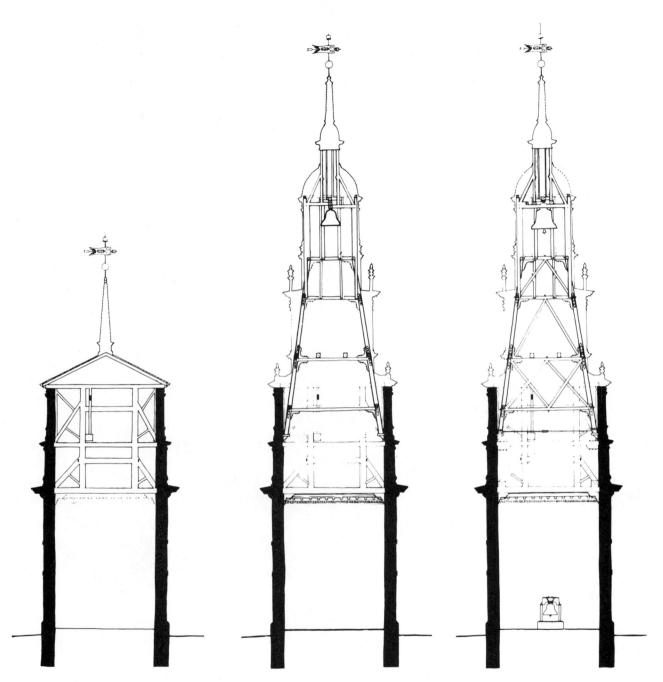

Fig. 15.13   The tower, constructed of rubble stone and veneered with brick, has survived the original steeple and its successors. At *left* is the tower after the original steeple was removed. The structure of the original steeple is unknown; the tower is shown here as it appeared temporarily capped with a pyramidal roof, a little spire, and a weather vane. It is likely that the original steeple structure was supported by the two level queen-post trusses which survived and are shown in the drawing. In the *center* is the tower with its 1828 steeple. The basic structure consisted of four corner posts on cast iron cantilevered brackets braced with wooden ships-knees. Additional wooden structural members facilitated the transition from a quadrilateral to an octagonal steeple stage, just below the cupola and bell. A 4,500-pound bell replaced the original Liberty Bell at the time of the steeple reconstruction. At *right* is the steeple as it was strengthened with diagonally placed timbers and tension rods to accommodate the enormous Centennial Bell which weighs a symbolic 13,000 pounds. (*Drawing by Gary Dysert*)

Fig. 15.14   Visible here is a portion of the original second floor framing. After the partial collapse of a plaster ceiling in 1961, the second floor was stripped of its relatively modern woodwork, flooring and plaster. As this material was removed, a remarkable succession of architectural changes and related structural remedial measures were revealed, such as an 8-inch sag in the floor framing which had been stabilized during the extensive work in the 1890s. Following an intensive investigation of the fabric, the National Park Service developed a comprehensive plan for structural rehabilitation. *(Photograph by Eisenman)*

restoration. This concept was central to the planning and logistics of the work. Of course, architectural research continued during the structural planning. And the interior finish work had to be coordinated with the basic structural program.

In June 1962 we opened bids for the structural rehabilitation; two of the bidders were members of the Carpenters' Company. The low bidder was the A. Raymond Raff Company. The contract was signed soon after, and thus began another relationship that I recall with a good deal of satisfaction. The contrac-

tor's Project Manager for the structural rehabilitation was Walter Riley, now president of the Raff firm, and a member of the Carpenters' Company.

The concept for the structural rehabilitation was relatively simple, but there were considerable problems due to the high priorities given to maximum preservation and visitor access. A series of shoring columns was installed in the cellar, penetrating through the Supreme Court Room and the Assembly Room (Fig. 15.15). From these shoring columns, it was possible to install platforms where needed on the second floor, and to help carry skid platforms for installing steel beams

Fig. 15.15   The structural rehabilitation of the early 1960s involved an extensive temporary shoring system (seen here in the Supreme Court Room). This system extended to the attic level and provided temporary shoring for the new grid of structural steel which was installed to support the roof and second floor ceiling. *(Photograph by Eisenman, 1962)*

293

in the attic. A significant fact working for us was that the old hard-burned brick walls, the stone foundation, and the soil bearing were all in good condition and needed only minor underpinning.

In October 1962, the first attic steel beams were installed through small holes in the roof; when completed they formed a system of major cross beams between the old wooden trusses, with a series of smaller beams located so as to provide anchorage for picking up the attic floor system with hanger rods, and to provide bearing for adjustable pipe columns to pick up roof load (Fig. 15.16). The system allows for easy introduction of additional members if needed in the future.

At the second floor level, two new systems were introduced: the first being two new plate girders to assume the function and position of the original wall trusses, that is, to be concealed in the longitudinal partitions (Fig. 15.17). Secondly, three new steel girders were installed over the Assembly Room and Supreme Court Room. This allowed the sagged

Fig. 15.16   View into portion of attic following structural work of 1962–63. A series of steel girders span the width of the building between the original roof trusses. The new girders were installed slightly above the original truss chords and floor framing, in order to preserve the entire old system *in situ*. A secondary system of smaller steel beams was installed between the major girders as needed. These secondary beams provided a convenient means of hanging the old attic floor framing at critical points, as well as a convenient and flexible framework for supporting the original wooden roof trusses, purlins and rafters. The latter was accomplished with adjustable steel pipe columns. The entire system is designed to accommodate future hangers or adjustable posts should the need for them arise. (*Photograph by Eisenman*)

Fig. 15.17 Steel plate girder being installed through second floor window. Two such girders replaced the 1890s trussed partitions and serve the same function as the original wooden trussed partitions, i.e., supporting the first floor ceiling framing. The building remained open to visitors at all times during the structural rehabilitation, although access was necessarily limited. *(Photograph by Eisenman)*

Fig. 15.18 1961 photograph of advanced deterioration in one of the queen-post trusses in the tower directly above the Liberty Bell. This is a most conspicuous example of the structural deterioration caused by alterations, abuse, changes in structural function, poor maintenance, and in some instances inferior original materials and workmanship. *(Photograph by Eisenman)*

framing to be raised back to level; eventually the floors and ceilings were restored to their approximate original levels. They had sagged as much as 8 inches.

The greatest sacrifice of the surviving original structural system was cutting the old hand-in-hand girders to allow the new steel members to be fitted within the thickness of the old framing. About twenty pieces were removed and accessioned into the park museum collection.

THE STEEPLE REINFORCED

During the course of the structural rehabilitation, we were working with Keast's office on the details for strengthening the steeple structure. By January 1963 the work was underway, as an addition to the existing contract.

Fig. 15.19 View into the second level of the tower (just above the Tower Stairhall) showing the 1750s queen post trusses preserved *in situ*. The trusses were supplemented with steel channels and plates during the 1962–63 rehabilitation by the National Park Service. The new steel members are supported by grillage beams embedded in the stone masonry walls of the tower. *(Photograph by Eisenman, 1963)*

295

LEE H. NELSON

Of the many instances of structural deterioration, few were as dramatic as the dry rot found in one of the 1750s queen-post trusses at the second level of the tower, directly over the Liberty Bell. This condition was concealed by later flooring. (Fig. 15.18); an extensive wooden shoring system was installed in the stairhall supported old trusses. We elected to preserve these trusses by sandwiching them between new steel channels, which were suspended from grillage beams in the old stone walls (Fig. 15.19). This was done to preserve the fireplaces in this Upper Tower Room, long ago used as caretaker's quarters.

Far more difficult was anchoring the immense wooden corner posts of the steeple (Fig. 15.20). This operation, with some risk attendant, involved increasing the very limited bearing area of the old cast-iron brackets and tieing the columns to the brackets to better resist overturning due to wind loads. The flanking wall areas were cut away to allow bearing for a steelbox girder, and holes were drilled through the cast-iron brackets to admit specially fabricated tension rods.

The old ships-knees and posts were drilled to receive steel shear pins, and steel bands were fabricated to fit the posts and prevent further checking and splitting. Other steel framing was installed at various levels to stiffen the awkward connections at the juncture of the octagonal portion of the steeple, to strengthen the hanging of the present 13,000-pound bell and to stiffen the cupola against any movement caused by ringing the bell.

The entire structural rehabilitation work described was completed by the late summer of 1963. It was followed by a much longer period of architectural research, physical investigation and restoration of the interiors.[10]

Fig. 15.20 The existing steeple structure consists of four enormous wooden corner-posts canted inward to accommodate the gradually decreasing size of the steeple. The 1962–63 structural rehabilitation provided new steel box girders under the corner-posts with adequate bearing in the walls and tie bars to relieve the unwanted outward thrust of the canted corner posts, and to reduce the possibility of overturning in the event of hurricane velocity winds. The ships-knee braces, shown here, were previously through-bolted to the posts. During the rehabilitation of 1962–63 they were supplemented with steel shear pins (half in the knee and half in the post) which provide for better transfer of stresses from the wooden cross braces. *(Photograph by Eisenman, 1974)*

NOTES

1. *Pennsylvania Archives,* Eighth Series, III, pp. 2154–55.
2. The hallway-to-courtroom arched openings are each about 7 feet wide, and the piers between the arches are about 3 feet wide.
3. Considering the early date, the window openings are surprisingly generous, and measure about 6 feet by 10 feet, with splayed jambs so that the inner openings are about 7 feet wide.
4. Nearly a century later, an almost

identical arrangement of framing was illustrated in Thomas Tredgold's well-known *Elementary Principles of Carpentry* . . ., London: 1820, plate IV, fig. 47.

5. Wall trusses or "truss'd partitions," are illustrated in such treatises as Batty Langley, *The Builders' Jewel* . . ., London: 1741, plate 83, *see* reprint of 1757 edition by Benjamin Blom, New York, 1970. They are also illustrated and priced in *The Rules of Work of the Carpenters' Company of the City and County of Philadelphia,* 1786, edited and annotated by Charles E. Peterson FAIA, Princeton, 1971, *see* plate XV and p. 17. The pricing rule states: "TRUSSED partitions framed, with posts and braces about 10 or 12 by 4 inches, either strap'd to a girder, or with a sill, per square, £1 s.10 d.-". Long-span trussed partitions were also used in the old buildings which flank Independence Hall, i.e., the County building (Congress Hall) has a trussed partition which is part of the 1793 extension, and Old City Hall has a 46-foot span trussed partition which is part of the original 1790–91 construction.

6. For a brief discussion of "cruck construction" and reference to further writings about the subject, *see* C. F. Innocent, *The Development of English Building Construction,* Cambridge, 1916, pp. 23–72 and Alec Clifton-Taylor, *The Pattern of English Building,* London: 1972, pp. 303–307.

7. The evidence of an octagonal turret is established not only by the mortise hole evidence of the eight cupola posts, but by the large octagonal opening still to be seen in the original roof sheathing. The latter evidence gives the plan dimensions of the cupola, if not the height. This evidence has been measured and recorded by the National Park Service.

8. There seems to have been little contemporary published information about the construction of steeples. One of the few such sources is Francis Price, *The British Carpenter: or, a Treatise on Carpentry* . . . London, 3rd ed., 1753; plate facing p. 34.

9. The excellent measured drawings of the attic roof and floor framing were prepared by student architects Dave Baker (University of Illinois), Blaine Cliver (Carnegie Tech) and Tom Frost (Princeton). They were assisted by Architect J. M. Everett.

10. Messrs. Gianopulos and Riley, present in the audience on March 29, were introduced and later they assisted as guides to the party visiting Independence Hall.

CHAPTER 16

Independence Hall: Its Appearance Restored

PENELOPE HARTSHORNE BATCHELER

Since 1955 Mrs. Batcheler has worked as restoration architect on Independence Hall and most of the other National Park buildings in Philadelphia.

The National Park Service, standing on the shoulders of many who had gone before, has worked for nearly 25 years at unravelling the complicated structural history of Independence Hall—the Pennsylvania State House, as it was first known. The eventual key to this story was the identification of the building's uses over the years. Research historians and architects examined the papers of each government or private body that had occupied the structure. An enormous amount of documentary evidence was accumulated from the surviving Pennsylvania provincial government papers, Charles Willson Peale's papers (from the time when he used the State House for his museum), the City of Philadelphia papers (as the caretakers and then owners of the building since 1815) and the papers of those who have previously sought to restore the building: architect John Haviland of the 1830s, Col.

Frank M. Etting in 1876, the Daughters of the American Revolution and their efforts of the 1890s, and the Philadelphia Chapter of the American Institute of Architects under the leadership of Horace Wells Sellers in the 1920s.

The restoration architects of the National Park Service, aided by repeated summer student teams, made measured drawings of the building and meticulously recorded the layers, so to speak, of physical evidence as uncovered. During this research and restoration process the general public was always allowed into some portion of Independence Hall every day of the year. The investigation of the fabric was therefore somewhat fragmented and had to be directed toward one room at a time. Actually this proved to be an advantage, as it permitted full concentration on the documentary references,[1] as they related to the minute details unfolding within the walls of each room.

As we were getting to know the building we also got acquainted with the techniques, foibles and individualism of its early builders. I would like to share with you a brief overall view of the National Park Service restoration, some of the workmen's imprints, and some of the experiences we enjoyed in restoring this grand old building.[2]

THE EXTERIOR

The committee appointed in 1729 by the Pennsylvania Provincial Assembly to superintend the building of the Pennsylvania State House was composed of lawyer Andrew Hamilton and medical doctor John Kearsley—both members of the Assembly—and Thomas Lawrence a member of the Governor's Council. Andrew Hamilton's persuasion placed the site of the building at the western edge of the growing capital city. The authorship of its design was probably shared in varying degrees by this committee and the master carpenters employed to erect the building, Edmund Woolley and Ebenezer Tomlinson.

Two unsigned drawings have survived, one of floor plans and the front facade (*see* Fig. 15.9) and the other of floor plans only (Fig. 16.2). A receipted bill of Edmund Woolley's for drawing front and side elevations and floor plans of both the main building and its wings and arcades indicated that yet more drawings once existed. (These documents were found among what is left of the Penn family proprietary papers in the Historical Society of Pennsylvania.)

Over and above authorship, what is interesting is the elaboration of the front versus the plainness of the rear facade. The front (100 feet long) has a balanced organization of openings, embellished with marble keystones, panels and belt courses; soapstone water table mouldings and corner (quoin) stones; and an acanthus-leaf carved modillion cornice of wood. Two of the window openings (those flanking the front entrance) are blind with closed shutters hiding discord with the interior plan, where the interior walls of the entrance hall abut the window positions. Other than this practical solution and the overall pleasing distribution of details, the organization of the front facade seems to have gotten away from them a bit in that the belt courses and corner quoin stones do not work together neatly.

There was clearly less emphasis on the rear facade. Its 1730s appearance sans tower, with molded brick water table, brick belt course, the neatly rubbed brick flat arches above the windows and *uncarved* modillion cornice are well executed and were all details being used repeatedly on the better domestic structures in the Delaware Valley.[3] One gets the feeling that they were more comfortable in handling the rear facade with its traditional and tried details, while at the front one senses the use of untried elements remembered from elsewhere or borrowed from some book.

Designed to receive the flanking arcades which connect the main building to the wings, the end gables are simple: their only refinement being the "tumbled" brick battlements (unusual to Philadelphia) and the arcaded chimney tops.

The surviving elevation drawing represents a design made prior to construction. As built, the front facade has two marble belt courses instead of just one as shown on the drawing. And the date A.D. 1732 shown on the drawing within the panel above the entrance was not placed here in actuality. Instead, a 1733 date, probably marking the completion of the roof,

Fig. 16.1 Wooden downspout conductor head, 1777. Four of these replaced leaden ones removed to make bullets for the Revolution. The total weight of lead stripped from the State House was 1636 pounds. The original flanges seen here were left. (*Photograph by R. Frear, 1975*)

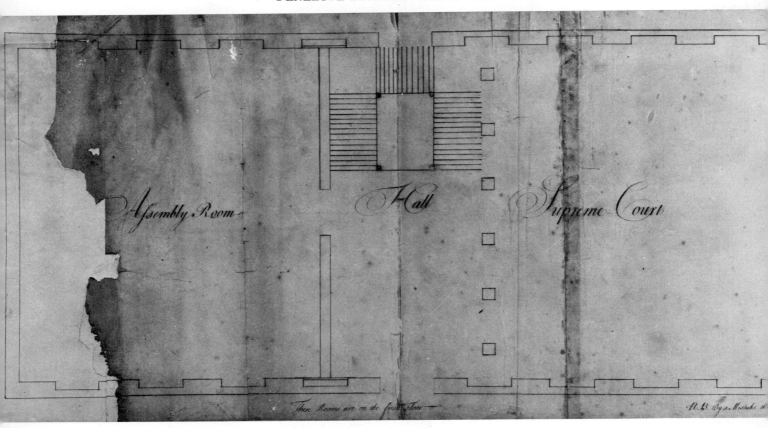

Fig. 16.2 Unsigned first and second floor plans, apparently drawn prior to construction, show the pre-tower distribution of rooms. The physical evidence of framing pockets and nailing blocks confirmed the position of this pre-1750 stair. *(Penn Manuscripts, Historical Society of Pennsylvania)*

was placed on the lead flanges of each of the four downspout conductor heads (*see* Fig. 16.1).

Less than 20 years after the main building was roofed the Pennsylvania Assembly embarked on the major addition of a massive tower containing a grand staircase (Fig. 16.3). Edmund Woolley was again employed as the master carpenter, this time working with a committee of the Assembly chaired by Isaac Norris, an influential Quaker whose interest in this great expenditure of funds was to invest in brick and mortar rather than to support the wars currently being waged by the British crown.

What is known today as the "Liberty Bell" was ordered from England to hang in the new steeple above the new grand staircase. Upon arrival and first test, that bell cracked. The Assembly immediately ordered a second bell

as a safety measure. The first bell was recast twice here in Philadelphia, to the end that the Pennsylvania Assembly found themselves the owners of two working bells. To justify the second bell a clock was ordered, the works of which were placed in the middle of the attic of the main building with the second bell striking the hours within a small turret above.

The curious design solution for powering this clock can probably be attributed to Isaac Norris. Among his papers the Park Service historians found a crude sketch of a highly unusual "grandfather" or tall case clock (Fig. 16.4). As necessity is the mother of invention—and tall case clocks were developed to house the weights and lines for eight-day works—Isaac Norris obviously chose the same route to develop a weight drop for the State House clock. A giant clock with a rusticated soapstone case was built against the west gable

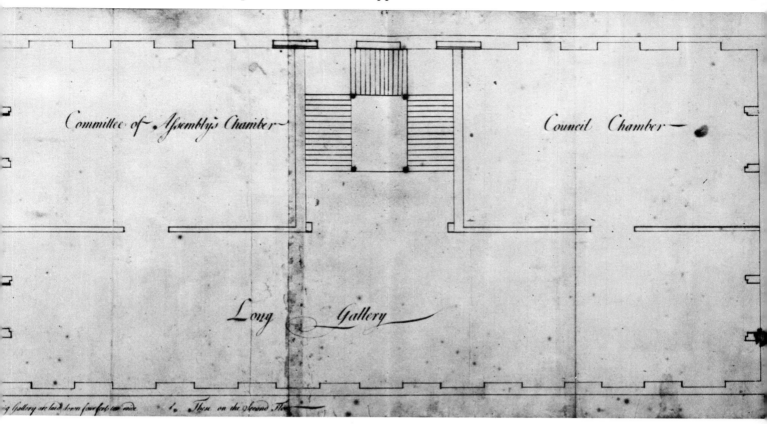

Committee of Assembly's Chamber

Council Chamber

Long Gallery

of the main building. The hood and face only were repeated at the east gable as there was no need for a second case for weights.

The distance of 50 feet to the clock works, over which power was transmitted to activate the two sets of hands and the bell, seems to have prevented the clock from almost ever ringing on time. But as a statement of the Pennsylvania Assembly's authority and stature, nothing could have meant more to the populace than that symbol of affluence aspired to by each household—a tall case clock. The National Park Service has reconstructed the west clock (Fig. 16.5); albeit it now contains air ducts headed for the second floor rooms.

On Walnut Street, in the south wall of the State House yard (now called Independence Square), there later was constructed a second giant symbol of affluence. Here was a triangular pedimented gateway enframing 20-foot high panelled doors and a fanlight (Fig. 16.6). Endless variations of such frontispieces were to be used in the finest domestic structures

built by Edmund Woolley and his brother members of the Carpenters' Company throughout the rest of the century.[4] It would be fun to reconstruct these giant gates to accompany the unbelievable giant clock.

Reconstructions are dangerous, however. Who could imagine the unusual relation of belt courses to quoin stones on the front facade mentioned before? And what twentieth century designer, unless he had graphic proof, would cut in half ionic capitals where they abut outside walls or window sash, as they do inside the State House at the tower stairhall Palladian window. And how remarkable it is that Edmund Woolley had a bricklayer working at the east end of the building constructing cellar windows with segmental arches, while at the west end another bricklayer used three-center arches! These men, perhaps one English trained and the other German, were apparently kept on all through the job despite their individuality. For in the attic, the supporting arch of the upper chim-

Fig. 16.3   The State House from the northwest after Charles Willson Peale's painting;
engraved by James Trenchard, *Columbian Magazine*, 1787. From about 1755 onward the
State House stood high and alone except for the wooden sheds at each end (note Carpenters' Hall cupola in the distance). The steeple shown here lasted only until 1781 when it
was taken down. Note Woolley's giant tall case clock. *(Historical Society of Pennsylvania)*

nies has a segmental arch at the east gable and
a three-center arch at the west gable.

This relaxed attitude toward variation in
detail came perhaps naturally to a generation
of builders which had to use handmade bricks.
The bricks in the State House walls vary from
4 ¼ to 4 ½ inches in width, by 8 ½ to 9 inches
in length. The exterior Flemish bond pattern
weaves to and fro as it rises. But the masons
were in control and started corners and masonry openings with their traditional closing
bonds of "queen closers" and headers alternating in courses with stretchers, or headers
alternating in courses with three-quarter
bricks.

Edmund Woolley's 1750 wood frame steeple remained above its massive masonry tower
only until 1781, when master carpenter Tho-

mas Neville contracted to take it down, its fabric having been declared unsafe. The tower
remained thus truncated (*see* Fig. 15.11) until
1828 when William Strickland designed and
built the present steeple with its four clock
faces. These clocks and their works replaced
the tall case clock and its mate which had been
modified as far back as 1812, when the wing
buildings and arcades were torn down for new
office wings designed by Robert Mills.

From 1828 onward the building's exterior
has largely remained unchanged. A Corinthian fanlight frontispiece was added, however,
in 1816 to ornament the front facade when
the city and county of Philadelphia were so
anxiously trying to rent space in the building
to help with the burden of its maintenance.
This frontispiece was removed and the origi-

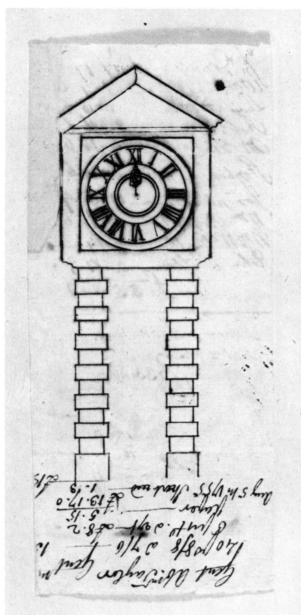

Fig. 16.4 Isaac Norris, speaker of the Pennsylvania Assembly during the 1750s, is thought to have drawn this "grandfather" clock built against the west gable. *(Norris Papers, Historical Society of Pennsylvania)*

nal doorway "restored" inaccurately in 1896, along with the reconstruction of the present arcades and wing buildings.

The National Park Service has corrected the 1896 front doorway, bringing back also the original pyramidal form of soapstone entrance steps and the 1780s layout of paving

along Chestnut Street. The copper roof deck and shingle roofing have been restored to their late eighteenth century appearance and heavy layers of paint have been removed from the exterior trim preparatory to repairs where the wood had rotted and the paint alone was holding the form. Repointing has also been done at the rough gneiss stone base. There have been numerous varieties of repointings of the brickwork done over the years, none of which are correct in appearance and many of which are of hard cement and are therefore irreversibly beyond correction.[5] When, for the Bicentennial, the Liberty Bell is moved from its exhibition spot inside the tower surrounded by the grand stair, the last touch to the exterior restoration should be accomplished. That is to remove two glass door panels for viewing the Bell and to replace them with solid wood panels, thus restoring the door of Edmund Wooley's 1750s doric frontispiece facing Independence Square.

CENTRAL HALL AND TOWER STAIRHALL

The woodwork of the central hall and around the grand stairway in the tower was completed by Edmund Woolley in the 1750's. Until that time the central hall had bare brick walls with a minimum of trim and a temporary stair at the south end. Woolley's panelling has survived except for a few minor details. In the central hall the walls are completely and richly panelled in the Doric order with pedestals, engaged fluted columns and mutule entablature (Fig. 16.7). The tower stairhall, on the other hand, was finished in the lighter Ionic order with decorative panelling flanking a Palladian window and only wainscoting to the half rail as the open newel stair climbs to the second floor (Fig. 16.8). The panelling of both these rooms was made up of several woods — principally long leaf yellow pine, Eastern white pine, Atlantic white cedar and carvings in gum or poplar. Such a variation in wood species indicates clearly the intention of painting the surfaces.

An account of the woodcarver, Samuel

Fig. 16.5   Independence Hall as it appears today with the reconstructed clock and stairway and door leading directly into the Supreme Court Room. Compare with the Birch
view, Fig. 15.11, previous chapter. *(Photo by R. Frear, 1975)*

Harding, itemizes in detail and thus dates the carved embellishments of both these rooms as well as those carvings which appear on the exterior of Woolley's now missing steeple and the exterior gable clock heads. Harding's account runs through the years 1750–1756 giving a very good idea of the pace of the

work, the product of his craft having been applied immediately prior to the painters' application of prime coats.

Indeed it is a pity that Harding's carving was meant to be covered with paint. Much is lost even with only a prime and two finish coats, but almost all its skillful relief—and

Fig. 16.6 Walnut Street Gate. By 1769 the Provincial Assembly had acquired all the square down to Walnut Street and enclosed it with a 7-foot-high brick wall coped with shingles. The garden was developed in the late 1780s by Samuel Vaughn with hillocks, serpentine walks and one hundred elm trees. Archaeological excavations have located sections of the red gravel walk as well as earlier house foundations along the street. *(Birch engraving, 1799)*

sometimes unique detail—was hidden under fifty odd coats when the Park Service undertook their removal. On the ceiling cornice above the grand stairway no paint had ever been previously removed (Fig. 16.8). A study was made of the sequence of colors used, and for the benefit of future restorationists a section of cornice along the south wall in the southeast corner was left with its full layering.[6]

For the tower stairhall cornice, Harding carved alternately acanthus leaf modillions ("Mundulyouns" as he spelled it) and flowers. Above the east and west windows the flowers are oval in form instead of round. This was done to continue the same rhythm of repeat in the shorter distance of cornice related to the windows. The moment of instruction to the carver to vary the shape of flower was brought to us 200 years later when we took down the flowers to soak off the thick paint. Here in white chalk was the sketch of the oval flower drawn on its back panel, perhaps by Harding himself directing his helper (Fig. 16.9)

One can understand finding lengths of ordinary moldings dropped behind finish panelled walls, but it was surprising and heady to find an extra length of the carved "water-leaf"

Fig. 16.7 Entrance Hall. In 1816 the front doors shown here were installed, along with a fanlight and exterior Corinthian frontispiece. The Philadelphia Centennial of 1876 added touches to the Hall. The archways at left, originally open to the Supreme Court Room, were filled in with closets in an attempt to heat the courtroom with stoves. *(Photo by W.H. Rau, 1896)*

moulding from the central hallway entablature. Unpainted, it was apparently just broken over the knee and tossed away — a fresh piece only removed by time from the carver's chisel and mallet!

None of the carved work removed so far bears the individual craftsman's signature. The tower stairhall cornice flowers in the corners of the room were all marked with the points of the compass, but the great work listed by Harding in his account goes unsigned by him or helpers. Not so the panelling. Thomas Neville, hired in the 1750s as a young carpen-

ter to help Edmund Woolley complete the central hall, penciled in bold script "T N." on a hidden rail back of the entablature. And elsewhere in the same room my architect colleagues tell me there were some penciled inscriptions describing one of the original workmen in colorful four-letter language.

Throughout the building one finds initials scratched into bricks, carved on framing members, or as plain graffitti carved and often dated right on the face of finish woodwork. Such initials are of interest, but even more so have been the dated inscriptions identifying a

Fig. 16.8   Tower Stairhall. The handiwork of Edmund Woolley and woodcarver Samuel Harding, survives almost intact. The height of the 50-foot ceiling had prevented the removal of the original paint, allowing a careful analysis by the author. A section of the original paint has been preserved. *(Photo by Frear, 1975)*

Fig. 16.9 Guide for workman. When removing the rosettes, the points of the compass were found penciled on the back side of those in the corners. This is a chalk sketch, perhaps by the master carver Samuel Harding, of the flower to be carved for this space. Normally most of the rosettes were full round. *(Photograph by the author, 1956)*

project and its craftsmen. Where we have found such records they are welcome documentation. We encouraged our own workmen to write their names and dates on hidden members. And we ourselves have left warning signs on new framing members to protect original evidence nearby. And to gain our own immortality we inscribed the dates and names of all architects, including students, and other National Park Service or contractual personalities involved in the project.

Back of and beneath the panelling of the central hallway and tower stairhall, portions of original brick paving were found, providing the size, pattern, and surface level used in the restoration of the flooring in these two rooms.[7]

Workmen's tools were also found back of this woodwork. An early iron gimlet, a small carpenter's square initialed "P. B.," apparently dropped or kicked into an irretrievable position to be found a century or two later by another generation of carpenters. Somehow, trivia of the former occupants also was found in hiding back of the central hall and tower stairhall woodwork: two swagger sticks, marbles, parts of shoes, nineteenth century tickets admitting visitors to the steeple for the view, and hundreds of oval glass waste pieces apparently left from the Peale Museum era when, for a fee, one could have his silhouette cut and framed under glass.

And to show that the debris of plastic coffee cups and sandwich wrappings of today's workmen have their early counterparts, many small animal and fowl bones, some peach pits and pieces of ceramic ware were found back of the woodwork. Each of the phases of renovation has left its debris, often dateable.

THE ASSEMBLY ROOM

The National Park Service task of returning the Assembly Room to its most historic period, 1776–1800, was difficult as this room had many times been subjected to the same well-intentioned ambition.[8]

The Assembly Room was first finished in the 1740s with the Ionic woodwork recorded at the end of the century in the painting and its engraving by Edward Savage known as

"Congress Voting Independence (Fig. 16.10)." In 1816 the room was modernized to a late Federal style, then in 1831 the architect John Haviland completely changed the room once more with his "restoration" to a fully panelled room in the doric order. The years 1896–98 brought further "restoration" of the east fireplace wall and alteration to the ceiling height and entablature.

It was essential to be able to differentiate the evidence of each of these eras. Fortunately wire nails of the 1898 period and the subtle differences in cut nails from the 1816 and the Haviland 1831 periods helped identify the woodwork. And upon the bare brick walls two

Fig. 16.10 "Congress Voting Independence" engraved by Edward Savage in the 1790s. Elements in the picture verified by physical evidence included the cove ceiling, dentil cornice, plastered side walls, window architraves without ears, the face of the east wall woodwork, its pulvinated frieze, and most important, the scallop shell frieze (Fig. 16.11) of the tabernacle frame centered on the east wall. (*Historical Society of Pennsylvania*)

varieties of plaster residue distinguished the 1816 renovation from the original plaster areas. Most important in detecting the placement of missing original woodwork was the survival of wrought iron anchors, some still containing the wrought nails which fastened the woodwork in place, and the detection of paint lines on the brick walls denoting the extremities of wood trim. The eighteenth century builders customarily erected and prime painted all their wood trim prior to plastering, thus the painters' brushes unwittingly deposited both the red iron-oxide prime paint and the light grey second coat on the bricks and mortar joints adjacent to the woodwork. The original boundaries of cornice, chair rail, window architraves, door enframement and the vertical plane and profile of the east wall panelling were clearly established by this type of evidence.

Proof that these were the original prime paints was found on the frieze board of the east wall tabernacle frame (Fig. 16.11). That particular board, adorned with the most splendidly carved scallop shell and high relief baroque entwined foliage, anonymous as it is, survived all the renovations. John Haviland incorporated the frieze in a pedimented tabernacle frame approximating the design of the missing original frame, but the proportions

Fig. 16.11 Carving of eastern white pine applied with hand wrought finish nails to a board of long leaf yellow pine. This piece of woodwork witnessed the signing of the Declaration of Independence. The darker shaded portions in the drawing indicate the exposed (or visible) parts of this frieze board as it was originally incorporated into the tabernacle frame. The lighter shaded portions indicate the additional exposed area of the frieze board as it was reused by John Haviland in his "restoration" of 1831. Arrows 1 and 2 point to the residue of the red iron oxide prime paint of the eighteenth century exposed areas; arrow 3 to the other exposed area with the full layering of paint colors still there. Haviland applied the board to the wall in his "restoration" with three heavy screws countersunk and hidden. *(Photo by Eisenmann, 1964)*

were such that some areas of the board still carrying the eighteenth century paint layers were covered over. Thus the original paint layers from the 1740's onward were protected for our identification. Some of the dimensions of the original tabernacle frame were evident from the presence and absence of the prime paint on the board, making possible the reconstruction of the frame.

Knowing the sequence of colors helped us identify other surviving fragments of original woodwork. A cornice dentil was identified. It had been wedged between an original floor joist and the brick wall, having fallen into this position during the 1816 removal of the original woodwork. It was protected from that date by an area of 1816 flooring laid above it.

Reflecting perhaps the 1816 intention to economically modernize the room for rental appeal, the carpenters and joiners of that renovation reused the woodwork they had just dismantled, turning it over and cutting their new mouldings on the reverse sides. Two splayed paneled window jambs of this 1816 period were found encased by John Haviland's 1831 carpenters back of two of their doric pilasters. The reverse sides of those 1816 jambs had the full paint layering used in the Assembly Room in the eighteenth century. Indeed these reworked fragments had lost any identification of their original use as part of the eighteenth century woodwork, but their painted surfaces had witnessed the political history we are commemorating and in themselves contained ample evidence of the original paint colors.

The structural members which were so cleverly preserved in place by the team of engineers and contractors as related in Mr. Nelson's preceding chapter, had a great deal to tell us of the Assembly Room and all the rooms. The underside of the girders displayed the boundaries of the flat ceiling with its plaster stains; the absence of lath and plaster marks indicated the projection of the missing wooden cornices and ceiling coves. The topsides of the girders and the extant original joists retained evidence of the width and butting ends of floor boards above. Pairs of nail

holes alone indicated tongue and groove boards. And the width of *butt* joint flooring was marked by ridges of dirt and water stains where the charwomen's wash water carried the dirt down between the boards. Each instance of this evidence and that found on the walls was meticulously plotted and recorded on the elevations and floor plans of the rooms.[9]

In designing the restoration details, a basic tenet was followed: where possible, develop missing details from those extant in the original fabric of the building. Where not possible, such as the detail for the marble trim of the Assembly Room fireplaces, a building in the Delaware Valley of the same period was turned to. In this instance a 1790 bill for repairs to the Assembly Room lists polishing its marble mantels and providing "2 New Keystones Moulded & polished." The grand house northwest of Philadelphia called Hope Lodge provided design precedent for just such moulded keystone facings. And a fragment of what is called "Pennsylvania blue and white marble" found lodged in a hole in one of the Assembly Room original fireplace jambs guided our choice of material.

Needless to say the Savage painting—and more specifically the Savage engraving—supplied the best guide to the appearance of the finished woodwork of the east and principal wall.

The west wall or entrance wall of the Assembly Room, is a story unto itself involving John Trumbull and his artist's license (or historian's ignorance) in portraying the room. A sketch made by Philadelphia antiquarian Joseph Sansom reminded Trumbull too late of the original appearance. The confirmation of this sketch was found by us on the wall itself.[10]

THE SUPREME COURT ROOM

There is one incident of our research on the fabric of Independence Hall which bridges architecture and an historic event. The Provincial and State Supreme Court Room across the hall from the Assembly Room had suffered some change in its architecture by the

Fig. 16.12   Assembly Room as restored in the 1960s. Anticipating reassessments of this restoration, Park Service architects made every effort to record each step with photos, drawings and written reports. *(Photo by Frear, 1975)*

age of photography. In 1896, a photograph records that it had lost the major portion of its west wall woodwork and windows (Fig. 16.13). The photograph also shows the surviving original doric entablature and tabernacle frame (with open scrolled pediment). Two years later even these were removed to make way for a modified cornice beneath a furred-down ceiling intended to cover the sagging of the framing. Much of this original woodwork was naturally discarded. Not everything, however, for when we removed the furred down 1898 ceiling all manner of carpenters' shims, wedges and hangers were found crudely fashioned from pieces of the original entablature triglyphs and the corona moldings! These were easily identified by the unmistakable pumpkin-colored paint found to be the origi-

nal finish paint on that part of the original woodwork hidden and protected by the modified cornice.

One board retrieved from this ignominious position turned out to be a board which, when glued up with others, formed the field panel of the tabernacle frame originally featured conspicuously on the west wall behind the judge's bench. The painted surface on this board stops short evenly at top and bottom, indicating the position of the applied frame. Tongues at each end indicate it originally fitted into grooves in the panelling rails.

A piece of iron was screwed fast with hand-filed screws dead center in the height of the board immediately atop the first finish coat of paint. It had a broken end and under the binocular microscope one could see all the subse-

Fig. 16.13 Supreme Court Room, looking west, 1896. The justices of the Pennsylvania Provincial and State Supreme Courts sat on a bench beneath the tabernacle frame just visible in the center of the west wall. This view reveals the "National Museum" installation of patriotic memorabilia. *(Independence Park Collection)*

quent paint layers covering the broken end. This piece of iron was most likely part of a hook. It could not have supported the Pennsylvania coat of arms, painted by George Rutter in the 1780s (now owned by Independence Park), as this canvas had always hung from the upper corners of its frame.

Remembering that the hook was mounted upon the first finish coat of paint, and that it is centered in the board, we are almost certain that the hook supported a solid carved wood coat of arms similar to the fragment of one still at Philadelphia's Christ Church. And we know that on July 8, 1776, a committee was appointed to take down the King's Coat of Arms and march it up to Center Square (to be thrown on a bonfire). In their enthusiasm they must have broken the hook.[11]

Except for the Park Service restored entab-

lature and dado, all the panelling of the Supreme Court Room north, south and east walls is original and dates from the 1740s. Its curved panels in the northeast and southeast corners cover the westernmost blind windows which flank the front entrance and those which, prior to building the tower, must have flanked the doorway to the yard in the rear.

To complete the space as a courtroom the Park Service had to reconstruct all the fixtures including the judges' bench (for which we had the original floor joist pockets) and two grand jury boxes (we also had the original joist pockets for each tier). For the petit jury box, witness stand, prisoner's dock and bar we had only documentary references. The back of the courtroom, open in the eighteenth century to standing visitors, is still used this way by today's visitors, and the archways to the Central

Fig. 16.14 Mending window jamb woodwork, Supreme Court Room. Approximately half of the original interior woodwork has survived in the building. The Park Service developed a team of skilled craftsmen for the restoration program; shown here is Carpenter Frederick C. Hoehn, Jr. *(Photo by Eisenmann, 1968)*

Hall continue to accommodate overflow crowds.

THE SECOND FLOOR

The executive office of the Pennsylvania State House—in the form of the Governor's Council Chamber—was on the second floor along with a room which served as a sometime Assembly committee room and, during the Revolution, for storage of the militia's arms. The third room was the Long Gallery across the front which took up the full length of the building (Fig. 16.16). It accommodated an occasional banquet and more often was simply a promenade for those waiting to see the provincial and state governors.

Over the years this floor of the building suffered the greatest architectural losses due to drastic changes in use. It underwent changes in room layout with resultant structural alterations. By the 1850s all the original

Fig. 16.15 The Pennsylvania Supreme Court Room restored and refitted. The authority of the Commonwealth of Pennsylvania is symbolized by the Commonwealth coat of arms painted in 1785 by George Rutter, now back in its original position. The door at the left exits to a set of steps believed to have been used privately by the judges (the door to the right is false). *(Photo by Frear, 1975)*

Fig. 16.16 The Long Gallery, ink and watercolor by Titian R. Peale, 1822. Portraits of the founding fathers border and ceiling atop cases of stuffed birds, shells and fossils. Five adjustable Argand chandeliers supplemented light from the nine windows. Peale had closed off the center doorway leading to the lobby. *(Detroit Institute of Arts)*

fireplace wall panelling had been removed, and the interior partitions were gone. The City of Philadelphia Select and Common Councils occupied each end of the building in large square rooms filled with heavy oak furniture. Fans hung from the ceiling and large ventilators drew the warm air up through vents on the roof.

Architect T. Mellon Rogers, engaged by the Daughters of the American Revolution in 1896, tried in his way to restore the original room arrangement and appearance, but his work was not well received (Fig. 16.17). In 1921–23 Horace Wells Sellers and a committee of the Philadelphia Chapter of the American Institute of Architects donated their time to documentary and physical research to improve upon previous interpretations. The

Park Service with its paid staff and time for research brought the second floor one step closer to the original appearance.

Amazingly, the old double hung sash survives in all the second floor windows, as well the original splayed and cased jambs, architrave trim and window stools or sills. The latter originally were integral with the chair rails. The chair rails had long ago disappeared, but beneath later layers of wall plaster we found one large fragment of original plaster with a crisp sharp edge where it had abutted the original chair rail. And still lodged in the remaining window sills were a few wrought finish nails which had fastened the chair rail.

The original paint colors for all the rooms were detected on the surviving window woodwork. And most fun was the discovery of a

315

Fig. 16.17 Long Gallery, T. Mellon Rogers restoration, 1896. The original room arrangement was approximated and original window sash and trim retained. But the sagging original flooring was covered over by sleepers and a new floor. The architectural detailing of the missing elements based on the then stylish Colonial Revival became a subject of hostile criticism immediately. *(1922 photograph, the Philadelphia Chapter, American Institute of Architects)*

dark brown finish color for the top surface of the chair rails and window sills with a 7/8 inch upturn of this color on the splayed jambs, trim and window frame, undoubtedly a protection to the lighter trim colors when the window sills were being washed. Again, in these rooms with the walls stripped to the brickwork, we were able to find red iron oxide prime paint lines on the brickwork not only indicating the plane of the end wall panelling, but at the entablature level we found profiles of the different entablatures precise enough to establish the levels of architrave, frieze and cornice.

In 1917 someone found in the attic of Independence Hall and deposited in the City of Philadelphia's "National Museum" collection, a small hand-carved acanthus leaf modillion bracket. The paint color of the bracket matched that found in the southeast and southwest second floor rooms.

From its wall profile the southeast room proved to have had only a simple crown moulding cornice. Thus we decided that the modillion had come from the Governor's Council Chamber—or the southwest—room where we had found profile evidence of an elaborate entablature, i.e., Corinthian, and bespeaking an office of the highest political rank.

As the Long Gallery apparently opened

Fig. 16.18   The Long Gallery restoration, completed in 1972. The exposed original framing and bare brick walls were examined for evidence of cove ceiling projection, floor board widths and door openings, fireplace hearth sizes and levels, and boundaries of wood trim. The original sash and window trim have survived and they provided evidence of the original paint colors of each room; the Long Gallery was found to have had a strong Prussian blue, a favorite color in the eighteenth century. A profile of the end wall entablature was found outlined by a red primer on the bare brickwork of the outside wall. *(Photo by Frear, 1975)*

through an archway to its lobby and in turn through another archway to the grand stair, we chose to keep the order of these spaces the same as the grand stairhall—the Ionic order. In the lobby we found the profiles of more of Samuel Harding's carved bow and bell flowers within the pilaster sunken panels of the archway leading to the grand stair. And reused in the stair to the garret are some fifteen of the original ballusters and one of the original carved tread brackets, placed at eye level for the comparison of eighteenth and twentieth century carvers' skills.

Without doubt this symbolic building will be subject to future restudy: we prepared not only many "evidence drawings," but an extensive photographic record of physical evidence and restoration progress.[12] And hopefully our Historic Structures Reports amply explain the care and attention which we expended on our assignment.[13]

It has been a privilege to work on this building—one which every participant felt and met with extraordinary interest and application. Those of us who have had an intimate view, such as no one will probably have again for some time, feel especially fortunate. This intimacy is perhaps best exemplified by one of my own experiences in seeing a hand-wrought rose nail head with what looked like a fresh blow of the hammer on its peak, having just been driven home. Next to it in the wood was an equally fresh round hammer blow. You could almost see the man wince 200 years ago!

NOTES

1. The Independence Park staff produced in-house Historic Structures Reports Parts I and II in 1959 and 1962. This work followed the excellent article by Edward M. Riley published as part of *Historic Philadelphia (American Philosophical Society Transactions, Vol. 43, Part I)*, Philadelphia, 1953. All documentary references are chronologically cross referenced on index cards now under the care of the Librarian of the Park.

2. Restoration work on Independence Hall was started by the National Park Service under the direction of the Resident Architects of the Independence Park. Charles E. Peterson held that post from 1950 to 1954. Charles Grossman then directed the work until 1959. From 1954 until 1962 the Historic Structures division of the Eastern Office of Design and Construction (located in Philadelphia) took over under the direction of Supervisory Architect Peterson and his assistant Henry A. Judd. Lee H. Nelson was the project director from 1960 and saw the work through until its completion in 1972. The Historic Structures Reports which have been produced so far have been microfilmed by,and are available to the public at, the American Philosophical Society Library.

3. Evidence (now only recorded in manuscript form and on NPS evidence drawings) indicates that the rear facade, prior to the addition in 1750 of the tower, had a centered entrance also flanked by two blind windows. Immediately inside was a stairway leading to the second floor within the south end of the central hallway. The stair started along the east wall, with a flight along the south wall and landed alongside the west wall. As far as we could tell it was relatively undecorated and stood within the bare brick walls of the as-yet unfinished entrance hall. It undoubtedly was lighted by a window above the centered south entrance. Edmund Woolley's account for adding the tower leads one to believe that a balcony originally projected in front of a window-door.

4. E.g., Charles E. Peterson, ed., *The Rules of Work of the Carpenters' Company of the City and County of Philadelphia*, Princeton, 1971, plate XXXIV.

5. Pre-1750s lime mortar "grapevine" tooled pointing was found in pristine condition back of the position of the tall case of the west gable clock, protected from 1812 to 1896 by the abutting Mills wings and, since exposure in 1896, by an attempted clock restoration of that year. For the record, a casting was also made of the original 1730s pointing of the main building. It was revealed when the grand stairway wainscot panels were removed for repair at the second floor landing.

6. There is an Historic Structures Report portion dealing with the paint colors of the central hall and tower stairhall.

7. As yet no comprehensive report has been written dealing with the history of the woodwork in these two rooms. There is a formal report dealing with the paved floor, front entrance and the small guard room in the southwest corner of the tower stairhall.

8. A good description of the Assembly Room's complicated history may be found in the *Antiques Magazine* article cited above. The Historic Structures Report on the Assembly Room deals principally with the immediate evidence for the restoration to its eighteenth century appearance.

9. The evidence drawings—as well as restoration drawings—are presently a part of the Park Service's Denver Service Center drawing files. Their title blocks reflect the periodic NPS departmental name changes for the administration of historic restoration and preservation work. But they retain their original numbers and are readily retrievable by numerical or subject matter listings.

10. In assembling this story we were grateful for the cooperation of the owner of the Sansom sketch, for the FBI analysis of handwriting confirming Sansom's hand, and for the aid of paper conservator Willman Spawn of the American Philosophical Society in dating the paper upon which the sketch was drawn.

11. This tabernacle field panel board, the Assembly Room cornice dentil and all fragments of the original woodwork or other parts of the building which have not been reincorporated in the restoration are all accessioned and catalogued into the museum collection of the Independence Park. These loose parts of Independence Hall are all stored in a special space created for their permanent security within the steeple of Independence Hall.

12. The photograph negatives have been entered in a numerical system established in the Philadelphia Eastern Office of Design and Construction Historic Structures Office and then transferred to the NPS Washington Park Preservation Office. Most of these photographs are duplicated by prints in the photographic file of the Independence Park.

13. As the deadline of our nation's Bicentennial approached, Historic Structures Reports normally required as prerestoration steps were waived to save time. The restoration of the Supreme Court Room fixtures, the second floor, the tall-case clock and the remainder of the building where little restoration was necessary have not yet been formally reported. The rationales for this work have been written and exist in manuscript form in the files of the NPS Philadelphia field office of the Denver Service Center.

CHAPTER
17

Restoring Jefferson's University

FREDERICK D. NICHOLS

Mr. Nichols, a Fellow of the American Institute of Architects, is Chairman of the Division of Architectural History, School of Architecture, University of Virginia, and is directing the restoration of its Jefferson buildings. He is author of *The Early Architecture of Georgia* (1957).

Thomas Jefferson began the actual plan of the University of Virginia at Charlottesville in 1816, when he was 73, but he had been thinking about its curriculum for 40 years. His goal was "A system of general instruction, which shall reach every description of our citizens from the richest to the poorest. . ."[1] Although he is now admired for its architectural design, he felt that planning the curriculum, choosing professors and a method of administration, and assembling a good library were as essential as the physical quarters for faculty and students, classrooms and library.

Believing that knowledge is power,[2] Jefferson was ready with a proposal in 1818 when the Virginia General Assembly approved a bill establishing a state university. The proposal, included in the so-called "Rockfish Gap Report"[3] of 1819, outlined ten courses: ancient languages, mathematics, physico-mathematics, natural philosophy, zoology and botany, anatomy and medicine, government, municipal law, ideology, and the fine arts.

Jefferson first laid out a square of about 700 or 800 feet for the campus, filled with grass and trees to reproduce the classical teaching grove, and set a mile away from the urban influence of Charlottesville. Around three sides of the square would be pavilions for the faculty, connected by students' rooms and colonnades serving as covered passages. He wrote to William Thornton, the amateur architect who had won the competition to design the Capitol in Washington, that the "Pavilions . . . shall be models of taste and good architecture, and of a variety of appearance, no two alike, so as to serve as specimens for the architectural lectures."[4] Thornton's reply suggested a pavilion at the corners of the square to mark the change of direction.

Jefferson then wrote to B. H. Latrobe, America's first professionally trained architect, who suggested that the square be changed to a rectangle to give it direction, and that a large building be placed to give the "Lawn" a focal point. Adopting these suggestions, Jefferson placed five pavilions on each side of the Lawn, spacing them closer together as they neared the Rotunda; this increased the apparent length of the Lawn, a subtle device

Fig. 17.1   The Rotunda, University of Virginia, shown before restoration, which, on the exterior, involved only the skylight portion. *(All illustrations courtesy The University of Virginia, Department of Graphic Communication Services)*

unique in America at the time. Jefferson never mentioned the derivation of the overall design, but the pavilion arrangement, the "falling terraces" and the Rotunda are similar to the great Chateau de Marly next to Versailles, where he and his friend Maria Cosway had picnicked.

On October 6, 1817, the cornerstone of the University was laid at Pavilion VII with James Madison and James Monroe, as well as Jefferson, in attendance. But not three weeks later Jefferson complained that "The dilatoriness of the workmen gives me constant trouble. It has already brought into doubt the completion this year of the building begun which obliges me to be with them every other day. . ."[5] He was forced to superintend the work, and was

beset by mounting costs: "I drew a plan . . . to demonstrate that it will not cost more than the sum allotted."[6]

Quality of craftsmanship was another problem. Previously, Jefferson had brought James Dinsmore and John Neilson from Philadelphia to work at Monticello. Now for the University he was forced to import column capitals from Italy[7] after two Italian stone carvers gave up and returned home. The poor quality of the local stone and incompetent local stonecutters are memorialized today in unfinished and rejected Ionic and Corinthian capitals preserved in the restored gardens.

The composition ornaments in the friezes of the orders were made by William J. Coffee of New York, who also made the interior

Fig. 17.2 The Böye Print, 1826, was admired by J.H.B. Latrobe (son of Benjamin Latrobe) as being the most accurate drawing of the University. This print was used to restore the Chinese railings throughout the Lawn.

frieze figures.[8] Jefferson wanted the students to live with the greatest examples of Roman architecture and "the orders" so that they would remember them on return home. He was to succeed more completely than he could have hoped; the Classical Revival swept the nation.

While Jefferson borrowed ideas freely, he usually adapted the design to make it his own. Pavilion II, although almost a literal copy of the temple of Fortuna Virilis in brick and white-painted wood, has a very different appearance from the stone original. He adapted Ledoux' villas for Madame du Barry at Louvecinnes and for Mademoiselle Guimard in Paris to create an entirely new effect at Pavilion IX. His design of the Rotunda, ostensibly a half-scale copy of the Pantheon on the exterior, exhibits the originality which ranks him as the first and greatest native-born architect. All drawings and specifications for the University are his and, except for the fact that he did not charge fees, he was a professional in every sense of the word.[9]

THE PAVILIONS

Because they were intended to teach the "classical orders" to students, each pavilion is different. Each was to house both the family of a professor and his classes, with a classroom and study on the ground floor, a kitchen in the basement and a drawing room, bedrooms and dressing rooms on the upper floor. Early on, professors' wives complained that they needed the lecture rooms as additional living space and were instead granted the space and ultimately two student rooms on either side of each pavilion.

On the upper floors the ceiling cornices were of the finest design, differing from room to room. Ornamented friezes appeared only in the drawing rooms. The ground floors had simple cornices or none at all.

Jefferson was always interested in the latest building materials and in heating devices.[10] For the ground floors he installed traditional wood burning fireplaces, but on upper ones he used the more efficient Franklin or Penn-

321

Figs. 17.3 and 17.4   The Bohn engravings, 1856: 17.3 is from the south; 17.4 from the
north, shows Monticello atop the mountain in the background. The addition to the rear
of the Rotunda was built by Robert Mills in 1853.

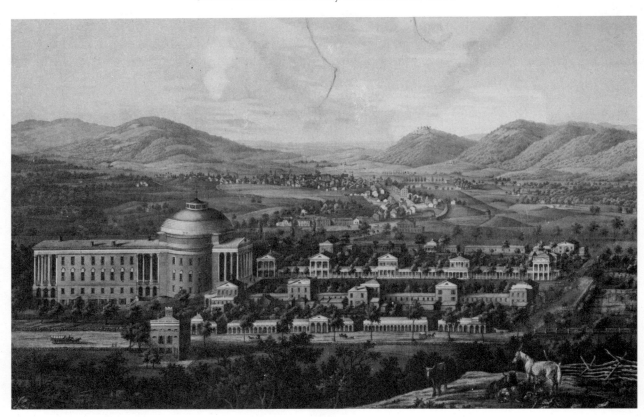

Fig. 17.5 Detail of the Bohn engraving showing the site of the Botanical Garden (far
right in Fig. 17.4).

sylvania stoves, obtained through Thomas
Cooper of Philadelphia in 1819. These were
cast iron and demountable, enriched with
classical motifs of swags and griffons. Two
of the original stoves are still to be seen
in Pavilions II and VII.

The rooms for students were rather small
even though Jefferson planned them for two
boys. But privacy was little known in the eighteenth century and this was no problem; the
rooms were still occupied by two students as
late as 1950. Each room had a single window
with Venetian blinds, for which Jefferson
spent $3,000 for the entire university. There
was a fireplace without a mantel in each room.

The most interesting architectural details in

the pavilions were the "rooflets," as Jefferson
called them. These were made up of small
gables and valleys so that from the ground
the roofs seemed to be flat. They were
covered with shingles and, unfortunately, they
leaked. In time, these roofs were covered with
an ordinary double-pitched roof and slate.

On each range, designed directly out of Palladio, were three "hotels for dieting the
students." Each had a large dining room with
fireplace and two rooms for the family which
prepared the meals. Jefferson planned for a
Spanish-speaking couple in one, French in
another, etc., so that the students would learn
the languages during meals. There is no evidence that this lofty plan was followed.

Fig. 17.6   Bohn engraving, 1856, taken from the south. The engraver ignored details of
the gardens and their walls.

THE ROTUNDA

Begun in 1823 when Jefferson was 80, the Rotunda is the finest building and the capstone of the entire University complex (Figs. 17.13 through 17.15). It was essentially completed in 1826, the year he died. The Rockfish Gap Commission, at Jefferson's suggestion, allocated the spaces within the Rotunda: the Dome Room should house the library, and the ovals should be for public examinations, religious services, instruction in music and drawing, a laboratory and similar studies.

The exterior design was adapted at one-half scale from the Pantheon in Rome—in Jefferson's view Hadrian's great classical temple was the finest of all spherical buildings. He simplified the design, put back the podium, or base, which had disappeared as the Roman street levels rose over centuries of constant resurfacing. The original eight columns were reduced to six. Interior changes were more significant: instead of a series of niches with columns lin-

ing the inner walls, Jefferson used a noble colonnade in the Composite order. It was fitting that this richest of all Roman orders should be used in the Dome Room—the largest and grandest of all Jefferson's rooms—as embellishment for the library which would impart the knowledge that its architect believed was the best guarantee of national freedom.

Jefferson even planned to use the dome for teaching purposes, the first planetarium in the New World. But he never perfected the mechanism which would have permitted the stars to be moved. Below the Dome Room there were two floors of three oval rooms each, with a free-form, dumbell-shaped stair hall.

THE GARDENS AND THEIR WALLS

Behind each pavilion were gardens enclosed by serpentine walls which Jefferson had probably seen on his tour of English gardens with John Adams in 1786.[11] In England the

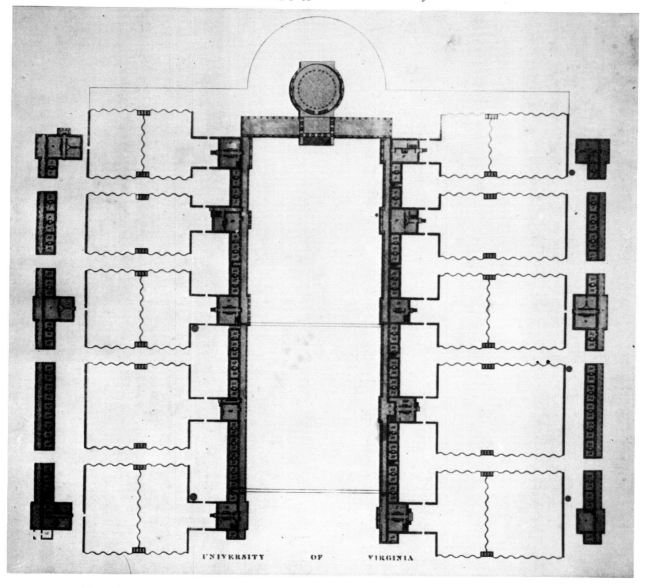

Fig. 17.7   In 1824, Jefferson used this drawing by Peter Maverick, a New York engraver, to send to students so they could choose their rooms, numbered, starting from left nearest the Rotunda, then right, left again, etc. The hotels follow the same system. The serpentine walls are clearly marked here. Student rooms, still in use, are between the pavilions and the hotels.

walls were used to catch and hold the warmth of the sun, hence forcing young plants. Jefferson thought they were beautiful, even though the mortar froze easily and they had to be rebuilt every few years. Being only one brick thick, they did stand up in strong winds because of their curvilinear design.

Midway between the serpentine walls were privies, or necessary houses, for students and faculty. Jefferson also left specifications for a meat house for each family. The gardens were not laid out in detail; each tenant was to plant as he saw fit. During construction of the University, a large vegetable garden was cultivated for the use of laborers and slaves.[12]

RESTORATION

Over the years the Lawn buildings remained in use, receiving haphazard care and

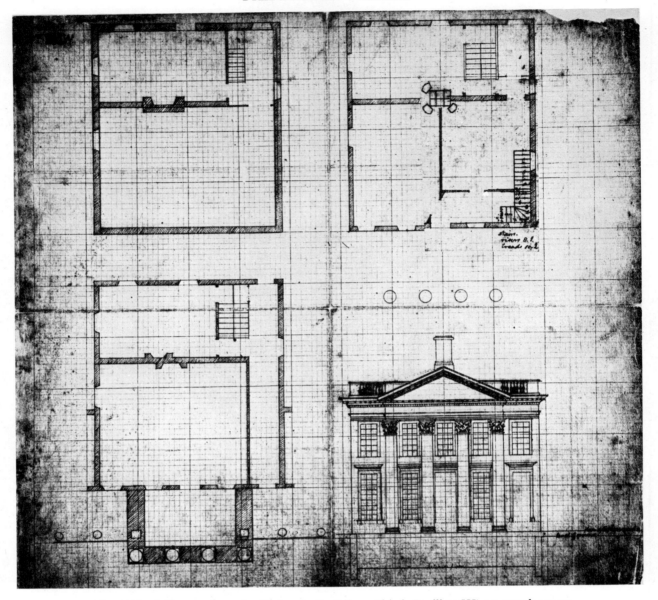

Fig. 17.8 Jefferson drew the University buildings (this is Pavilion III) on graph paper marked in tenths, which he had learned to do in Paris. He showed the elevation and floor plans, complete with the Franklin stoves ordered from Philadelphia in 1819, to heat the second floor. The order is Corinthian, Palladio.

many alterations. It is hard to understand today why they were treated so casually, but until 1952 there was very little interest anywhere in scientific restoration. One factor was that Jefferson was regarded as a great statesman abroad but a neighborly amateur architect at home. Local tales persist about his forgotten chimneys, the leaky roofs and the missing stairs at Monticello. His brilliance as

an architect, first established by Fiske Kimball,[13] has not yet been fully explored.

Another factor was that the University collected only token rents from the Lawn tenants, with the understanding that the professors keep up the interiors. Each tenant thus abandoned all work as retirement approached and gradually the successors could not catch up. Most of the additions had been made be-

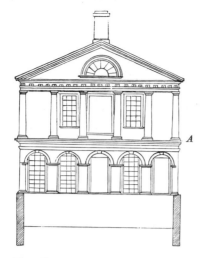

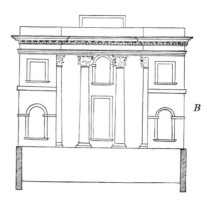

Fig. 17.9a Pavilion VII (now the Colonnade Club) was the first building erected at the University. The cornerstone was laid on October 6, 1817. (Doric Order of Palladio.) Jefferson drawing.

Fig. 17.9b Pavilion VIII (Corinthian order from the Baths of Diocletian, Rome) showing the portico *in antis*. Jefferson drawing.

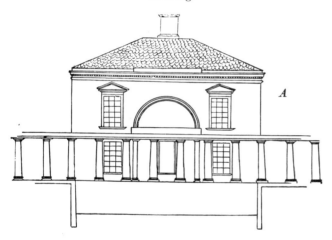

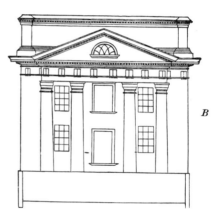

Fig. 17.10a Pavilion IX (Order from the Temple of Fortuna Virilis, Rome) with the exedra, a half dome which Jefferson had admired in Ledoux' work in Paris. Jefferson drawing.

Fig. 17.10b Pavilion X (Order from the Theater of Marcellus, Rome) with the large parapet recommended by Jefferson. This is the only Roman order without a base. Jefferson drawing.

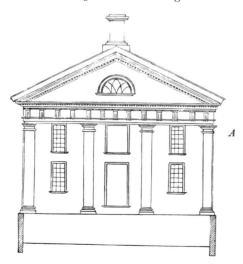

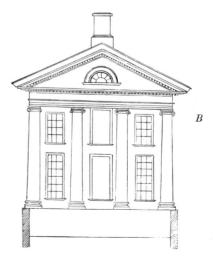

Fig. 17.11a Pavilion IV (Order from the Temple at Cori, south of Rome). Jefferson drawing.

Fig. 17.11b Pavilion II (Order of the Temple Fortuna Virilis, Roman Forum). Jefferson.

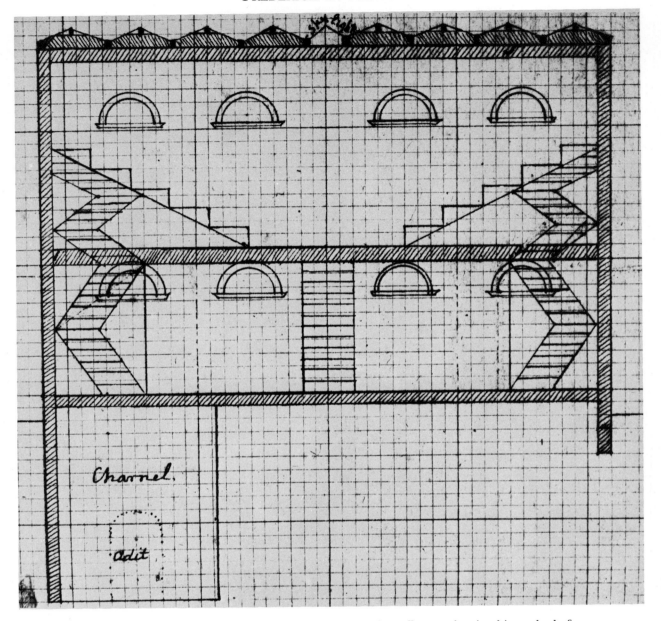

Fig. 17.12   Section of the Anatomy building, drawn by Jefferson, showing his method of building flat roofs or "rooflets" with a series of small valleys. The Anatomy building was destroyed in 1938.

fore the Civil War and subsequent modernizing of kitchens and baths was difficult.

At the suggestion of Charles Smith, Chairman of the Art Department, and Vincent Shea, Comptroller, the author was appointed in 1952 to renovate Pavilion II. Decisions made then set the pattern for the entire restoration. It was decided to keep all the post-Jefferson additions and put the kitchens, baths

and other modern services in them. We were not creating museums, but restoring the exciting series of Jeffersonian interior spaces, some of which had been completely obliterated.

Fortunately for the restoration work, the University had an aged but excellent plasterer, Luther Dorrier, who took molds of all the ornaments in the drawing room friezes and reproduced them in plaster for use where

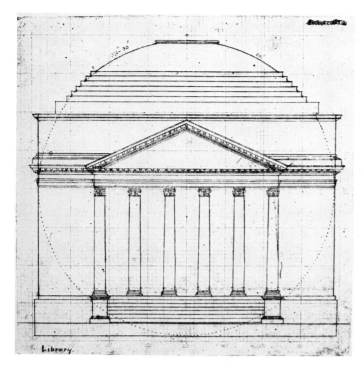

Fig. 17.13   The Rotunda, as drawn by Jefferson. Dotted lines represent his spherical concept.

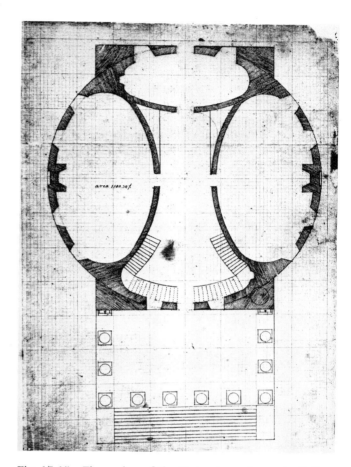

Fig. 17.15   Floor plan of the Rotunda as drawn by Jefferson. Note the penciled lines which he added after discovering that the acute angles would make awkward spaces. The "window" openings on the exterior of the fireplaces were actually fake windows to preserve the external rhythm.

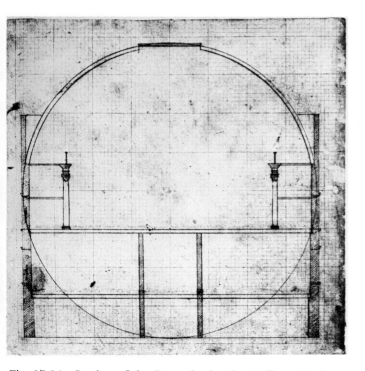

Fig. 17.14   Section of the Rotunda showing Jefferson's plans for three floors with a Corinthian colonnade in the Dome room.

necessary. Most of the original ornaments, made by William J. Coffee of New York City, were a terra cotta substance, a kind of hard plaster. Some, probably the original models from which the others were made, were metal. Coffee had worked for Jefferson on earlier buildings.

Throughout the Lawn, the Chinese railings had long since been replaced by a variety of other styles, usually in iron. These are now being restored to the original design and will add the final unifying touch which Jefferson had specified.

Pavilion II

Pavilion II, which had been the Medical School in the mid-nineteenth century, had undergone many changes and was in very bad condition. The splendid proportions of Jefferson's interior had been totally lost; a large hall had been added across the front, a large arch cut between the front rooms, a Greek Revival mantel substituted for the original in the school room, and the free-standing stair had been pushed against the rear wall. A one-story addition at the rear contained a kitchen, bedroom and bath.

The first floor plan was returned to the original with a small entry in the center of the front. The small stair to the basement was removed, along with vestiges of wallpaper printed with a rusticated stone design. Fortunately, the old cornices had merely been moved during previous alterations, and they were carefully replaced in the original locations. Three generous windows were provided when the rear wall of the schoolroom, which had been added to form a pantry, was removed. The earlier design of his typical fireplaces was used for the mantel, and the arch closed up. The restored room is one of the most exciting of all Jefferson's spaces.

The study, or dining room, on the north was made smaller, making the fireplace wall asymmetrical. Some community protest resulted, such is the power of symmetry. The new wall was placed on the old headers where it was shown on the drawing, and the free-standing stair, one of the most dramatic of all

Jefferson's designs, was replaced. A modern kitchen, bath and dining room were provided in the rear wing.

A Federal mantel with pilasters and a Gothic Revival coal grate were found in the basement. As the original fireplace had been plastered up, these were used in the study. All trim was painted the original white, with dark brown baseboards, and the doors were painted and grained. The existing wall colors, perhaps selected by tenants rather than the architect, were yellow, gray, Wedgwood blue and buff. In the basement kitchen there was whitewash with some evidence of an iron oxide paint thinned with buttermilk for economy.

On the second floor, one original Franklin stove used by Jefferson was placed in the drawing room, borrowed from Monticello. The other two stoves are American Empire design with brass ornaments. A large dressing room was converted into two baths.

Pavilion VI

One year later Pavilion VI was restored. It required much work, having been used as classrooms for the Romance language department. The small vestibule and the fireplaces on the main floor had been removed to make a center hall and a large room had been added in the rear; fortunately, the garden doorway was reproduced in the new room and the original left untouched. Upstairs, the two bedrooms with angle fireplaces on the north were untouched, but the chimney had been removed in the drawing room to enlarge it, and the Doric cornice had been patched to cover the change. This had happened about 1900, when the French government had given a series of frescoes to the University in honor of Jefferson. Painted by Robert and Marthe Saint Hubert, they show such scenes as Layfayette being greeted by Jefferson in 1824 and cover all four walls. An inlaid floor had been put over the original. The decision was made not to return this room to its first phase but to retain the gift of the French government.

All the other Jefferson rooms were restored. The drawing room mantel, which had been taken to the Observatory by the wife of an astronomy professor, was returned to its original location. During repairs to the stair, we found that Jefferson had used iron rods, exactly like the wooden ones, to reinforce the balustrade at every third or fourth step.

Another unusual structural detail is found in Pavilion VI and elsewhere on the Lawn. At Monticello, the floors were insulated by brick nogging laid over strips nailed to the sides of the joists. At the University, the same treatment was used except that red clay was used for insulation instead of brick.

A modern kitchen and pantry were installed on the main floor, leaving space for a sunny dining room overlooking the garden and Monticello mountain.

Pavilion III

Pavilion III, which had been used by the Graduate School, was next. As with IV, it had not been added to and we had to provide a dining room in the old basement kitchen. We removed a pantry and restored the separate door to the school room from the colonnade. The principal change on the main floor was reversing the stair to the basement; previously the basement entrance was under the landing and one had to duck to enter.

While the school room has the usual Jefferson plain fireplace, the rear study has an Empire one with pairs of pilasters—the coal grate is not original. The arch separating the front hall from the stair hall is directly out of *Palladio Londinensis*, by William Salmon, London 1734, a type popular in Virginia until the Revolution but very *retarditaire* by 1819. In the basement, small French windows had been added in the 1850's during the Tuscan Revival, and we restored the original double sash. A handsome outside batten door with a small pilot[14] door is unique on the grounds.

On the second floor, one bath had been installed in the original dressing room on the northwest corner, so a second bath was added in space taken from the northeast room. The cornice in the drawing room is similar to the one in the hall at Monticello, from Desgodetz's book of the Temple of Vespasian in Rome.[15] As it was only possible to secure four

Franklin stoves, the plain fireplace here was allowed to remain.

Pavilion IX

At Pavilion IX, the spacial form of the exedra, or semi-dome with a screen of columns, at the entrance portico had been destroyed by use of a fixed blind, or jalousie, to cover the top of the portico, which we removed. On the drawing for this house is the name "Latrobe" written in pencil in Jefferson's writing. It probably means that Latrobe had suggested the form of the house and the low roof, but Latrobe himself never seems to have used this motif.

On the interior an arch had been cut in what is now the dining room, and a curved dummy wall had been installed in the drawing room to make the east end symmetrical, as the curve of the semi-dome intruded into the room. A Victorian study had been added to the southwest corner of the house. The kitchen had been moved to pantry space on the main floor and in the mid-nineteenth century a two-story gallery had been added to the garden side. During the restoration we divided the gallery to expand the kitchen and faced the exterior wall to match the existing privacy trellis. Painted black, the trellis seems to disappear behind the white columns of the garden facade and the kitchen extension is little noticed.

Baths were installed in the addition off the gallery, but the drawing room with its Doric cornice, and the pair of bedrooms were intact. Here, as in all the pavilions, we installed freestanding movable wardrobes as no closets had been provided by the architect. Although closets had been popular in Virginia before the Revolution, clothes presses had replaced them during the late eighteenth and first half of the nineteenth centuries.

Pavilion IV

Pavilion IV originally was the only one entered directly through the drawing room. An entrance hall and a bath had been put into the north end of the room and we left this arrangement to preserve the much needed bath.

Early in the twentieth century a Colonial Revival portico had been added to the garden side, and this was removed. A new kitchen was installed in the basement, and both it and the dining room open on a broad brick terrace.

We installed two baths in one large dressing room on the second floor. The drawing room has a "cushion" frieze in the entablature, undoubtedly the work of a *retarditaire*, or old, carpenter, as it is in the style of Virginia 100 years previously. Also in this room is a rococo Victorian Franklin stove, dated 1857.

Pavilion V

A kitchen wing with a large bay window had been added to the rear of Pavilion V, early in the twentieth century. Long before, a large wing had been added by one Professor Tucker with the provision that at the death of his mother-in-law, for whom it was intended, it would revert to the University. The family turned out to be unexpectedly long lived, so at their deaths it was ruled that all occupants had to leave a pavilion when the professor reached retirement. There is now one family in Pavilion V and one in the Mother-in-Law annex.

A new kitchen was installed and a lavatory placed at the end of the entrance hall. Repairs were made in the woodwork, whose most unusual feature is the pilastered arch in the hall, separating it into two areas. One rear window had to be opened up. Upstairs, repairs were made to the woodwork and to the two small fireplaces in the north bedrooms. The plumbing was modernized.

The outstanding features of Pavilion V are the stair which, like others on the Lawn, is slightly tapered to make the winders less dangerous; a beautiful dentil cornice in the upper stair hall; and a splendid Doric entablature in the drawing room. The school room, with its noble proportions, is one of the handsomest on the Lawn; it has six windows, two triple-hung to the floor.

Pavilion VIII

Because Pavilion VIII has been the President's Office continuously, it has not yet been restored. It does need repairs because an old house cannot be used as an office for a long period of time without considerable damage. The pavilion has a portico *in antis,* originally containing only a central bridge from the second level to the upper terrace. The entire level, however, had been filled with a floor which will be removed during restoration. Two rooms had been added on each floor, but the Jefferson rooms are largely intact.

Pavilion X

About 1870, a large room had been added to the rear of Pavilion X in the "Bracketed" style, and an entrance hall built. Aside from some repairs, little work was needed on the main floor Jefferson rooms. A new kitchen and lavatory were put into the addition, and a French door and stairs were provided to give direct access to the garden. In the basement, those brick floors which had not been cemented over were cleaned and repaired. On the upper floor, the middle bedroom on the north was opened up; a sewing room at one time had been put in part of it.

Pavilion VII

Now the Colonnade Club, Pavilion VII had been expanded in 1846 and in 1912, when the reading room and several bedrooms were added off the garden at the height of the Colonial Revival period. The south end of the schoolroom had been partitioned off into a useless narrow hall, and when the partition was removed the handsome Jefferson room was revealed. The old cornices, which break to allow floor-to-ceiling triple hung windows, were moved back to their original locations.

On the upper floor the dressing room was rebuilt off the front bedroom. The drawing room, with its octagonal end, was restored and the door reopened to the portico. The Wedgwood blue color and the Franklin stove were restored.

Restoring the Students' Rooms

Over the years the students' rooms had undergone various changes. Closets had been added; and mantels with shelves installed in style ranging from Greek through Eastlake to the Colonial Revival. These additions were removed in 1960; the walls replastered where necessary and the fireplaces finished with backbands. There were no cornices, but we added simple picture moldings to prevent the students from driving nails into the plaster. Two wardrobes, one containing a sink, were installed. After much deliberation, very plain furniture of solid walnut was chosen. As it is an honor and a privilege to live on the Lawn, even though the baths are in the basement or in other buildings and there is room for only 100 students, they have taken great care of the furniture. Today it is still in excellent condition, and shows no signs of dormitory fatigue.

Restoring the Hotels

Each hotel contained two rooms, a large dining hall and basement kitchen. The most serious changes had occurred on the West Range in the early twentieth century when a central campus heating system was installed and large pipes were run directly through the basements. The kitchens are otherwise intact, and it is hoped that before long a more modern system can be installed, freeing the kitchens for restoration.

Hotel C, long the home of the Jefferson Society, a debating group, was restored as one large room to suit the Society's needs; but if the Society should move, the partitions in C can easily be replaced.

On Christmas Eve, 1920, Hotel A burned, but only the floors and interior partitions were destroyed. It was restored in 1965 for the 40th anniversary of the *Virginia Quarterly Review*. Work included rebuilding the central chimney with its three fireplaces on the main floor. Fortunately the great kitchen chimney in the basement had survived, so it was easy to reconstruct the upper part.

As usual with Jefferson's houses, the basement had full windows on the south due to a change in grade. Once the partitions and fireplace were restored, a plaster arch was placed under the chimney over the hall, a detail to be found in most of the pavilions. This allowed central chimneys instead of external ones which would have spoiled the exterior cor-

nices. The interior woodwork needed little restoration, but the west doors had been badly hacked, as if someone had once intended to plaster over them. The great interior cornice had not burned and there was enough of it to complete the large room. But the other cornices had to be run in plaster, as the full Tuscan entablature in wood today is violently expensive. Most of the flooring is old, and this is the only completely restored hotel to date.

Hotel B, or Washington Hall, was gutted and enlarged in 1870 by the Washington Debating Society. Even though the diminutive portico and the pedimented windows are Jeffersonian, they also indicate the style of the Second Empire in France which was popular at the time. This hotel still needs restoration.

At Hotel E, only the three main rooms on the first floor were restored by repairing the great Tuscan cornices and the fireplaces. The fourth room had been stripped of its woodwork and used as a kitchen when the building had been a residence. The lower floor had been converted to a bar and dining room. It now houses the Colonnade Club Annex and could be restored.

About 1950, Hotel F was converted into offices as Levering Hall. This is the only two story hotel, with a single room on the main floor with a fireplace on the west wall and a stair in the northeast corner. Upstairs were rooms for the family. It has not been restored to the Jefferson design. Behind this building is a former summer kitchen, known as the Cracker Box, which had been converted into a garage in the early twentieth century. Although built after 1826, this kitchen has the finest eighteenth century type chimney left in Albermarle County. It is of buttress design, with large brick ramps building up to the flue.

Restoring the Colors

Unfortunately, Jefferson left few records of colors for his buildings. Even Monticello has required extensive research; its dome retained bits of white paint, and the brick below the watertable was once painted white to emphasize the podium[16] although it may be that a white grout had been used, or a substance called Roman Cement, in order to waterproof the basement wall.

As the old dome at the Rotunda had disappeared in the fire, it could provide no physical evidence, but evidence for a white dome turned up in several Jefferson letters.[17]

All of the exterior woodwork seems to have been white, except for the door frames of the students' rooms which were originally sanded and painted to look like stone. Jefferson believed this treatment would wear better than non-sanded woodwork. The blinds were dark—almost black-green, and the University now uses black with a small bit of cadmium yellow to produce this color. By avoiding the use of blue to produce the black-green, the result is a color which will not fade.

Again, the interiors were generally white with some exceptions. Buff was used in Pavilion II, Wedgwood blue with a lighter blue ceiling in several others, some strong clear yellows and several instances of gray. In Pavilion II, traces of pink (made with iron oxide red) were found in the ground floor kitchen.

All interior woodwork was white, and all interior doors painted and grained. As painting and graining is almost a lost art, a painter had to be trained in this craft for the restoration. All craftsmanship is under the direction of Alton Leake.

Restoring the Rotunda

The Rotunda did not long remain as Jefferson built it. In 1853, pressed for space, the University called upon Robert Mills to design an annex on the north end containing a meeting hall, classrooms and laboratories. Mills as a young man had been invited to Monticello to study drawing and architecture with Jefferson, and had later worked for Latrobe. It was in the meeting hall that fire, caused by faulty wiring, broke out in 1895, destroying the annex, the dome and interior of the Rotunda.

The faculty voted to restore the Rotunda to its original splendor but, unfortunately, were unable to prevent architect Stanford White[18] from transforming it into his own version of Pantheon.[19] To start, White eliminated the sec-

ond floor to make a more convenient library, designed a Renaissance dome sheathed in copper and decorated the interior in plaster reinforced with iron, with railroad rails for lintels. To make the structure fireproof, he used tin for all exterior trim and lined the damaged brick walls with Guastavino tile.[20] He also built a north portico, which Jefferson had planned but not built due to lack of funds. The Rotunda was used as a library until 1938, but had no purpose thereafter and was allowed to deteriorate.

In 1952 a committee under the chairmanship of Frank Berkeley Jr. obtained permission from the Art Commission to restore the Lawn buildings. In 1955 President Colgate W. Darden Jr. requested that Pavilion VIII, his office, be enlarged to accommodate the Board of Vistors; instead, it was suggested that the Rotunda be restored and the Board be given the use of one of the splendid oval rooms. In 1959, when Edgar Shannon became president, Mr. Darden asked him to pursue the restoration. Funds for preliminary plans came from a state grant and in 1966 came funds for restoration and repair of the Jefferson buildings from the estate of Cary Langhorne. These were matched in 1972 by a special HUD grant and Ballou & Justice of

Fig. 17.16 Photograph, believed to be previously unpublished, was taken by an unknown photographer just after the great fire of 1895, but before the parapet collapsed.

334

Fig. 17.17 Stanford White rebuilt the Rotunda without the top floor, attempting to more faithfully recreate the Pantheon and provide a "more useful" library. He used iron railings, fire resistant wall tile and plaster reinforced with iron. He also included Poe with the great classicists Homer, Bacon, Tacitus, Milton, etc.

Richmond were chosen as architects, and Robert E. Lee, contractor.

We were fortunate in having all the drawings and memoranda necessary—these are the most documented buildings in America—to recreate Jefferson's original designs as far as practical. Investigation of the fabric revealed no surprises except for some of White's work. He had left few drawings or photos, and by mid-century no one could remember what he had actually done.

We made no attempt to obliterate all of White's work. The walls, which are Jefferson on the outside and White inside, were left intact, and little structural work was required for the foundations and walls. White's exterior railings, originally concrete but converted to marble in 1939 by the WPA, and his terraces remain. The iron ring supporting the dome walls was also kept, but the oculus was rebuilt to Jefferson's design, this time in steel with triangular skylights. The dome will be covered, as Jefferson's was, with tin painted white. He had thought that tin, new to Virginia in 1810, would last 100 years but the roof soon leaked and a second layer of tin was installed over the first.

The returns of the porticos on the north and south, and chimneys on the east and west act as buttresses for the dome. In order not to interrupt the rhythm of the windows, Jefferson had installed fake windows behind the

Fig. 17.18   The Rotunda railings by Stanford White, originally done in concrete, were reproduced in marble by WPA workers in 1939. This and other White details have been retained in the current restoration.

chimneys, backed with painted black boards; we are doing the same. Inside, steel and concrete floors were installed and the main stairs restored. An elevator is being placed in one of the two small circular stair openings (we found evidence, charred stair ends embedded in the brick, that only one of these openings had ever been fitted with stairs). The walls have been replastered except where the Guastavino tile had been used, and the new columns are hollow wood with caps of plastic. If we had been "authentic" and used carved locust caps, they would have cost an additional $120,000 and also might have caused some confusion for future architectural historians.

The doors are painted and grained, and the rooms painted white. Jefferson had planned to use a sky blue dome with gold stars and a moon but he never completed the drawings for the machinery which would have allowed its use as a planetarium, so we made no attempt to follow his original plans.[21] The brick in the basement was restored in the typical herring-bone pattern.

Air conditioning, sprinklers and electrical systems will bring the Rotunda to current standards, and a large pantry under the stairs will allow the University to continue the custom of serving banquets, a tradition begun by Jefferson in 1824 when he gave a dinner there for Lafayette. The lighting has not yet been fully designed, but there will be adequate electric illumination supplemented with argand lamps to recreate Jefferson's experiments with this type at Monticello.

Fortunately, quality labor and craftsmanship are still available in Charlottesville. There has been no pressure to introduce new and cheap building methods, so plasterers, extinct elsewhere, still carry on father-son businesses. Painters and carpenters also have not been spoiled by assembly line methods, although for the fine woodwork the specialists of the Knipp Company of Baltimore have been called in.

Fig. 17.19 The Oculus being restored to Jefferson's original design, with triangular skylights framed in steel rather than the original wood. Photograph taken November 18, 1974.

The entire Rotunda project cost $2.3 million in contrast to Jefferson's expenditure of $55,000 for the Rotunda and $250,000 for the entire University complex through 1826.

Fig. 17.20 The Jefferson Neoclassical dome has replaced White's Renaissance ogee dome. The covering will be tin painted white, which research indicated was the original.

Work will be finished and everything complete by Founder's Day (Jefferson's birthday), April 13, 1976.

CONCLUSION

The restoration of the Jefferson buildings at the University of Virginia has not been done with the intent to duplicate the originals. Instead, we have preserved the use of the buildings by keeping additions required by subsequent generations and further modernizing them for the benefit of present and future occupants. The Lawn is not a museum. It is a vital part of the University and must continue to serve its occupants.

The thrust of the restoration has been to recapture all of Jefferson's masterly architectural spaces. Jefferson, who studied with the avante garde architects of his day while he was the U.S. minister in Paris, was intently interested in form, which resulted in what we today call "spatial experience." He seized upon classical Roman forms not because they represented any "ideal form for a democracy" but because they were in themselves ideal, perfect and beautiful models. He did not hesitate to adapt them to his own purposes.

In my view, the Rotunda is one of the greatest neoclassical buildings in the world. It represents the goals of the visionary French architects in which a perfect sphere is enscribed in a cylinder and made useful by two floors which provided six oval rooms. Until the restoration of the Rotunda, only a handful of specialists knew of its importance in this light. Now, the Rotunda will not only be more useful to the University and more beautiful, but it will be available to teach a lesson to all Americans. To some, it will be only a "spatial experience." To others, it will be a reminder that centuries before Buckminster Fuller, Jefferson built a dome in the "Delorme method," using short pieces of wood formed into ribs pinned together.

This view of Jefferson as an avant garde form giver will be explored in a new book by the author.

Fig. 17.21   The University today, seen from the north. What was to have been an open vista to the south has been closed off by White's Cabell Building. The variety of gardens can be seen clearly in this winter photograph.

NOTES

1. TJ to Joseph C. Cabell, January 14, 1818. P. L. Ford, ed., *The Writings of Thomas Jefferson*, New York, 1892–99, Vol. 10, p. 102.

2. William Thornton to George Tich-
nor, A. A. Lipscomb and A. E. Bergh, eds., *The Writings of Thomas Jefferson*, Washington, 1903, Vol. 17, p. 396.

3. The Rockfish Gap Commission was set up when a bill was passed for the

establishment of a state university. It not only had the Governor of Virginia as a member, but also Madison and Jefferson. It first met on August 1 at the Rockfish Gap in the

Blue Ridge mountains, a central location in the state.

4. TJ to Thornton, May 9, 1817. Lipscomb and Bergh, 17:396.

5. TJ to J. C. Cabell, October 24, 1817. Lipscomb and Bergh, 19:250–1–2.

6. *Ibid.*

7. The imported capitals, ordered through the American consul in Leghorn, were Carrara marble from the famous quarry near Florence. The page and plate numbers of an example can be found in Giorgio Leoni, ed., *The Architecture of A. Palladio,* London, 1770.

8. Other workmen were Daniel R. Clarerly of Richmond, painting and glazing; John Garman, stone work; John M. Perry, who sold most of the land to the University and submitted bids for brickwork and carpentry; William Phillips, brickmaking and laying; Matthew Brawn, who did brickwork through S. J. Harrison; and Joseph Antrive, plasterer. Francis L. Berkeley, Jr., Constance Thurlow, Anne Freudenberg and John Casteen, comps., *The Jefferson Papers of the University of Virginia,* Charlottesville, 1973.

9. Monticello was the first house in the United States to be built from complete working drawings.

10. He owned the books of Count Rumford, and installed Rumford fireplaces at Monticello. *The Writings of Count Rumford,* Boston, 1870 with a Memoir by G. W. Allen.

11. One architectural historian has counted 45 serpentine, or "crinkle crankle" walls as they are called in England, in Suffolk alone. While he loved the great landscape gardens of England, Jefferson abhorred its architecture. He preferred the buildings of France, especially those that were being built at the time he was there.

12. Restoration of the Lawn, the gardens and their walls was begun in 1948 by the Garden Club of Virginia. Alden Hopkins was the designer, with Edmund S. Campbell and Edwin Betts as consultants. Decorative gardens were placed behind the pavilions and the kitchen gardens have fruit trees and grass beds to suggest vegetable plantings next to the hotels. The restoration was completed in 1965 at a total cost of over $250,000. For an account of Jefferson's interest in landscape, see Nichols, "Thomas Jefferson, Landscape Architect", program for the Restored Gardens Presentation, University of Virginia and the Garden Club of Virginia, 1965.

13. Fiske Kimball, *Thomas Jefferson: Architect,* Boston, 1916.

14. A double door with one leaf so hinged to be used alone.

15. *The Ancient Buildings of Rome,* Portico of Antoninus & Faustina at Rome, English Translation from the original French, London, MDCCXCV, Vol. I, plates III–IV.

16. In conversation with architect Milton Grigg, June 2, 1975.

17. January 26, 1809, from Edmund Bacon (ms., Massachusetts Historical Society, Boston). "N. B. Burill tells me you was to send some spanish Brown and oil for painting which he wants. He says he has used 300 pounds white lead on top of the house that he did not give one coat. He also says he will need 50 to 100 pounds dry lead, Burill says it will take 60 gallons of oil . . ."

January 30, 1908, to Edmund Bacon (ms., Huntington Library, San Marino, California). "I shall immediately order the oil & paints from Philadelphia."

February 15, 1809, to John Taggert (ms., Massachussets Historical Society, Boston). ". . . I must trouble you again on the subject of paint & oil by asking the favor of you, to send for me to Messrs. Gibson & Jefferson at Richmond, 60 galls. linseed oil & 100 lbs. of dry white lead, by the first vessel going to Richmond, after your river shall be open, & to note to me the amount which shall be duly remitted . . ."

March 24, 1809, to George Jefferson (MHi). ". . . my packages from Washington must be now near arriving . . . there will also be from Mr. Taggert of Philadelphia a cask of linseed oil & keg of white lead . . ."

18. The MacDonald Brothers of Louisville had been called in earlier, but White soon replaced them.

19. White also destroyed Jefferson's open vista to the south by building Cabell Hall and its flanking buildings.

20. The Guastavino Brothers of New York had developed this, based on ancient Roman brick 1 inch thick, 12 inches long and 5 inches wide but made of terra cotta.

21. Bills were found only for white lead and Spanish brown used on the doors.

CHAPTER
18

Training Architects for Conservation in Britain

DEREK LINSTRUM

Dr. Linstrum, a Fellow of the Society of Antiquaries and an Associate of the Royal Institute of British Architects, is Director of Conservation Studies, University of York. He is the author of *Sir Jeffry Wyatville, Architect to the King* (1972) and many shorter studies.

"They order," so we have been told, "this matter better in France." Some of us might feel tempted to agree when we hear of ninety-five students attending lectures at the Palais de Chaillot in Paris as part of their work for a Diploma of Higher Studies for the "Knowledge and preservation of the ancient monuments of France." Indeed, the French system, which originated in the Ecole des Beaux Arts, has more than once been held up as a model for the training of specialist architects in that branch of the profession. Ironically, the whole subject of architectural education in France has been in a critical situation since 1968, and there is no assurance that this system of specialisation will continue; yet, whether or not it is serviceable as a model seems to depend upon three essential conditions which exist in France.

First is a recognised necessity to qualify by passing a reputedly difficult examination in order to be able to compete for one of a limited number of appointments as Architecte-en-Chef when it becomes vacant; this provides a definable incentive which is lacking in Britain. Secondly, it depends upon an acceptance of the general principles of a traditional method of restoring historic buildings. Thirdly, it requires the support of the resources and experience of an established state department. The post of Inspecteur-Général des Monuments was created as long ago as 1830, and the Commission des Monuments Historiques 7 years later.[1]. One has only to remember that from 1834 to 1860 Prosper Mérimée, the distinguished holder of the appointment of *Inspecteur,* as well as being the author of *Carmen,* was occupied with extensive journeys and voluminous, detailed reports of France's historic buildings, and that he was followed by Viollet-le-Duc who was no less actively restoring most of them,[2] to become aware of the long tradition upon which the system of training in France has been based. When Viollet-le-Duc wrote in 1866 that famous definition of the restoration of a building as "not just to preserve it, to repair it and remodel it, [but] to reinstate it in the complete state, such as it may never have been in at any one moment,"[3] he was at least laying a firm foundation for suc-

340

ceeding generations, although I would hesitate to assert it was still being taught without some modification by my friends in the Palais de Chaillot.

The English tradition is less easy to define, but undeniably different. If we go back to 1715 we find Nicholas Hawksmoor advising, "What ever is good in its kinde ought to be preserv'd in respect to antiquity, as well as our present advantage, for destruction can be profitable to none but Such as Live by it."[4] And again, he recommends a practical attitude when he uses an argument that has been repeated many times since 1715; an old building may well be "Strong and durable, much more firm than any of Your New buildings because they have not ye Substance nor Workmanship." In 1723 James Gibbs made a similar observation about the medieval tower of All Saints, Derby: "the great old gothick Tower . . . is a very good one, and stands firm and strong, which was the reason it was not pulled down."[5] And his design for the new body of the church was said to have possessed a "plainness . . . [which] tenders it more suitable to the old steeple.[6] To this early eighteenth-century common sense and respect for the old, we can add Sir John Vanbrugh's plea for Old Woodstock Manor[7] because, in its partly ruined state, it was evocative and picturesque; perhaps it is not being too fanciful to suggest that an English attitude to historic buildings, based on a practical view tinged with a touch of romanticism, was beginning to manifest itself at about the time of the inauguration of The Philadelphia Carpenters' Company.

The setting up of the Inspectorate of Ancient Monuments in France did not pass unnoticed in England, where the concern for the treatment of historic buildings was essentially a private endeavour stimulated by enthusiastic historians and topographers. There was no official recognition of this concern, and no comparable appointment in England to that of Mérimée in France. In 1841 a Select Committee discussed setting up a similar scheme, and 3 years later the Minister responsible noted that his attention had been drawn "to the rapid destruction of many ancient ruins and other Architectural memorials of our History which is going on—partly for the sake of improvement of Properties, partly from ignorance of their value and other causes, but by no means more common or more generally mischievous than misjudged *reconstruction* under the name of *restoration* and *repair*."[8] However, no official action was taken and the Royal Institute of British Architects, founded in 1834, appears to have been oblivious of the implied criticism of some of its members.

Nor was the situation clarified by the attitude of John Ruskin, who wrote privately from Italy, where he was incensed by the restorations he saw, "You will ask what I would have, if I would neither have repairs, nor have things ruined. *This* I would have. Let them take the greatest possible care of all they have got, and when care will preserve it no longer, let it perish inch by inch, rather than retouch it."[9] Four years later, in an influential public utterance in *The Seven Lamps of Architecture,* Ruskin enlarged on this attitude toward restoration in words which sank deep into the English consciousness: "Neither by the public, nor by those who have the care of public monuments, is the true meaning of the word *restoration* understood. It means the most total destruction which a building can suffer . . . Do not let us deceive ourselves in this important matter; it is *impossible,* as impossible as to raise the dead, to restore anything that has ever been great or beautiful in architecture . . . The thing is a Lie from beginning to end . . . But, it is said, there may come a necessity for restoration! Granted. Look the necessity full in the face, and understand it on its own terms. It is a necessity for destruction. Accept it as such, pull the building down, throw its stones into neglected corners, make ballast of them, or mortar, if you will; but do it honestly, and do not set up a Lie in their place."[10]

The foundation of the Society for the Protection of Ancient Buildings in 1877 helped to define an attitude which was essentially that of Ruskin. The publication of William Morris's famous manifesto which drew attention, among other facts, to the undoubted truth that "those who make the changes wrought in our day under the name of Restoration . . . have no guide but each his own whim,"[11] was not in fact as revolutionary as some have claimed. It was the ratification of an English attitude which was the result of experience and Ruskin's writings; but it set down a series of principles which were followed by many practitioners in ideal situations; it sensibly advocated as a general rule repair rather than restoration, but it carried the generalisation to untenable extremes and failed to recognise there is a time for each. Nevertheless, it helped to propagate basically worthy principles and methods and, perhaps more important, its foundation can be seen historically as the beginning of the wider public concern for buildings which has become such a remarkable feature of the 1960s and '70s.

The National Trust for Places of Historic or Natural Beauty[12] was founded in 1895 as an organisation of private citizens who would act as trustees for the nation in the acquisition of buildings and land worthy of permanent preservation. In 1882 the Ancient Monuments Protection Act marked the acceptance by the State of formal responsibility for selected monuments, although it was not until an Amendment Act was passed in 1913 that a branch of the then Ministry of Works was formed with responsibility for the treatment of monuments and historic buildings. And in 1897 the then London County Council began the great inventory of London buildings which is still incomplete, an action which ultimately led to the setting up of the Historic Buildings Division of the Architects' Department. It can be seen from this that many of the bodies which play a leading part in building conservation in England today were in existence by 1913, but there was no special training for architects. Was it necessary?

Perhaps that can be answered in part by considering a well known, but not untypical, provincial practice at the turn of the century, that of Walter Henry Brierley of York.[13] If no one has heard of him, that is no matter, but it was in such an office that the treatment of old buildings was undertaken. In this particular practice, from the 1880s to the 1920s there was a varied programme. There were new country houses, and alterations and additions to old ones; there were new churches, and repairs and changes to old ones; there were new banks, schools and commercial offices — and there were what we would categorise as ancient monuments and historic buildings. At that time the treatment of ruined abbeys and castles, which were slowly being taken into guardianship by the State, was beginning to abandon the traditional picturesque approach, and techniques were being developed which became the basis of the later work of the Inspectorate of Ancient Monuments; Brierley is believed to have provided a model for some aspects of the official codification. He was one who, following the general principles of the Society for the Protection of Ancient Buildings, believed in repair and the retention of existing fabric wherever possible; he also had a profound knowledge of the construction and materials of historic buildings. He loved the qualities of old, natural materials which were, after all, those he used in his new buildings; and there was, as in the work of his contemporaries, a close affinity between his rebuildings of eighteenth-century country houses after severe damage by fire, his restorations of dilapidated timber-framed halls, barns and houses, his encouragement of the revival of the stuccoists' craft, and his designs for new buildings. It is doubtful if Brierley thought of himself or was regarded, in the present sense of the word, as a specialist in the treatment of historic buildings. It can be assumed that pupils in practices such as his learned to appreciate the qualities of traditional building materials and were quite conversant with the practicalities of traditional methods of conservation. Perhaps even more important was the sense of continuity in the life and use of a building which the repairs and alterations to country houses could give to those who were working on them.

The heyday of such practices coincided with the beginning of architectural education in England. In Leeds, for example, day classes for architectural students were inaugurated in

1903, and the prospectus shows that a great deal of time was taken up with lectures on history and traditional materials and construction. A major requirement was a set of measured drawings and the whole system was intended as a preparation for designing and building in a variation or pattern of one historical style or another. While the tradition was still alive there was naturally no need to consider special training for the treatment of historic buildings in use. The Inspectorate of Ancient Monuments evolved its own method of consolidating the largely ruined buildings which were its responsibility; the thoroughness with which ecclesiastical buildings had been dealt in the nineteenth century left comparatively few problems for the early twentieth.

It was in the 1920s and '30s that the change began: that commercial buildings occupied practices more than residential and ecclesiastical; that Corbusier had his visions of *Plan Voisin;* that there came into being those ideas of obsolescence and total renewal which materialised in the 50s and 60s. Many schools continued to devote a great deal of the curriculum to history, traditional construction and measured drawings, but the advanced ones dreamed of brave new worlds of white towers and green spaces occupying the sites of existing cities. Such visions affected even the plans for historic cities, and when Professor Adshead prepared his proposal for York in 1949[14] he recommended demolishing a wide belt of the city outside the walls and leaving the medieval pattern intact within a plain of lawns, in which were to be isolated schools, welfare centres, social centres, etc. These were the influences that were passed on to another generation of students; but while whole areas of cities were being razed completely the situation was changing, first as it affected some types of isolated historic buildings and then in a total environmental concern.

The virtual destruction of the old social pattern by taxation during the 40s and 50s, added to the neglect of the war years, meant that new uses had to be found for an incalculable number of country houses if they were not to be demolished. Many were, but the National Trust took some into its care; local authorities and institutions saved some by turning them into museums, colleges and schools; and some were bought by private trusts for conversion into flats or homes for the elderly. These changes alone meant that from the late 40s historic buildings were now becoming the concern of some sections of the architectural profession in public and private practice for the first time. This pattern of change has slowed down but it is still continuing; the cumulative maintenance of these buildings is obviously becoming an increasing responsibility. To this class of historic building has now been added the redundant church, with its own problems of conversion and maintenance, and the changed official attitude to our architectural heritage is also adding to the profession's responsibility.

The annual allocation of Government grants for historic buildings, which began in a small way in 1953, is currently around £2 million and, in addition, since 1962 local authorities have had powers to make grants for the repair and maintenance of buildings of architectural and historic interest. These are both dependent, generally speaking, on a contribution of about 50 per cent from the owner, which means that from this source alone something like £4 to £5 million worth of conservation work is being initiated this year and that most of it will be undertaken by private architects. There is also the work, difficult to ascertain, which will be necessary without grants, and the adaptation to new uses of old buildings. In addition, there are 34 cathedral architects and 43 diocesan surveyors with responsibility for ecclesiastical buildings. When it is realised there are now 183,444 listed buildings of architectural and historic importance in England alone, excluding churches and cathedrals, it can be seen that the number of practices in which conservation work occupies a part of the annual programme, though difficult to determine, is obviously considerable.

Yet there is some doubt about whether conservation work is sufficiently rewarding financially as a basis for a successful practice. The Society for the Protection of Ancient Buildings founded its Lethaby Scholarship in 1930, enabling a young architect to spend a period working in different practices and acquiring experience in the repair and treatment of old buildings. The previous holders of this scholarship were recently asked about their subsequent careers, and the replies showed that although some have been able to find appropriate appointments in public or private offices, some have accepted that an occasional job of this nature is all they can expect, and some have given up the idea of conservation altogether. On the other hand, both the Inspectorate of Ancient Monuments and Historic Buildings and the Greater London Council Historic Buildings Division have great difficulty in recruiting staff.[15]

These reports, on the one hand of difficulty in sustaining a practice in conservation work, and on the other of the apparent unavailability of architects to do this work, are at first sight contradictory; but they both seem to contain one common factor I have already identified—a lack of incentive.

There are a few dedicated private practitioners who make an adequate living out of specialising in conservation, but most of this type of work is undertaken in general practices which combine new buildings and conservation. It has been suggested[16] there might be a new professional man, the architect-conservator, trained as such from the beginning of his professional education, but there seem to be certain serious objections to such an idea. First, where would such a man fit into the present system? Unless he had an official appointment in a greatly enlarged Inspectorate of Ancient Monuments and Historic Buildings, and unless he was given an opportunity to acquire a recognised status and a salary reasonably equivalent to his expectation in practice, it is difficult to see how a prospective architect-conservator would be induced to undertake the necessary training. Secondly, there is not, nor can one see the likelihood of, a system like that in France which requires a recognised additional qualification before an

architect is eligible to apply for a position in the Inspectorate. Nor is there any requirement of proven ability or qualification before an architect undertakes a commission which includes an element of conservation. This is one of the incentives that is at present obviously lacking. Thirdly, there is no guarantee that a practice will have continuity of conservation work, a difficulty which practically excludes the idea of the architect-conservator outside an official department. Fourth, when conservation in practice is becoming more closely integrated with other architectural work, especially in urban redevelopment, is it sensible to consider setting up an isolated role for a specialist?

But, it might be asked, does the repair and protection of historic buildings necessarily call for training additional to the 7-year programme now obligatory before registration as an architect? The answer to that must be that there is, at present, no opportunity for an undergraduate to pursue under appropriate supervision an option which would enable him to work with students of art history, social history and archaeology, or to gain a useful experience on site with the techniques of building conservation and traditional crafts and materials. It is open to question whether such a course would be appropriate in an undergraduate school, but it might be suggested that some opportunities are not exploited as fully as they might be within the present system.

From information kindly given to me by the university and polytechnic schools of architecture it appears that architectural history, which must be regarded as one of the basic elements in a knowledge and appreciation of buildings, is seldom if ever related to projects. In some schools the time allocated to the subject has been severely curtailed and in hardly any is it included, except occasionally as an option, after the third year. The once-obligatory production of a set of measured drawings has almost entirely lapsed, although a certain amount of "investigation" is required as part of some projects. Inevitably perhaps, a concentration on present-day standard methods of construction and building materials has superseded a study of traditional types in some schools. Indeed, the meaning of *traditional* seems to have undergone a change and is now thought by some students, according to their answers in a recent examination, to signify the speculative builder's alternative to pre-fabrication. I doubt if anyone would claim that undergraduate training at present gives a young architect sufficient theoretical or practical knowledge to undertake work on historic buildings without further study and observation. What it can do is to stimulate an interest in such work and encourage some students to prepare to develop in that direction. I would like to return to undergraduate education later, but will now look at the current concern in my country for our architectural heritage and see if it suggests an associated direction for conservation studies.

Change has always caused protest; there have always been people who have made in their journals such entries as:

> When my hostess brings in my dinner, talk she will; and news of a new street, or of an old building demolished, comes as regularly as the salt. Her pet hobby is a remorseless railway which is to cut through all obstruction, chopping up my native town. . . . And she has told me that the old Fleece is to come down. . . . 'tis a subject on which I dare no longer dwell; but indignation at the thought of some new-fangled abomination rising upon the ruins of the old Fleece should choke me quite'.[17]

Substitute motorway for railway and the entry would be a currently common attitude although it was written in 1868. The lament goes on through the decades and there is a familiar note in such observations as this from *The Times*, which draws attention to the break-up of villas around London, with often considerable acreage, "when the speculative builder descends on them, cuts straight roads through them and plants as many houses as he thinks he can let." That was written in 1904, topical though it sounds; but it is well known that the melancholy Chekhovian twilight, in which nothing was heard but the whispers of regret

and the thud of axes striking on the cherry trees, has been shattered by the eruption of organised public opinion.

The formation of the Georgian Group[18] in 1937 and the Victorian Society[19] in 1958 as offshoots of the SPAB; the reorganisation of the Ancient Monuments Society[20] in 1952; the founding of the Civic Trust[21] in 1957; and the growth of something like 2,350 local protest and amenity societies since the end of the Second World War, have all introduced a formidable element in conservation issues: one which represents a wide social and professional spectrum. The 1944 Town and Country Planning Act introduced the listing of buildings of architectural and historic importance, a procedure that was more fully set out in the 1947 Act and then embodied in the Civic Amenities Act of 1967 and the Town and Country Planning Act of 1968. By strengthening the legislation concerning historic buildings, both in isolation and as groups; by making available various forms of grants to owners of historic buildings of all kinds; and by setting up the machinery of advertising proposed demolitions and compulsory acquisitions, inviting objections from societies and individuals; and holding public inquiries presided over by independent inspectors appointed to recommend to the Secretary of State for the Environment, these Acts have effectively provided for public participation in conservation issues. They have also inevitably made professional responsibility for historic buildings far more complex than a question of the treatment of the fabric.

I use the word *professional* rather than *architectural* because — while there can be no substitute for the architect's skill and knowledge in the conservation of the building, nor in the designing of alterations and extensions to an existing building or replacements in a street or group — he cannot now be working in isolation. The implementation of the various legal acts has given the official planner considerable control, but it is not an essential part of the planner's training to identify the architectural or historic character of a building. The estate agent advising the private developer about the scheme for a site which might be a designated conservation area or might include a number of listed buildings occupies an important place in urban and rural issues, since the principle of conservation often needs justification in an economic or commercial sense before it will be considered as an alternative for which grants might be available. The quantity surveyor and the building surveyor are employed in preparing evidence at an inquiry into a proposed demolition of a listed building, and there is a growing practice to involve an economist's cost benefit analysis to add weight.

The architect who proposes the retention and rehabilitation of an historic building on the grounds of its importance in architectural and visual terms, or the architectural historian who believes in its importance in historical terms because of its design, its designer, or its techniques, has to face such a statement as: "Rehabilitation is the preferred method of renewal if redevelopment costs exceed the sum of rehabilitation costs plus the present value of the next renewal cost outlay, plus the present value of the annual difference in the maintenance costs of a rehabilitated structure compared with a new one."[22] He cannot avoid an assessment of social and technical considerations, or of what has been called "the moment of truth for every building . . . when the site value is sufficiently higher than the building's value to make redevelopment economic."[23] That is indeed the time when the value of a building is questioned; and it has been well said recently that "in a typically British way there has been no attempt to define *why* we believe there is value, or to quantify *what* we accept as valuable heritage."[24]

I suspect that is not a peculiarly British sin of omission; but a form of training at any level for architects in conservation which does not acknowledge the problem as more than one of techniques and philosophies, important though they are, is not accepting the realities of the 1970s. This seems to me one of the arguments against the idea of the architect-conservator unless it can be linked with a complementary additional qualification. A result of the increased legislation and system of grants,

and of the pressure of local and national amenity societies, is the new importance attached to the appointment of conservation officers with special responsibility for the recommended designation of conservation areas, the control of development and new designs within them, the implementation of grants, and the building up of a good relationship[25] between a local authority, building owners, and the public.

One profound difference between the undergraduates of the 1970s and those of previous generations is their preoccupation with social and environmental issues. If I may use Lord Esher's phrase, they "feel the wounds of the earth in the right part of their emotional anatomy,"[26] and they earnestly wish to see these wounds healed by the application of appropriate political and social panacea. The environmental implications of this attitude are obvious. In the forefront of their targets are comprehensive redevelopments and speculations which can destroy human scale and values as well as buildings.

Nor can we overlook the very real and potentially dangerous growth of a lack of confidence in the architect's ability to design buildings as part of a human environment—a disillusionment which is affecting the architectural profession itself as well as the public and which is certainly in evidence in undergraduate schools.

Modern architecture has not only lost the confidence of the public; architects themselves have lost the way . . . People want architecture which is warm and comforting to the senses, architecture which is pleasant to live with, which caters for man as he is and not for man as an abstraction, architecture which is seen to be appropriate to its purpose, bearing in mind the habitual responses of people who have been brought up in a living society, not processed in a laboratory.[27]

Who can dispute that is not an accurate interpretation of current public opinion? Or that it is not an essential factor in the present vociferous opposition to the destruction of old

346

buildings which would have been disregarded a decade ago? There is also the widely advocated policy of designing new buildings according to the theory of "long life, loose fit, low energy," which brings with it the idea of building in future adaptations and even conservation. All these current tendencies condition the climate to receive the principles of conservation and to accept that they are likely to become an integral part of practice, suggesting a strengthening in that direction of the undergraduate curriculum.

Many schools are already including occasional programmes which require an assessment of the character of an area of existing buildings, the adaptation of some and the design of substitutes for others. The thoroughness with which such studies are undertaken and criticised obviously depends on the experience of the supervising tutors; but even allowing for an element of topical trendiness, such programmes serve to introduce students to some current issues.[28] The steps already taken in this direction by some schools are reflected in the increasing number of conservation subjects chosen as final options or dissertations. I believe a realistic assessment of the place of conservation work in general practice and the present tendency to examine the potential use of old buildings would confirm the importance of revising the emphasis on new buildings and techniques in the majority of undergraduate courses; but I will conclude with a review of the recent developments in postgraduate training in Britain.

In 1959, the Conference on Training Architects in Conservation[29] was founded because "the constituent bodies, all of whom have some practical responsibility or special concern for historic buildings, were deeply concerned at the increasing need to make better provision for professional training in conservation." The first result of their work was the setting up of a postgraduate course[30] in the Institute of Archaeology in the University of London, leading to a Diploma in the Conservation of Historical Monuments. Students spent four days per week during term and five days per week during vacations throughout

two whole sessions, following a course of practical training in an approved office; and concurrently during term one and a half days per week in academic study at the Institute. The course was established with the help of the Gulbenkian Foundation and lasted until 1969, when the cessation of the financial support effectively increased the fee to a figure beyond the reach of most students.

That was a serious casualty, but the London course led to the formation in 1968 of another organisation, the Association for Studies in the Conservation of Historic Building,[31] which consisted originally of staff and the thirty students who had taken part in it during the 8 years it was in existence. Subsequently, it has extended the membership by invitation to those specialising in disciplines involved in conservation, and there are now 128 members. Both these organisations, COTAC and ASCHB, have produced excellent reports and recommendations on the training of architects (and others) for conservation; the latter arranges meetings and discussions and circulates a newsletter to its members. The Royal Institute of British Architects has been a party to various discussions about forms of training and awards. At present a COTAC/RIBA group is examining the situation nationally, bearing in mind the relatively small number of potential students and the need to rely on a limited number of practitioners as visiting lecturers. The group is also considering if there is a need for providing training in the London area in addition to the two courses already in existence.[32] One of these, a one year, full time, postgraduate course leading to an MSc. is held in the Heriot-Watt University in Edinburgh;[33] the other is in the University of York.

If one had a free choice of the appropriate English city in which to study conservation, York would have few competitors. 1903 years old, it contains within its medieval walls, important survivals of most phases of its history, but especially of the Middle Ages and the eighteenth century. Like all historic cities, it has serious conservation problems of individual monuments and whole areas of traffic and obsolescence, of changing patterns of social and economic conditions, all adding reality and immediacy to conservation studies. Nor could there be a more appropriate building in which to set up such studies than the historic King's Manor, a notable restoration and adaptation for a new use of an important monument close to the medieval walls and gates and within sight of the magnificent Minster. Yet the choice was not consciously made. The present pattern of studies grew out of the annual programme of short courses for practitioners in the building professions which was initiated shortly before the foundation of the Institute of Advanced Architectural Studies in 1951.[34] In 1963 the Institute became a part of the University and in 1966 moved into its present historic building. From the beginning, each year's programme included a

number of short courses concentrating on various aspects of conservation. It was out of the success of these that the idea developed of offering a one year course of study for which a diploma would be awarded. With the blessing of most of the organisations concerned with conservation, and with very generous support from the Radcliffe Trust, the Pilgrim Trust and the Department of the Environment, we accepted our first course members in 1972.[35]

During the planning of the course of studies we were able to establish friendly relations with Rome, Ankara and Columbia, and since then we have become more closely acquainted with our friends in those centres as well as with professional groups concerned with conservation and education in Amsterdam, Paris and

Fig. 18.1   Restored Buildings at King's Manor, York. A diploma course in historic buildings conservation began here in 1972. It is part of a program of the Institute of Advanced Architectural Studies of the University of York. *(Photograph by Crown Commission on Historical Monuments, 1970)*

348

Delft.[36] Each autumn we organise a two-week course for overseas conservation architects; through welcoming visitors from something like thirty countries, we have established friendly relations all over the world. This international fortnight is also the time when we receive our diploma course members who, according to the regulations we drafted after considering various alternatives and the patterns of the courses in other countries, can be from any profession or discipline which is engaged in the conservation of buildings, sites and cities, provided he has had at least 4 years experience after qualifying or graduating.

The decision to make it a midcareer course was partly because of the Institute's experience in organising short courses of that nature, and partly because of the belief that at that stage in his career a man knows what he wants from a course of studies. He has had sufficient time in practice to come to a conclusion about the direction in which he wishes to develop; he has had sufficient experience on which to build additional theoretical and practical knowledge which should be useful to him when he returns to practice; and he should be able to discuss and work out his course for the year with the Director of Studies. We are not attempting to train a course member to be a skilled practitioner in one year, but we are trying to complement his experience and knowledge with an opportunity to study, discuss, research and observe conservation in action.

Because of its multidisciplinary character, the pattern of the course of studies is variable.

Fig. 18.2   The Travelling Summer School for Restorationists visits the Queen's House, Greenwich. In the summer of 1972 a party of young professionals funded from the United States toured active restoration projects in Northwestern Europe under the auspices of the Rome Centre. (*Photograph by Jack E. Boucher, USNPS*)

A number of formal seminars, regarded as essential, concentrate on a single theme: for example, masonry, decorative plasterwork, timber-framed buildings, landscape, museum conservation. Each of these, except for the last which is the valuable contribution of conservators from the Victoria and Albert Museum, is organised in consultation with an expert in the subject who is brought in from outside the Institute; he recommends the speakers and arranges visits or practical demonstrations on sites or in workshops. The full-time and short course programmes are planned at the same time, so that the full-time course members may attend the short courses on conservation subjects.[37] There are also organised visits, arranged in consultation with the course members, which are intended to supplement the seminars by offering opportunities to visit architects' offices and discuss work on the drawing-board with the staff, to see historic buildings under conservation, to hear about the routine problems in historic cities, and to try their hands in specialist craft workshops. There is a spring visit to a foreign city to meet architects, planners and historians and to see conservation in practice; in 1973 we went to Amsterdam and learnt about Dutch legislation and its implementation, the financing of conservation work and the part played by developers, the techniques used in restoration, the architects' attitudes toward the treatment of historic buildings. In 1974 there was a similar programme in Paris.[38]

The terms are longer than the normal university term, 13 weeks each, to make the most of the time away from practice. During the first and second terms the course member pursues, in the free time when there are no seminars, short courses and visits, his own agreed course of study and also presents a seminar based on his own experience or interest. The third term is devoted entirely to the preparation of a dissertation, which is submitted to external examiners and for which the Diploma in Conservation Studies is awarded.[39]

Whether such a one-year pattern is what is needed or practical for those who could bene-fit from it, or whether a sandwich course would make it easier for a practitioner to attend; whether it might plant offshoots in other centres or cooperate with other universities; or whether there is sufficient demand for such a course at all, will be clearer after another year or two. But now the diploma course has been established and the short courses continue to attract, the implication of these activities in helping to make York a centre for conservation studies is becoming clearer.

The Institute set up a Design Unit in 1966 and a Research Unit in 1971. These function separately, but with obvious links which now extend to conservation studies. Designs for adaptations of old buildings or for new buildings in conservation areas relate closely to the diploma course. A recent commission from the Department of the Environment to prepare a feasibility study of an important historic building taken into guardianship strengthens the links between study and practice. Similarly, recent research completed or in process by short-term appointed Research Fellows investigates the use of existing building stock, the criteria adopted in the designation of conservation areas, and the implications of legislation for conservation.[40]

Perhaps the most encouraging and striking development has been, not the inauguration of the diploma course in itself, but the way in which that has acted as a magnet and drawn together several separate activities and ideas to suggest a coherent pattern for development as a centre for conservation studies in the widest sense. As the options become clearly defined and the links already made are strengthened, it should be possible to incorporate several variations of advanced training and study to suit what is needed.

The city of York has a long tradition of being at the centre: Roman military headquarters and an Anglo-Saxon seat of learning; a Medieval centre of religious life and a Tudor centre of commerce; the eighteenth-century social capital of the North and the nineteenth-century centre of railway communication. It would be tempting providence to continue the analogy but, although one might agree with that idiosyncratic Yorkshire parson Laurence Sterne—that they order this matter better in France in some respects—perhaps there is a slight chance that the old city in which he spent much of his life might have something to offer in conserving English matter in a manner of its own.

NOTES

1. L.Réau, *Les Monuments Détruits de l'Art Francais,* Paris, 1959, II, pp. 123 ff.

2. P. M. Auzas. *Eugène Viollet-le-Duc,* Paris, 1965.

3. E. Viollet-le-Duc, *Dictionnaire Raisonné de l'Architecture Francaise,* Paris, 1869, VIII, p. 14. 'Restaurer un édifice, ce n'est pas l'entretenir, le réparer ou le refaire, c'est le rétablir dans un état complet qui peut n'avoir jamais existé à un moment donné . . . . Nous avons dit que le mot et la chose sont modernes, et en effet aucune civilisation, aucun peuple, dans les temps écoulés, n'a entendu faire des restaurations comme nous les comprenons aujourd'hui."

4. *Explanation of Designs for All Souls College, Oxford by Nicholas Hawksmoor,* Oxford, 1960. This reproduces Hawksmoor's MS explanation which accompanied his designs for additions and alterations to the college buildings. It also contains this comment: 'What I am offering at in this Article is for the preservation of Antient durable Publick Buildings, that are Strong and usful, instead of erecting new fantasticall perishable Trash, or altering and Wounding yᵉ Old by unskillful (knavish) Workmen."

5. Sir John Soane's Museum, London. *Gibbs MSS.*

6. J. Gibbs, *A Book of Architecture,* London, 1728, p. viii, pls.26, 27.

7. D. Green, *Blenheim Palace,* London, 1951, pp.92 ff, 303 f.

   There is perhaps no one thing, which the most Polite part of Mankind have more universally agreed in; than the Value they have ever set upon the Remains of distant Times. Nor amongst the Severall kinds of those Antiquitys, are there any so much regarded, as those of Buildings: Some for their Magnificence, or Curious Workmanship; and others, as they move more lively and pleasing Reflections (than History without their Aid can do) On the Persons who have Inhabited them; On the Remarkable things which have been transacted in them, Or the extraordinary Occasions of Erecting them.

8. J. M. Crook and M. H. Port, *The History of the King's Works,* London, 1973, VI, p. 641.

9. H. I. Shapiro (ed.), *Ruskin in Italy,* Oxford, 1972, p.119 ". . . . they are perpetually at work chipping & cleaning, & putting in new bits, which though they are indeed of the pattern of the old ones, are entirely wanting in the peculiar touch & character of the early chisel . . . . Take them all in all, I detest these Italians beyond measure. I have sworn vengeance against the French, but there is something in *them* that is at least energetic, however bad its principle may be. . . ."

10. J. Ruskin, *The Seven Lamps of Architecture,* London, 1849, pp.179 ff.

11. The manifesto is printed in full in J. Harvey, *Conservation of Buildings,* London, 1972, pp.210 ff. The Society for the Protection of Ancient Buildings, 55 Great Ormond Street, London WC1N 3JA, continues to play an important part in conservation. Chairman: The Duke of Grafton. Secretary: Mrs. Monica Dance.

12. The National Trust, 42 Queen Anne's Gate, London SW1H 9AS. Chairman: The Earl of Antrim. Secretary: J. D. Boles. Historic Buildings Secretary: St. J. Gore.

13. University of York, Institute of Advanced Architectural Studies, unpublished dissertation, C. Carus, "Walter Henry Brierley," 1973.

14. S. D. Adshead, C. J. Minter and C. W. C. Needham, *York: A Plan for Progress and Preservation,* 1949.

15. In 1973 the Greater London Council initiated a training scheme which offers a two year appointment and an opportunity to study and attend appropriate short courses in educational institutions. Two appointments were made, but it is worth noting there were only eight applicants for what seems a generous and useful opportunity for a young architect who is unable to embark on a full-time course of study.

16. Harvey, *op.cit.,* pp.81 ff, 202 ff.

17. D. Atkinson, *Old Leeds,* Leeds, 1868, p.4.

18. Georgian Group, 2 Chester Street, London S.W.1. Chairman: I. O. Chance. Secretary: Miss E. V. de B. Murray.

19. Victorian Society, 29 Exhibition Road, London S.W.7. Chairman: Sir Nikolaus Pevsner. Secretary: Mrs. Jane Fawcett.

20. Ancient Monuments Society, 33 Ladbroke Square, London W.11. Chairman: W. A. Eden. Secretary: Mrs. Jennifer Jenkins.

21. Civic Trust, 17 Carlton House Terrace, London S.W.1. Director: M. H. Middleton. Secretary: F. J. Barroll.

22. A. H. Schaaf, "Economic Feasibility Analysis for Urban Renewal Housing Rehabilitation," *Journal A. I. Planners,* 1969, XXXV, No.6.

23. G. Shankland "Conservation through Planning," P. Ward (ed.), *Conservation and Development in Historic Towns and Cities,* Newcastle, 1968, p.79.

24. G. Ashworth, "Conservation," RTPI Conference Paper, 1973.

25. These appointments, which might be filled by an architect or a planner, ideally require a knowledge of and feeling for historic buildings as thorough as do those which concentrate on the building fabric. They represent an important new figure in conservation, concerned with it in a wider sense than an architect-conservator would be.

26. L. Brett, *Parameters and Images,* London, 1970.

27. B. Allsopp, *Towards a Humane Architecture,* London, 1974, p.4.

28. It would be possible to extend such studies to incorporate, for example, a detailed investigation of a building, using observation and recording of the fabric as well as whatever documentation is available; the same building could be related to other examples of that date, type or construction as exercises in the application of historical knowledge. There could also be such considerations as a cost-study of alternative uses and treatments, a design for infill buildings made with full appreciation of their context, and an introduction to the techniques of building and traditional crafts by visits to work in progress on similar buildings and practical exercises in workshops. Indeed, one can see a whole year of a course could profitably be occupied by such a comprehensive programme.

29. Conference on Training Architects in Conservation, 19 West Eaton Place, London S.W.1. Chairman: Sir Ed-

ward Muir. Secretary: Donald Insall.

30. The Director of the course was W. A. Eden, then Surveyor of Historic Buildings, London County Council.

31. Association for Studies in the Conservation of Historic Buildings, Institute of Archaeology, 31–34 Gordon Square, London W.C.1. Chairman: Ashley Barker. Secretary: Miss Corinne Wilson.

32. The joint report of COTAC and RIBA, prepared under the chairmanship of Dr Bernard Feilden, was approved by both bodies in June 1974.

33. Heriot-Watt University, Department of Architecture, Lauriston Place, Edinburgh EH3 9DF. The Director of the course is Peter Whiston.

34. Since its foundation, the Directors of the Institute of Advanced Architectural Studies have been Dr William Singleton (1956–60), Professor Patrick Nuttgens (1962–69) and Professor Robert Macleod (1969–74). The development of conservation studies in the Institute owes much to their initiative and enthusiasm.

35. I would like to mention especially the great help we have received from the Radcliffe Trust, and it adds an additional gloss to the Institute's gratitude and prestige to know that we share a common source and a titular association with those well-known architectural monuments in Oxford, James Gibbs's Radcliffe Library and James Wyatt's Radcliffe Observatory. As well as providing us with this noble historical background and funds to operate the course, the Trust has established a research fellowship in conjunction with the course. Nor can I omit to record the work it has done in providing grants for training apprentices in the conservation crafts in the cathedral workshops at Gloucester, Hereford, Lincoln, Salisbury and York, and in supporting the York Glaziers' Trust in training and research.

Other organisations which encourage craft training for conservation are: The Council for Small Industries in Rural Areas (CoSIRA), 35 Camp Road, Wimbledon Common, London SW19 4UW, which runs courses for thatchers, ironsmiths and furniture restoration; The Orton Trust, 9 Cheyne Walk, Northampton NN1 5PY which organises short courses for masons; The Ernest Cook Trust, Fairford Park, Fairford, Gloucestershire GL7 4JH, which encourages the work and training of craftsmen through grants to appropriate Institutions; The Worshipful Company of Carpenters, which has an old-established building crafts training school, 153 Great Titchfield Street, London W1P 7FR, for carpenters, joiners and masons; The Ironbridge Gorge Museum Trust, Church Hill, Ironbridge, Telford, Salap, which is in the process of setting up a one-year course in building conservation techniques.

36. One of the five weeks spent in Europe by the Travelling Summer School for Restorationists (organised by Professor Charles Peterson) in 1972 was the responsibility of the Institute of Advanced Architectural Studies.

37. During the years 1972–74, these have included such topics as urban conservation, the recording of historic buildings, the maintenance of historic buildings, open-air museums, timber-framed buildings, and urban archaeology.

38. Reports of these tours may be found in the newsletters of the Association for Studies in the Conservation of Historic Buildings, nos. 20 (1973) and 24 (1974).

39. There has already been a wide range of subjects, and some of the dissertations submitted are valuable studies which have greatly extended the course members' knowledge of the contributions made to the whole problem of conservation by other professions and disciplines. In some cases they have already proved useful models for further investigation and of practical value.

40. Completed research reports are normally available to the professions and libraries. Immediate plans include the compilation of a critical bibliography of publications in English about conservation, an investigation of the state of the traditional crafts and training requirements, and a study of historic colour and paints. Within the university there is specialist research in other departments, such as medieval stained glass and landscape reclamation, and there are useful reciprocal links with a Decorative Arts course in the University of Leeds and York Archaeological Trust.

CHAPTER
19

Canada Prepares for a Great Program

JACQUES DALIBARD and MARTIN WEAVER

MESSRS. DALIBARD and WEAVER are with the Restoration Services Division of the Department of Indian and Northern Affairs, Ottawa, Mr. Dalibard as Chief and Mr. Weaver as Head of Preservation Training. The former is a member of the Royal Architectural Institute of Canada, past President of the Association for Preservation Technology, and Chairman of the Canadian ICOMOS Committee. The latter is Secretary of the Canadian Group of the International Institute for Conservation and author of two UNESCO studies on the conservation of Iranian monuments.

THE PROSPECTUS: 1970[1]

At the moment Canada has no facilities for the training of restoration and preservation specialists except the "on the job" trial and error approach—and there are plenty of errors. I won't try here to justify the need for a formal course in restoration and preservation in Canada. Other countries have already recognized this need. At this moment, even though it has the best opportunities and resources in the country, the Department of Indian Affairs and Northern Development alone could employ 30 to 40 specialists in this field.

In 1969 and 1970 a number of meetings took place involving representatives of Carleton University and of the Department of Indian Affairs. The discussions turned around the need for, and the types of training possible for, such work. A graduate degree program at Carleton University, combined with an apprenticeship within the Department of Indian Affairs, was considered. This combination would probably be ideal, but the school of architecture at Carleton is very young and does not anticipate having a graduate program before 1973. This is too long to wait, thus an alternative approach has been considered.

The Federal Government, through its Bureau of Staff Development and Training of the Public Service Commission and through its Departmental Staff Development and Training Divisions, has already developed a number of courses for its special needs. The Department of Indian Affairs and Northern Development is considering the possibility of establishing a new course for restoration and preservation specialists in the Department. There are precedents for this approach. In

353

France, for example, the training of restoration architects is largely handled by the "Service des Monuments Historiques" itself. Courses in the history of building construction and the preservation of ancient monuments are given by the architects of the Service.

Such a formula would be, of course, very advantageous for the Department of Indian Affairs. It would be possible to meet more closely the departmental needs by selecting people in the discipline especially lacking at that moment. However, it is believed that this system will also eventually benefit the provincial organizations or others in need of such specialists because of the way the candidates for the course would be recruited.

The Department would encourage a number of people already on its staff to undertake this training program. In addition young, promising architects, engineers, historians, etc., from as many provinces as possible, who want to attend the course, would be engaged by the Department for a period of two years and given a chance, during that time, to enroll and to serve an apprenticeship. Because they would be paid, they would, in effect, be receiving a scholarship. It is expected, of course, that the Department would benefit from their learning, but it is also hoped that a number of these students would eventually work for other organizations, provincial or otherwise, or even practise privately. Candidates would preferably already have university degrees, but this would not rule out especially talented ones without such degrees.

There are restoration specialists of many kinds: historians, architectural historians, archaeologists, architects, landscape architects, engineers, and others. In devising a curriculum, it would be important to consider the number and the proportion of the different disciplines represented among the candidates. I won't attempt at this time to define the curriculum, but I will mention the format of the proposed course. It will be given as a series of one-week seminars, each covering one aspect of the field such as "Repair of Old Stone Buildings," "Repair of Old Wooden Buildings," "History of Building Technology,"

"History of Preservation and Preservation Legislation," "An Introduction to Historical Archaeology," "Furniture and Furnishings," "Historic Gardens," and so on. These seminars would be given by experts in these subjects.

There are many reasons for this approach. One is that it reduces the need for a large permanent teaching staff. Another, that it is then possible to bring in a number of the best experts for a short time, while it would be difficult to retain any of these people away from their normal occupation for a longer period. It is hoped that these experts would indicate the breadth of their fields, give two or three detailed lectures on some significant aspect of their subject, suggest further possible studies, and provide an intensive bibliography. A number of the lecturers would be drawn from the Department itself. The staff of the Historic Sites Service — historians, architectural historians, archaeologists, furniture curators and administrators — and that of the Technical Services Branch of restoration architects, engineers, landscape architects, etc., provide an immediate resource.

The apprentices would perform useful work in their chosen disciplines, but they will be given the opportunity to sample other aspects of the field. For example, an architect could spend some time on an archaeological dig and in the artifact lab, prepare an historical report on the historical research section, select furniture with a curator, work on an inventory in the design office, or on construction in the field. All these aspects of preservation and restoration are routine in the Department and it would be easy to incorporate apprentices in any given activity. In addition, the students would be given a chance to widen their knowledge by looking at the work of other organizations and by examining historic buildings and sites in Canada and in the United States on field trips.

Even though Ottawa is not itself a city of outstanding architectural or historical interest, it has in the different governmental departments the best facilities available in the country. The Department of Forestry can be a valuable asset in studying wood, its deterioration and its treatment; the same is true for the Department of Energy, Mines and Resources for the study of stone. The National Research Council has helped the Department of Indian Affairs develop recording techniques using photogrammetry and rectified photography. This collaboration could continue in teaching those techniques. The National Archives and the National Library are, of course, basic resources for carrying on any kind of historical research in Canada. The Library of the Department of Indian Affairs and Northern Development has collections relating to preservation and restoration and this could, of course, be augmented considerably when teaching facilities are developed within the Department.

One of the main problems in establishing such a course outside of a university is that of academic recognition. Obviously the Government cannot award a degree, but while waiting for a university to take on the job, a diploma or a certificate based on merit would be given by the Department to successful candidates. In the final analysis what really matters is the quality of the course. If the standards are high, prestige will follow and in turn prestige would accrue to those who take the course. It is hoped that such a course will start in the fall of 1971.

Progress Report: 1974[2]

A little over two years since the announcement at Quebec, we are able to report that the course has now been started and is well underway. The first lecture was given on February 12, 1973.

It is interesting to reflect that the course closely follows many of the guidelines discussed in 1970. The Restoration Services Division has actually developed a pool of trained conservators to facilitate plans for the decentralization of certain federal government functions. The trainees will be scattered across Canada to be closer to the field projects.

As far as we know, our program is unique in the world as an international government course at the Master's degree level. In varying ways it is now delivered to seventy participants who include not only the restoration architects and engineers of the Restoration Services Division but also historical landscape architects, archaeologists, conservators and others involved in the conservation of historic monuments under the care of the Federal Government of Canada.

At the end of the first year, following a number of reviews of the way that the course had been operating, the program was revised and expanded to the form described here. The entire course is recorded on video and audio tapes by the divisional staff and kept under their control. This not only means that lectures can be reviewed at any subsequent date, but also these tapes can be reproduced in specially edited versions and programs for other users. Thus parts of the course are available on tape to such groups as regional staffs, as-found recording teams in the field, and the drafting section of the Restoration Services Division. In this way it should be possible not only to improve individual staff advancement potentials but to generally upgrade expertise and performance—not only in the Division but in the whole Department.

This is a very ambitious program. It is hoped that the course participants will eventually establish a sound reputation for the course not only in Canada but in the world. Various international bodies have already shown interest and some have expressed a willingness to cooperate in the supply of lecturers and material. The program is organized under the following major subjects: a) philosophy and theory of conservation and restoration technology, b) conservation legislation and administration, c) historic centers, d) historic sites, e) historic landscapes and gardens, f) historical studies of period building crafts and practical work.

The philosophy and theory section of the course will be devoted to such subjects as the evolution of international conservation under restoration standards and philosophy, basic conservation and restoration philosophy in relation to works of art in historic monuments and sites. The best way to explain the content of the course is to offer samples taken from the tapes. The first excerpt is from a typical lecture on restoration philosophy.

MARTIN WEAVER: *Restorations*

You see here one of the major, really major, problems of the conservation of monuments. These two pictures are of the facade of Notre Dame in Paris. [Figure 19.1] is a very early photograph, taken in 1853 and interesting here because it shows that facade before Violett-le-Duc did a bit of conservation and restoration on this monument. [Figure, 19.2] shows the facade in 1937 after Viollet-le-Duc had had a go. Now it has been fashionable to take a kick at Viollet-le-Duc for the terrible things that he did to French monuments in the name of conservation but some of the problems which he faced in conserving these monuments were stupendous and, in many cases, if he hadn't taken a crack at some of these problems we

Fig. 19.2 Notre Dame in a 1937 view, long after the Viollet-le-Duc restoration. The architect did his best under the circumstances but, although we now concede that his work preserved this great French monument, his over-energetic methods would not be acceptable today. Les Pierres de France. *Achille Carlier.* Paris, Mar.– May 1937. *Notre Dame Photos*

Fig. 19.1 Notre Dame in a prerestoration photograph taken in 1853.

wouldn't have a lot of these monuments today—certainly not in this complete form. It's arguable, perhaps, that the restorationist went a little too far. Just to demonstrate that this sort of problem is an almost impossible one—even if you're a genius—here is a genius having a go at Milan Cathedral. The drawings by Leonardo Da Vinci show his attempt to get around the problem that the original builders of the Milan Cathedral had left him with—a lot of the building couldn't stand up any more [Fig. 19.3]. So here he was designing new vaulting with all sorts of special keying to make sure it didn't fall down and that it didn't exert thrust where the structure was not able to take it. So Viollet-le-Duc was doing his damndest and probably going a little too far, and Leonardo

Fig. 19.3   Milan Cathedral, with drawing by Leonardo Da Vinci, who attempted to restore it with new vaulting and keying. Detail from *Codex Atlanticus*, folio 310, verso b.28x 23.6 cm. Ambrosiana Library, Milan. *Leonardo and the Age of the Eye*, London: 1970. *(Illustration from Ritchie Calder).*

da Vinci was doing his and probably hitting just about on target.

GEORGE MACDONALD: *Wood*

The technology section contains a number of lectures on wooden structures and their historical development. Dr. George Mac-Donald of the National Museum of Man in Ottawa speaks on the timber buildings of the Canadian West Coast Indians (Fig. 19.4):

A typical Bella Coola house had a crest figure on the top. Some of the figures were mobile; they could wave or, if birds, flap their wings. The Bella Coola went in for that sort of thing in a great way, especially to welcome guests to their feasts. The houses had heavy gables, some about four feet deep, and decorative carvings on the posts and beams. Some were highly decorated with painting.

CHARLES E. PETERSON: *Brick*

Under the technology heading, Professor Peterson speaks on brick construction:

I've heard lots of people rationalize about why frame walls were "nogged" with brick and it makes sense to say it was good insulation. True. It also would contribute to fire-proofing. But there is only one contemporary document that said why it was done, and it was not for those reasons at all. Alexander Hamilton, who was a great figure in our Revolutionary period, married the daughter of Phillip Schuyler, a merchant

Fig. 19.4   The Bella Coola Indians, a Canadian West Coast tribe, decorated their wooden houses with carved wood figurines as well as totems. The figurines could be made to wave, beckon or flap, a sign of welcome to visitors. The heavy gables are typical of the area. *(Collection of the National Museum of Man, Ottawa)*

and a man of big affairs at Albany and who ran a lumber yard and sent the lumber down the river in sloops (or in rafts, I don't know which) to Manhattan, where Hamilton was to build his house. He said to be sure to fill in the frame with brick "to keep the vermin out." (By vermin he meant rats, mice and squirrels. I have found some examples of houses that got into trouble because they didn't keep the vermin out.) But this is the explanation to the man who built

it—to keep rats out. They could otherwise build nests in the walls, you see.

HARLEY J. MCKEE: *Stone*

Professor Harley McKee has given a series of lectures of historic stone working and quarrying methods in the U.S.A. A sample on stone dressing technique:

I have given a sequence of "bankering-up" or "squaring-up" a block of stone. I

358

don't know if any of you are old enough to have taken manual training in high school and had to take a hand plane and square-up a block of wood. Well, there's more of an art to it than meets the eye, and so there is in taking a rough piece of stone and working it out into nicely dresssed surfaces with all of those surfaces perpendicular to each other.

But I want to add a couple of things with respect to the tools that were used. According to Euclid's geometry the first two straight edges determine a plane, but how are you going to know that the third edge is also in that same plane? This was done with the aid of straight edges. Usually the stone-dresser had a couple of straight edges of about two inches in width, and he would lay them on the surface and then sight. I guess he had to have three and if the third one lined up with the others he knew that they were in the same plane. The Egyptians had a very complicated process for doing that — it's not completely understood and it's not relevant to us. I would suspect that the method of using straight edges and sighting, though it is probably a fairly old one, is just a kind of ingenious thing that smart people in early times would have figured out.

KENNETH HUDSON: *Industrial Housing*

In the category of historic studies the English broadcaster and writer, Kenneth Hudson, talks about industrial housing as a record of social history (Figs. 19.5 and 19.6):

These houses are in the English railway town of Swindon, which is 60 miles west of London. It's a railway town—Brunel's Great Western town—where he had his enormous carriage and locomotive works. These houses were built in the late 1830s and early 1840s in the open fields. They were built (you simply see here one street) with great urgency because there was nothing but open fields and housing was needed for the workers so that they could function.

Now these houses look perfectly pleasant; they are in good repair. You can leave it at

Fig. 19.5   Railway workers' housing in Swindon, England, after restoration. Built in the 1830s and 1840s, the houses look pleasant by modern standards, but investigation of social history shows otherwise. (*Photo courtesy Kenneth Hudson*)

Fig. 19.6   The backs of the Swindon tenants' houses, before restoration, show very short clothesline lengths for the actual number of tenants in each house, usually six lodgers in addition to a family of eight. Laundry and privy facilities would have been insufficient, especially during frequent outbreaks of cholera and typhoid. Such social history adds to archaeological evidence. (*Photo courtesy Kenneth Hudson*)

that. You can simply say this is a very interesting piece of town planning, if you like, worker's housing of the mid-nineteenth century. Or you can start digging and, I think that if you dig, the thing becomes about ten times as interesting. For example, you can look at these little houses in the front and you can say, "aren't they nice?" They're very simple, and then you look at the size of them and you find that what they've got is two rooms up and two rooms down.

Now when you look at the backs, it looks a little less garden-city like, you see. But if I say to you that the most significant thing in that picture is this bit of rope here and the bit of rope over there, you can start off by saying "this is just stupid." But suppose I explain it this way: the tenants' lists for these houses have survived in the British Railway records and they show that it was perfectly normal in the 1840s and 50s to have fifteen

people in those four rooms. The sort of thing you would find would be eight in a family plus six lodgers. It was a condition of having tenants at all because this is railway housing, and unless you could persuade families to take the single men, you were losing half your labour force. Right.

You then discover that the problems of water supply were very, very bad and there were bad cases of cholera and typhoid and God knows what else. You then look at the little outside privy, and the little outside wash house. Supposing you have an attack of cholera in these houses. What then? You've got fifteen people packed inside and that's the length of clothesline you've got to deal with the washing, and people with cholera and typhoid are not nice. You've got an appalling human problem and a sanitary problem in there.

Now it does seem to me this is one of

Fig. 19.7   Bits and pieces of the Fountain of Trajan at Ephesus were put together in their true "context" but without proper vertical relationship. This illustrates the danger of creating a misstatement of what had existed previously. (M.E. Weaver photo)

these cases where archaeology does add another dimension to the record. You can perfectly easily go and see these tenants' lists, you can read up about problems of epidemics, water supply in Swindon, but so far as I'm concerned the thing does not gel—it doesn't come alive until you've seen those small back yards. Then everything comes together and you've got the sense of what the history was really about.

HUGH MILLER: *Ancient Sites*

In the section devoted to historic centers and sites and historic landscapes and gardens, a typical lecture is that given by Hugh Miller of the United States National Park Service. Here he speaks on the preservation of the ancient sites of Ephesus and St. Jean in western Turkey:

The problem of three dimensioning, I think, is probably best shown in the Fountain of Trajan at Ephesus where the philosophy of "don't do anything you don't know" was carried to extremes [Fig. 19.7]. The bits and pieces were put together in true context and yet, because there was not enough material to determine their actual relationship vertically, they were set on small concrete piers. We have here an ugly dwarf that is not a lie but is so much of a misstatement that it would be better if it didn't exist at all.

The other problem of philosophy is to what extent is restoration appropriate. Here at the Temple of Hadrian we see reconstruction, some casts of the frieze which are now in museums, and some reconstructed portions [Fig. 19.8]. It is very well done, and yet there is this whole question of how honest is it as a reconstruction to put back new materials and to put back casts? This is a philosphical question that I won't get into at this time.

Fig. 19.8 How honest are reconstructions? At the Temple of Hadrian we see reconstruction, partially from casts of parts of the frieze now in museums. This type of restoration raises grave philosophical questions. *(M.E. Weaver photo)*

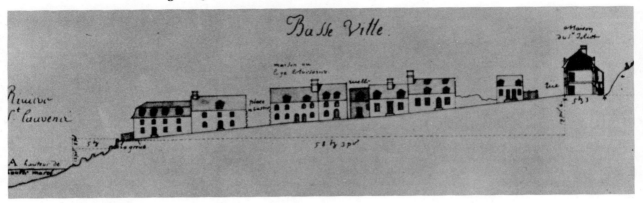

Fig. 19.9 Elevations of Sous le Fort Street, Quebec, drawn in 1685, suggest that the buildings did not have steep roofs. Plans and other drawings of the period suggest otherwise. *(Map Division, Public Archives of Canada)*

And then there's the whole problem of reconstructions themselves, the rebuilding of structures that no longer exist from parts that no longer exist: from fragmentary evidence. In the Park Service we have a rather clear policy on this. The most important thing is that the reconstructing is not conjectural, and very often as one gets more involved in the reconstruction the more it becomes conjectural. This is the snare and delusion of the whole fun and games of reconstructions.

A. J. H. RICHARDSON: *Quebec*

Rather than give a full course of architectural history, a series of study papers will be given covering a whole range of subjects from the fields of architectural, economic, social and industrial history. Here Jack Richardson of the architectural history section of the Department of Indian and Northern Affairs speaks on the development of domestic architecture in Quebec and the development of that city.

Here is an elevation of Sous-le-Fort Street in 1685 [Fig. 19.9]. The elevation suggests that the buildings were not too steep-roofed but, unfortunately, the elevation doesn't agree with the plan—the good, very neat plan that goes with it. The other pictorial representations of the period don't look at all reliable, so we have to fall back on the photographs of the oldest buildings that survived into the late nineteenth century and the few that survive today. They're all steep-roofed—all that can be clearly traced back to about 1720.

Most of these houses were one-chimney buildings. So we can understand Marie de l'Incarnation when she said "the excessive

Fig. 19.10 Detail from Vue de la Ville de Quebec, ca. 1720 Capitale du Canada. Nouvelle France, *Carte Marines*, 1727. Watercolor. *(Original in the Newberry Library, Chicago, Edward A. Ayer Collection)*

Fig. 19.11 The small scale of early Quebec is shown by the Jacquet house, now a small part of a restaurant dwarfed by its neighbors. In 1713, it was one of the largest houses in the upper town. *(Picture Division, Public Archives of Canada)*

cold doesn't permit us to make large spaces—there are times when our priests are in danger of having their fingers and their ears frozen." Only the great religious buildings rose above this general level. The smallness of scale comes home to us when we look at the great manuscript plan of 1713 in the Depot des Fortifications des Colonies in Paris, and find the Jacquet or Desjardins House [Fig. 19.11] at St. Louis and Jardin Streets. This is the smaller part now of the restaurant *Aux Anciens Canadiens*, which is now dwarfed even by the oldest of its neighbors. It was in 1713 one of the biggest houses in the upper town. We have another view of about 1720 and the houses were still mostly two stories.

MARTIN WEAVER: *Tools for Working Timber*

The course will be partially integrated with the division's craftsmen training scheme. The general idea is that course participants will be able to see historic building crafts processes taking place and should in some cases have an opportunity to actually work with the tools and materials themselves. A lecture related to this subject was given by Martin Weaver to investigators of the Canadian Inventory of Historic Building. The subject was a history of timber working tools and tool marks.

The reconstruction done in the 1930s at Port Royal [Fig. 19.12] used timber that had been gang-sawn but it's a little dubious as to whether the original timber was gang-sawn. From this reconstruction you get the idea of how important the actual tool marks are on timber; they can give you a totally wrong impression of how the original boards were made.

At Fort Wellington—built between 1830 and 1840—one finds saw marks on the old lumber [Fig. 19.13]. There is a little bit of doubt about this example; it could have been produced by the *pit* saw—so named because the bottom sawyer literally stood in a pit, and the big timber was just laid across it at ground level [Fig. 19.14]. The top man, being the most senior man, stood up above and merely operated out in the open.

The poor chap below, who got all the sawdust and muck dropped on him was usually an apprentice. Now the difference between the sort of marks that the pit saw produces and the reciprocating saw—it's a machine powered saw—is really very slight, because in its essentials the blade is identical. In fact in many cases you'll find that a pit saw blade has been taken and put into a sawmill. The only difference is that you'll get a slight see-sawing backward and forward. The marks won't be exactly parallel. Now in this particular case it is rather difficult to tell whether those marks are really parallel or whether they are see-sawing a bit.

KENNETH R. ELDER: *A Field Report*

The normal office and site work of the course participants is integrated into the course, in that such work will form the subject of research papers for lectures and seminars and finally theses or dissertations. Practical

Fig. 19.12   Reconstruction at Port Royal done in the 1930s points out the importance of tool marks on boards. A band saw may have been used in the reconstruction, to imitate a reciprocating saw. *(Photo, S. MacKenzie, D.I.N.A.)*

work can take many different forms, including small projects such as one at Fort Wellington where the dismantling of a deteriorated building was used as a subject for the preparation of training material.

You may wonder how the Riel Rebellion and the Battle of Batoche can be connected with our practical work. Kenneth R. Elder of the Restoration Services Division carried out an investigation of the *Louis Riel* House in St. Vital, Manitoba, in June 1972 and completed the field work in September of the same year. A large report was then prepared containing all the results of the examinations and correlations of the many samples gathered:

Wherever possible, attempts were made to establish a proper chronology of the various parts of the building, while at the same time discovering what the whole scheme of

interior decoration was at the time of Louis Riel's occupation. Lack of records on this comparatively recent building forced us to extract as much information as possible from the building itself; this meant going down to a microscopic examination; even of pieces of cloth found in the chinking. Although the house is small, it took a controversially long time to examine it in sufficient detail to provide significant answers. It is too easy in the restoration of such buildings to apply a roughly correct period skin to the bones of the original house while the surviving archaeological evidence for the actual state of the original is ignored and even destroyed. I feel that it cannot be overstressed that archaeology is not simply concerned with the remains beneath the surface of the ground. Any evidence which remains in or on the structure itself is archaeological evi-

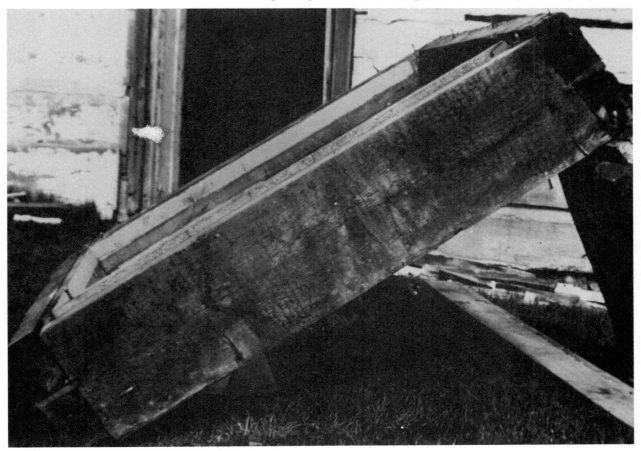

Fig. 19.13   The marks on these old planks at Fort Wellington could have been left by a
pit saw. *(M.E. Weaver photo)*

dence. It does not matter if the scrap of paper or cloth, etc., is overlaid by paint or wallpaper rather than earth—it is archaeological evidence and should be treated as such.

General Observations

Elder's methodology might well be the subject of a research paper for the course. The course functions via a number of different vehicles: the basic vehicle is the formal lecture program which consists of a number of speakers giving specially prepared papers on subjects under the course's seven main divisions, typical portions of which appear above. Those who wish to participate in the course on an intensive basis will be required to form themselves into a number of research seminar groups to meet outside office hours.

Where course participants feel that they have not received sufficient information on a particular subject in the program, or where they feel a new subject should be introduced, the informal seminars will act as a means by which they can correct the situation themselves. Following a short course on synectics, creative problem solving, and the use of video equipment, the groups will be expected to elect their own chairman on a rotating basis from among their own members. Members of the group will be called upon to carry out research on a specific subject and to prepare a paper which will be presented to the rest of the group.

Alternatively, the group may obtain the services of external lecturers or researchers via the course administration. A coordinator will be allocated to each group to ensure any assis-

Fig. 19.14 Pit sawing was named for the fact that the top man stood at ground level while his helper, usually an apprentice, stood in a pit and took the brunt of sawdust and muck dropped on him. Identifying marks left by various types of sawing operations are important in restoration work. Illustration from *The Book of Trades* (1st ed. 1804) Part 1, facing p. 80.

ments in this area as and when the need arises.

Because much of this work could be done by correspondence, the number of participants in the intensive course would not necessarily be limited by the scarcity of tutors or specialist advisers available within the Department. When a course participant feels that he or she is ready to submit a dissertation for the diploma of the course—and if the course administration agrees—the work will be circulated in much the same way as a doctoral dissertation. After it has been read, the paper will be presented as part of a formal lecture program and will be judged by a panel consisting of the

tance needed and to report back to the course administration, so that the overall aims of the course can be effectively maintained. General study tutors or "moral" tutors will be available to any course participant who requires their services. But as a rule it is suggested that participants should consult external tutors or specialist advisers. As the name implies, such advisers would be consulted for their expertise in specific areas of interest to the course participant—possibly on a corresponding basis. The course administration will endeavor to make the necessary contacts and arrange-

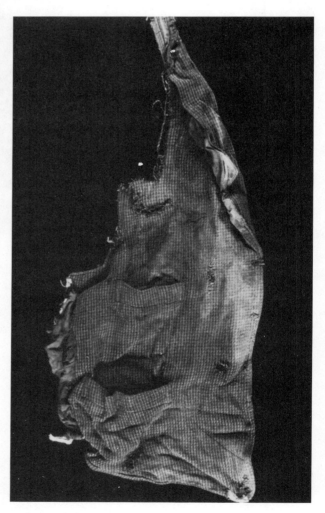

Fig. 19.15 Part of a vest, found in the house of Louis Riel, helped date and therefore reconstruct the Riel House in St. Vital, Manitoba, in 1972. Such seemingly irrelevant material as this must be considered as archeological evidence. *(Photo, S. MacKenzie, D.I.N.A.)*

Director of the Course, general study tutors, and such additional jurors as the participant may nominate. In any final dissertation the course participants will be expected to demonstrate that they have a full understanding of the complete restoration or conservation process for a monument or site and that if necessary, with appropriate technical and administrative assistance they could plan and control such a process. Normally, participants would be expected to base their dissertations on actual projects on which they have worked or are currently working. However this in no way precludes anyone from preparing a dissertation on an entirely different subject, although this might very well turn out to be a handicap.

Qualifications

It is expected that the full course would last two years but the participants may submit their dissertations at any time after they have attended the course for 6 months. Those whose theses are of the required standard will be awarded the Department's diploma in the Conservation of Historic Monuments and Sites. Participants and auditors who do not wish to submit a dissertation but have taken the course—either in person or by video tape—will receive the Certificate of Attendance. A submission of dissertations before the end of the two year course is arranged to permit individuals who have already attended relevant courses to gain the diploma when they are ready. There is no point in insisting that everybody go through an entire course if they are only repeating lectures and reading which they have already covered.

Timing

For those taking the intensive course, the whole of one working day is set aside each week, currently Fridays. With the 7 ½ hours available, 4 hours are allocated to formal lectures and the remaining time for individual research and reading. As has already been stated, the informal research seminars will be out of office hours and therefore could take place at any time, depending on the wishes of the members of the groups. However, some of the guest lecturers from the formal lecture program will be employed to their advantage by the seminar groups to explore subjects in greater detail or to enter into group discussion. This would mean that certain seminars would have to be scheduled when lecturers would be available.

Postscript: 1975

In the months which have elapsed since the Philadelphia presentation, the course has been reexamined. In a Status Report prepared in October 1974, the development of the course was reviewed, and problems and solutions were discussed. Part of the report is summarized here because of its interest in the present context.

In the second trial period, a larger number of guest lecturers were brought to Ottawa from within Canada and from the U.S.A. and Europe.

Despite requests for the presubmission of lecture papers and illustrative material, few lecturers complied adequately, and there were a number of unsuitable lectures delivered. This raises the whole question of quality control when the lecturer arrives suddenly—possibly from abroad—delivers a lecture, and then takes the next plane out. It is obviously difficult to obtain satisfactory results in this way.

After experiments of this nature it was decided to adopt a one week seminar approach where course participants are presented with groups of lectures on major aspects or themes of conservation studies—the lectures being delivered by carefully selected leading Canadian and international experts speaking on their own specialties from presubmitted and edited papers. The lectures will be accompanied by general discussion periods and working, problem-solving sessions.

In the following week, the material would be reviewed and a shooting script worked out with the lecturers for a one hour television programme. The original lectures, possibly augmented with additional material, will then be recorded in a commercial studio at the end of the week to produce a "broadcast quality" colour programme for distribution. The final product would be ready one or two days later.

The seminar approach also lends itself well to the training of staff for regional offices. The Departmental decentralization plan for restoration design and other similar functions has forcefully pointed up the scarcity of trained professionals in this field. Consequently, training regional staff members is one of the highest priorities.

The black and white videotapes made thus far have proved to be excellent for base material but lack the quality desirable for extended viewing by trainees. It is clear that one needs colour and an appropriate format to show detail, and properly edited material, if one intends to distribute videotapes and other audiovisual programmes. Acceptable standards for videotaped programmes are dictated by the standards which everybody sees on his home set. Even the ever-present advertisements and soap operas offer a high standard of technical competence, and people unconsciously expect this standard of any programme.

During what we can term our experimental period, we have tried to encourage small discussion groups to work within the main framework of the course. Our idea was to encourage course participants to assist in the planning and selection of training material, and to fill in the gaps which they felt existed in the formal programme. Apart from one successful interdisciplinary group, this idea appears to have failed largely because participants wanted to be continuously organized and directed.

We are currently holding a series of sessions wherein course participants from various disciplines are themselves presenting illustrated papers to their colleagues, on subjects which are usually related to their current work. This is working well and serves two main purposes. First, information on current practice in the "conservation/restoration" field is being disseminated among disciplines. Second, we are provided with an index of the course participant's proficiency without having to set up exams or indulge in other academic procedures.

The synchronized slide/tape medium has been used by some of the participants for their presentations; this vehicle is also used by staff to explain Divisional work organization and method for internal and external use (e.g., special presentation on Restoration Services Division's "as found" recording and photogrammetry programme, given to the 1974 Athens, Greece, International Conference on Photogrammetry: is was repeated, with Spanish translation, at a later conference in Mexico City).

One serious problem has been the lack of appropriate texts and manuals for this type of course. Now that the Restoration Services Division is to be given the responsibility of maintaining standards and guidelines for Federal conservation and restoration projects, it is clear that our training course will play a very important role in supplying the necessary manuals and guidelines.

One of the new duties of the Restoration Services Division will be to prepare conservation zone studies, as guideline prototypes and as specific area schemes to satisfy departmental requirements. The division has already prepared studies in the latter category on Dawson City, Yukon Territory, and on the Artillery Park area of old Quebec City. Both studies have been prototypical and will inevitably be used as precedents or models for later work of this nature. We have realized that the approach to such studies, the methodology used, and the form and content all have a strong educational value to all involved in the conservation of monuments and sites. With the current plethora of individuals and groups involved in the conservation of any historic urban or rural zone—the mixture of professionals, politicians and interested laypeople at different levels of experience and expertise—it is essential that some means be provided to inform all parties on the various philosophies and techniques which are employed in the conservation process. It is hoped that the material gathered for our course will be used to this end.

A very much broader educational purpose can thus be seen to lie beneath the surface of the course. In cases where the Federal Government has already been involved in urban conservation projects in partnership with municipal officials, professionals and lay groups, the problems of explaining all the diverse aspects of the conservation process — and of Federal roles in that process — would have been greatly reduced. The use of video-tapes and other audiovisual programmes prepared initially for the course would have simplified the educational process.

Eventually, the existence of this educational resource in Canada can do much to foster the conservation of Canadian heritage by reaching the people who actually live in and near the historic buildings and areas.

NOTES

1. This prospectus is drawn from a paper presented at the Second General Meeting of the Association for Preservation Technology, Quebec City, 1970. For a parallel statement on the American situation, *see* Charles E. Peterson, "Training Architects," *Journal of the Society of Architectural Historians*, Vol. X, No. 4 (December, 1951. p. 33; and Grant C. Manson, "Training Architects for Restoration," *Ibid.*, No. 2 (May, 1955) pp. 28–29.

2. This paper was adapted from a video tape presentation by Jacques Dalibard at the Carpenters' Company Symposium at Philadelphia, March 29, 1974.

CHAPTER
20

The New York State Historic Preservation Program

JOHN G. WAITE

MR. WAITE began as a student architect with the Historic American Buildings Survey in Philadelphia and was Senior Historical Architect in charge of the restoration program in the New York State Office of Parks and Recreation until November of 1975. He is now a partner in The Preservation/Design Group in Albany.

The state historic preservation program in New York began in 1850 when Washington's Headquarters (Jonathan Hasbrouck House) in Newburgh was acquired. With the Hasbrouck House, New York State became the first unit of government in the United States to acquire an historic site for preservation purposes. This system was expanded in the late nineteenth and early twentieth centuries to include other significant properties relating to the colonial and Revolutionary War history of the State. Much of the impetus for the acquisition of these additional properties stemmed from the general reawakening of interest in the nation's history brought about by the Centennial of 1876.

Most of these sites, which were acquired by the second quarter of the twentieth century, were houses located in the Hudson and Mohawk valleys. Included in this group were such well-known buildings as Philipse Manor Hall in Yonkers, the Knox Headquarters near Newburgh, the Senate House in Kingston, Schuyler Mansion in Albany, Fort Crailo in Rensselaer, Johnson Hall in Johnstown, and the General Herkimer Homestead near Little Falls. The site system also included other properties such as the Bennington, Oriskany and Newtown battlefields and the Saratoga battle monument which also related to the state's Revolutionary history. Although some of the buildings are of considerable architectural significance, they were acquired by the state primarily because of their historical associations.

Before 1966, the state historic site program was placed first in one agency and then another. Unfortunately none of these agencies had considered historic preservation to be óne of its major program responsibilities, and the sites did not receive substantial funding. In that watershed year for historic preservation, 1966, the National Historic Preservation Act was passed at the federal level. It called for an expanded National Register of Historic Places and a unique partnership of the federal government with the states to carry out the purposes of the act. On the state level in 1966, legislation was passed creating the New York State Trust. Consisting of a policy-making board of seven citizen and *ex officio* members, the Trust was established administratively within the Division of Parks of the Conservation Department as the first New York State agency exclusively responsible for historic preservation activity. The chairman of the Trust became the State Historic Preservation Officer for New York under provision of the National Historic Preservation Act and, for the first time, a state agency was established with responsibility for coordinating historic preservation activities on all government levels as well as in the private sector within the state.

In 1972, a new Office of Parks and Recreation was established under the leadership of Commissioner Alexander Aldrich. The New York State Historic Trust was reorganized as the Division for Historic Preservation within OPR. The state legislation was recodified, giving the Office of Parks and Recreation even more responsibility as the state's primary historic preservation agency. A State Board for Historic Preservation was established to advise the Commissioner on historic preservation matters, and the position of Deputy Commissioner for Historic Preservation was established with the primary responsibility for directing and administering the program of the Division for Historic Preservation. This change in legislation provided a stronger legal base for the preservation program and facilitated the use of OPR resources to assist the program.

Under the Historic Trust, the state historic sites system was expanded by the addition of a number of very significant properties during the late 1960s. These properties expanded the geographical and thematic base of the state historic site system and include Olana near Hudson, Hyde Hall near Cooperstown, Lorenzo in Cazenovia, and Schoharie Crossing, an Erie Canal site at Fort Hunter. In addition, several historic properties in state parks, such as Mills Mansion at Staatsburgh and Clermont at Germantown were transferred into the historic sites system. Interestingly, most of these sites are significant because of their architectural and environmental significance rather than their connection with a single historic event. At the present time the site system consists of three dozen properties, including the Schermerhorn Row Block in New York City which was acquired in 1974 as the future site of the New York State Maritime Museum.

Since the establishment of the Historic Trust, the primary thrust of the preservation program has been to establish professional staffs at all levels to implement the programs mandated by legislation, including physical upgrading of the historic site system. This process was begun by Mark Lawton, the first Director of the Historic Trust and now First Deputy Commissioner of the Office of Parks and Recreation. It has been significantly expanded and advanced by Frederick L. Rath, Jr., Deputy Commissioner for Historic Preservation.

Under Deputy Commissioner Rath's leadership, an organizational plan was prepared for the Division for Historic Preservation, which called for the establishment of six bureaus to carry out the historic preservation mandate. These bureaus include the following: National Register and Statewide Survey; Historic Archaeology; Preservation and Restoration; Conservation and Collections Care; Research and Publications; and Historic Sites.

For the Bicentennial, a program called "Heritage '76" has been presented to and accepted by the Governor and Legislature. This is a long-range program of historic site management which provides for the repair and restoration of the state-owned historic sites and their contents. The major historic sites have been designated Preservation Centers. Under the program, the public is invited to visit the site while restoration, archaeological and conservation work are occurring. Information now available on safeguarding the historic, architectural, and archaeological resources is shared with the public. Visitors are encouraged to seek advice for their own preservation problems from the professional staffs of the Division for Historic Preservation.

PRESERVATION AND RESTORATION OF HISTORIC SITES

The preservation problems inherent in the system of historic sites taken over by the Historic Trust were almost overwhelming. In addition to lack of maintenance and preservation work at the sites which had been owned by the state for some time, many of the buildings which had been recently acquired from private owners were in a badly deteriorated condition. In many cases, the buildings were not even weathertight and major structural damage had occurred. While some buildings were sound, others were in a state of collapse. Many buildings with superb interior decoration and furnishings were the victims of badly leaking roofs. Virtually every site had inadequate, and sometimes dangerous, mechanical systems.

Where preservation and repair work had been carried out in the past, few if any records were kept. There were no central files of plans and specifications for the previous work. In some cases, restoration techniques based on unsound building practices had resulted in cures which often did more damage than the original maladies. These unsound techniques included the use of dissimilar metals on roofs, which resulted in severe cases of electrolysis; sandblasting and siliconing historic masonry, causing major deterioration; repointing historic masonry with modern hard portland cement mortar; and the unknowing removal or replacement of original architectural elements.

Very little historical research had been carried out for sites, and no analysis existed on the physical fabric of the buildings. As far as could be determined, no historic structure reports had been prepared for any property. Even basic architectural tools, such as accurate measured drawings and photographs, were lacking except in a few isolated cases.

It rapidly became apparent that it was necessary to design an entirely new system to implement preservation and restoration work on the three dozen state-owned sites, located from Long Island to the Great Lakes. Before the establishment of the Historic Trust, there had been no professional architectural staff or trained craftsmen on the state payroll responsible for the preservation of historic buildings.

The system that was established has evolved into the Preservation and Restoration Bureau of the Divison for Historic Preservation. Included on the staff of the Bureau are restoration architects, historic landscape architects, architectural historians, draftsmen, preservation technologists, restoration supervisors and craftsmen. The Bureau is responsible for the preservation and restoration work at the state-owned sites, as well as providing professional assistance to other state agencies, units of local government and private groups.

The procedures established by the Bureau call for a carefully defined and researched process for the restoration of historic structures. The first step is emergency stabilization of the building by making it structurally sound, weathertight and vandal proof using simple and nondestructive methods to "mothball" the structure. After this basic stabilization has been completed, fumigation and cleaning up are carried out if required. Extreme care is required in cleaning out a building to insure that nothing of historical significance is discarded or destroyed. At this stage, everything that could be of importance is salvaged, labeled and then stored for future reference.

The next step in the preservation process is the preparation of an Historic Structure Report by architects and architectural historians on the Bureau staff. The Historic Structure Report is the basic research document for all work on the building and it consists of three sections. The first is an historical analysis of the structure, its construction, the land on which it stands and the people associated with it. The second section is a detailed architectural analysis of the building that utilizes measured drawings and photographs. This section includes an analysis of the construction history and physical development of the building, as well as a detailed compilation and dating of the historic fabric of the entire period on a room-to-room basis. The third section consists of a series of detailed recommendations for the treatment of the building. Based on the first two sections, the recommendations for the treatment of the building form a logical and consistent program for future preservation and restoration work. The Historic Structures Report may be supplemented by additional reports regarding specific historic building materials and processes, as well as modern stabilization techniques. For grounds restoration, an Historic Landscape Report is prepared, following the general format of the Historic Structure Report.

Plans and specifications for general preservation and restoration work are prepared by the Bureau's architectural and landscape architecture staff. Private consultants or engineers from the New York State Office of General Services prepare contractual documents for the installation of new mechanical systems, parking areas and access roads, and public facilities in conjunction with the Preservation and Restoration Bureau and the engineering staff of the regional park office where the project is located. These kinds of construction work are usually carried out by private contractors although some of the state restoration crews include plumbers and electricians. Outside consultants are also used on specialized jobs such as ruins stabilization and slate and sheet metal roofing work, as well as projects of especially large scale. This construction work is also carried out by·private contractors.

However, the vast majority of preservation and restoration work on the state-owned historic sites is carried out by restoration crews on the state payroll. The first state restoration crew, consisting of five men, was established at Hyde Hall in Otsego County in the fall of 1970. Since that time, workshops have been established at seven strategic locations across the state to service as many state historic sites as possible. At the present time the seven workshops are located at the following historic sites: John Jay House, Westchester County; Olana, Columbia County;Peeble's Island, Saratoga County; Hyde Hall, Otsego County; Lorenzo, Madison County;Sackets Harbor,

Jefferson County; and Crown Point, Essex County. The crews based at these seven workshops are presently working on twenty historic sites extending from Long Island to the shores of Lake Ontario. Each crew is responsible for sites within a hour's driving time of its home base. The map in Figure 20.1 indicates the locations of the workshops as well as the other historic sites.

One of the most encouraging aspects in the establishment of a system of state restoration crews has been the fine craftsmen who have been attracted to the program. It has been possible to recruit very capable and experienced craftsmen to lead the restoration crews in every area of the state where work is occurring. The Department of Civil Service and the Division of the Budget have been cooperative in establishing permanent positions, with the title of Building Restoration Specialist, for team leaders. This title is the first civil service restoration craftsman title established by New York State, and its level reflects the levels of responsibility and skill required for the head of a restoration crew. To qualify for the position, at least 6 years of restoration experience and a proficiency in at least two trades is required. Under the supervision of the Building Restoration Specialists on each crew are foremen, carpenters, masons, plumbers, electricians and laborers.

A decision was made to establish a basic core of permanent positions for each field crew, consisting of the Building Restoration Specialist, and the necessary foremen and critical mechanics. To this permanent core, temporary and seasonal positions are added so that the necessary tradesmen may be hired to accomplish a particular project for which capital construction money has been appropriated. This permits the hiring of appropriate tradesmen to carry out special projects such as roofing, plumbing, and masonry restoration.

It has been possible to establish a ladder of job titles in five stages, ranging from laborer to Building Restoration Specialist. This makes it possible to hire workmen with varying amounts of experience and skills at an appropriate level. Although concern has been ex-

Fig. 20.1   Map of New York State indicating location of restoration workshops and State historic sites where construction is underway. *(All illustrations in this chapter courtesy of the New York State Office of Parks and Recreation)*

pressed that it is almost impossible to find restoration craftsmen today and that specially subsidized programs should be established to train these men, it has been the experience of the Division for Historic Preservation that such men do exist in all areas of New York State and are willing to work, provided there is a viable program to employ them. It has also been found that there is a sincere interest among young men to enter the restoration field. A number of graduates of high school BOCES, or two-year technical college construction programs, have been recruited for the restoration crews. Today there still are many craftsmen interested in doing highly skilled preservation work in which they can

take pride. State government can offer them job security, year round employment and recognition. During winter when outside work is impractical in New York State, the crews work inside, fabricating millwork, repairing woodwork, plastering and doing sheet metal work.

Each of the workshops has been established with its own specialty. For example, the facility at the John Jay House is a basic carpenter shop, and in addition the crew specializes in heavy structural repairs. At Olana, there is also a carpentry shop, but the specialization is finely executed finished woodwork. At Hyde Hall is another carpentry shop specializing in finished work, as well as a sheet metal shop. There the crew was fortunate in acquiring tools from two nineteenth-century tinsmith shops which the workmen now use for restoration work. A masonry workshop along with a carpentry shop will be located at a former Cluett, Peabody & Co. bleachery building on Peeble's Island, at the junction of the Mohawk and Hudson Rivers. The island has been acquired by the Office of Parks and Recreation as a state park and now houses a central storehouse for the Preservation and Restoration Bureau's building materials. Historic building materials and architectural fragments will also be stored at Peeble's Island as a study collection.

Those workshops with specialties provide those services on a system-wide basis. Therefore sheet metal flashings might be fabricated at Hyde Hall for sites as far away as Sackets Harbor or Olana.

Each restoration crew is assigned a draftsman who prepares measured drawings of the building as it is opened up for restoration, as well as as-built drawings after restoration is completed. The basic measured drawings of each building consisting of plans, elevations and sections are prepared by the architectural staff from the main office, and the work of the field draftsmen complements these drawings. A technician is also assigned to each crew with the responsibility for collecting and labeling historic building fragments, as well as keeping the payroll records of the crew and ordering materials and equipment.

The basic truck used by the restoration crews is a double-cab pick-up truck which can carry six men. The trucks have either 6- or 8-foot cargo boxes with aluminum shells and ladder racks, which make it possible to carry an entire crew as well as its materials and tools. The aluminum shell protects the contents of the cargo box from the weather and vandals. The restoration crews also have access to equipment and trucks assigned to the regional park offices.

The crews are under the direction of two Building Restoration Supervisors, one for the upstate area and one for downstate. The men in these positions must have extensive restoration experience and knowledge. They are responsible for directing the work of the restoration crews and coordinating that work with the main office and the regional park offices. The Restoration Supervisors are also responsible for implementing training programs for the restoration crews. These programs consist of actual on-the-job training, as well as classroom sessions. Demonstrations of appropriate restoration techniques are often made in the field. The Restoration Supervisor for the downstate area is Anton Vermey, and Douglas Clinton serves the upstate area. Both have many years of restoration experience and were among the first restoration specialists hired by the Division for Historic Preservation.

Throughout the entire restoration process there is close communication and interaction with the Archaeology Bureau of the Division for Historic Preservation. Beginning with the basic research and examination of the building prior to restoration, the staff members of the bureau, under the direction of Paul R. Huey, Sr., Historical Archaeologist, work closely with the architects and architectural historians of the Preservation and Restoration Bureau. The research programs and work schedules of both bureaus are closely coordinated and a joint laboratory is being developed. The laboratory will be used for the treatment of archaeological artifacts, as well as for the analysis of paints, mortars, stuccos and other materials used in restoration.

THE "HERITAGE '76" BICENTENNIAL PROGRAM

A 4-year Bicentennial program for the restoration and development of the state-owned historic sites has been submitted to the Governor and Legislature. This program, beginning in fiscal year 1974–75 and extending through 1977–78, calls for an expenditure of $19.5 million. It includes the restoration of buildings and landscapes, the installation of adequate parking and comfort facilities, and the construction of visitor orientation centers for the major sites.

Under this plan, the major sites are designated Historic Preservation Centers. These will serve the surrounding communities by providing information and assistance on historic preservation problems ranging from the care of antique furniture to methods of surveying historic areas, to the restoration of historic buildings. As restoration work is carried out on the State-owned sites, the work will be made accessible to the public. Instead of being closed to the public during restoration, the sites will be open and visitors encouraged to view the work in progress. The state sites will serve as building preservation laboratories for the development of new preservation methods and technology. This information will be disseminated through visits to the sites, seminars, publications and technical consultations to local groups by the Preservation and Restoration Bureau staff both through the Albany office and the Historic Preservation Centers. This will provide a mechanism for direct assistance to the people of the State.

In October, 1974, the first public seminar on preservation was presented by the Division for Historic Preservation in conjunction with a local historical society. At the day-long session, members of the restoration crews and staff from the Albany office presented a program on preservation and restoration techniques and answered questions regarding technical problems people had with their own buildings. The seminar was very successful and it is anticipated that additional programs will be scheduled in a number of areas across the state.

As the program has been developed, administrative relationships and channels of communication have been developed between the Albany office staff, field crews, regional park administrators and site personnel. To define these, a manual of procedures has been prepared which includes statements defining the preservation and restoration policies and philosophy of the Division for Historic Preservation. Also included is a detailed outline of the preservation and development procedures for State-owned historic sites.

In order to support a series of restoration crews, much coordination is required to in-sure that the necessary research is carried out and that the required restoration drawings and engineering services are available at the appropriate time. As the restoration program becomes more extensive, the purchase of building materials and equipment becomes critical. Materials which are not ordinarily purchased by the state, such as treated wood shingles, hand-made brick, lead-coated copper, and pressure-treated lumber, are ordered in bulk and stored at the Peeble's Island warehouse.

Although some new trucks and scaffolding have been purchased for the restoration crews, the program has expanded so rapidly that it has been necessary to supplement these with surplus equipment and trucks from other state agencies. The problem of getting adequate and reliable trucks has been a major one and hopefully will be solved by the purchase of new vehicles.

Critical path methods are used for programming and managing the restoration and construction program to insure that the crews and their equipment are used as efficiently as possible. The financial accounting of each job is kept on a computer in the Department of Audit and Control. These accounts are monitored each month and expenditures for labor and material are noted. Careful and accurate expenditures records are kept of each job; after completion, a complete cost analysis is made for selected projects. This information is used by the Bureau staff in Albany for programming and budgeting purposes.

After the program for the restoration of the state historic sites has been completed, it has been recommended that the architectural staff and restoration crews be made available for work on other state-owned historic buildings which are not designated historic sites. The state owns thousands of structures, many of which are eligible for listing on the National Register of Historic Places. The vast majority of these historic resources are used for non-museum functions. Another possibility that has been suggested is the use of these forces for the restoration of buildings owned by units of local government and private organizations as a kind of special grant-in-aid program to complement the existing cash grant programs administered by the Office of Parks and Recreation. The crews would undertake specialized work that is not easily or economically carried out by private contractors.

CASE STUDIES

Schuyler Mansion

The house of Philip Schuyler was constructed in 1762 on an estate called The Pastures on the outskirts of Albany. Schuyler was the scion of a Dutch family prominent in the city beginning in the seventeenth century. Although Schuyler came from a Dutch-influenced background, the house he built reflected the latest Anglo-Georgian architectural taste of the period. In 1912, the house, now enclosed by the expanding city, was acquired by New York State as an historic site and subsequently restored.

During the 1950s, the brick walls were sandblasted, given a coating of silicone, and repointed with portland cement mortar. Serious problems with the exterior brick walls became evident in the late 1960s. Some were caused by the unusual geological conditions of the Albany area, consisting of layers of clay which tend to slide away from each other. One corner of the house was settling and moving down the hill away from the rest of the building. Steel tie rods were inserted through the house to control the problems of the differential settlement and movement. There was also severe deterioration of the surfaces of the brick walls which had been sandblasted, siliconed, and repointed a dozen years earlier.

A restoration crew specializing in masonry was established at the site to begin work on the problem of the deteriorated masonry. The hard portland cement mortar was not plastic and therefore did not accommodate the movements of the building, causing the bricks

Fig. 20.2 Schuyler Mansion, east elevation. The wood balustrade has been removed from around the edge of the roof. Evidence of deterioration of the hexagonal entrance vestibule can be seen in the upper left corner under the vestibule cornice.

themselves to fracture as settlement occurred. The cement mortar was carefully removed and replaced with a more plastic lime mortar which was compatible with the original construction and better accommodates the building movements.

The crew found that what appeared to be a minor pointing problem in the hexagonal entrance vestibule in reality was evidence of major deterioration (Fig. 20.2). The vestibule was originally roofed with tinplate pans, including a built-in gutter. During the 1930s, when the tinplate gutters were replaced with ones of copper, the tinplate roof was retained. Wherever the copper and tinplate met along the perimeter of the gutter, electrolysis occurred, causing the tinplate to deteriorate; this in turn permitted water to freely penetrate the interior of the wall. This caused the destruction of the inner construction of the masonry wall and rotted out the wood framing members for the ceiling and wall. To correct the damage, the upper fifteen courses of brick had to be numbered and removed along with the decorative wood cornices and the metal roof covering. The restoration crew from Hyde Hall was called in to stabilize and replace the damaged wood framing. Wherever possible, original fabric was retained and new structural elements were installed alongside the old. The decorative plaster ceiling inside the vestibule was stabilized in place and a new roof of lead-coated copper pans was installed along with a new gutter of the same material. The pans of the new roof were of the same size as the original and were installed in the same manner.

It was also found that the vestibule was sliding away from the house and had to be restrained by the installation of new steel tie rods. In order to accomplish this, the brownstone steps and stoop had to be dismantled, the underpinnings rebuilt, and the finished surfaces stabilized and reset (Fig. 20.3).

Not all of the restoration work involved masonry. The decorative wood balustrade around the main house was badly rotted and some of the sections had collapsed. The balustrade, which was found to date from at least

Fig. 20.3 Schuyler Mansion: detail of stairs and landing to entrance vestibule. Because substructure had disintegrated, steel tie rods and plates were inserted to prevent the vestibule from sliding away from the main part of the building. The brownstone blocks have been numbered, removed and replaced in their original locations.

the beginning of the nineteenth century—and possibly earlier—was carefully removed and taken to the Hyde Hall shop. There an exact duplicate was fabricated from redwood. The new railing was painted and installed on the house and the entire original balustrade was placed in the Preservation and Restoration Bureau's study collection.

Future work at Schuyler Mansion will include the restoration of the interior of the house, installation of new mechanical and detection systems, and rehabilitation of five adjacent nineteenth-century rowhouses as a visitor orientation center and offices.

Lorenzo

Lorenzo was constructed in 1807–08 overlooking Cazenovia Lake for John Lincklaen, director of the Holland Land Company, which did much to encourage the settlement

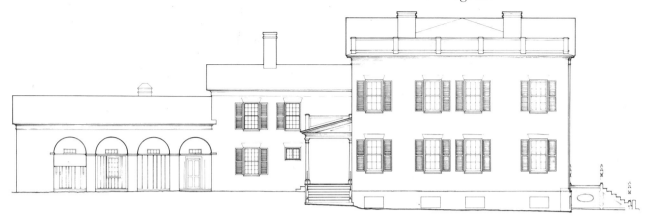

Fig. 20.4   Lorenzo, east elevation. Work on the two-story section of the building in the center revealed an original built-in gutter along the eaves. The wood gutter was restored and protected by a sheet neoprene liner. *(Drawing by John Shank)*

of central and western New York by the sale of lands to settlers after the Revolutionary War. Handsomely sited on the landscaped estate, Lorenzo remained in the ownership of the original family until 1967, when it was acquired by New York State. The house has survived intact with its original furnishings and collections, as well as its historic landscaping and outbuildings.

The first work done by the restoration crews was the restoration of the back porch, which had collapsed. The original framing for the porch was reinforced with modern lumber and the old sheathing was retained. The original porch roof profile was restored and covered with wood shingles to duplicate the historic appearance.

After the completion of the porch, the crew began work on replacing the roof on the house. Lorenzo was covered by a wood shingle roof; however, when acquired by the State the house had a badly deteriorated sheet metal roof. In removing the sheet metal roofing, it was found that much of the original moulded wood cornice, containing a built-in gutter, was badly rotted (Fig. 20.5). As much as possible of the original cornice and gutter was stabilized and retained; those sections that were not salvageable were replaced by new materials identical in appearance to the originals. The built-in gutter was restored and was lined with sheet neoprene to prevent future rotting of the cornice. A great deal of new flashing work was

required and the details were executed to be as inconspicuous as possible, and yet be weathertight in the harsh central New York climate.

A woodworking shop has been installed in the historic carriage house and work has been started on the restoration of that building. The restoration of the roof on the main house is continuing and future work includes extensive stabilization work on the exterior brickwork. Deterioration of the masonry caused by

Fig. 20.5   Lorenzo: detail of the plate and rafters where the roof joins the wall in the center section of the building. At left is a solid section of the original plate which had not rotted. At the right, where the original piece had rotted, is a new plate built up of modern dimensional lumber. The two sections are connected with a scarf joint. An original rafter is on the right. A new rafter has been nailed alongside the original, which had deteriorated where it joined the plate.

379

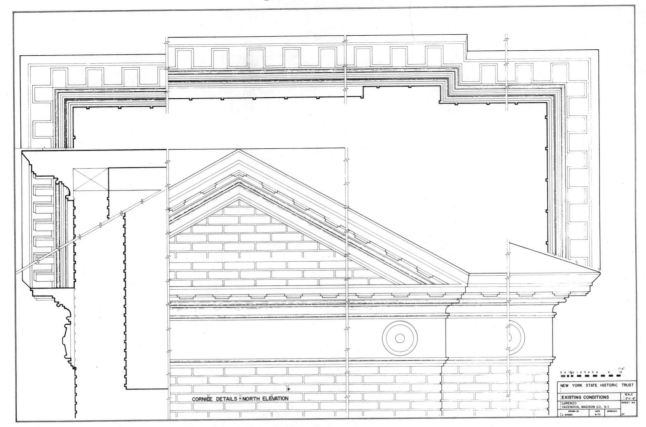

Fig. 20.6 Lorenzo: detail of the cornice and pediment of the front (north) elevation. A leaking roof caused deterioration of this section of the building. The drawing, by John Shank, serves as the basis for its restoration.

defective built-in gutters has occurred in a manner similar to the damage on the Schuyler Mansion vestibule. Also scheduled is an extensive restoration of the interior surfaces and the replacement of the existing inadequate heating, plumbing and electrical systems. All of the outbuildings will be stabilized and restored. An historic landscape report is being prepared for the very significant grounds of Lorenzo to guide the preservation and development of the estate. A major design problem is determining a location for the visitors' parking area where it will not be a major intrusion into the historic environment and yet be viable for the operations of the site.

Hyde Hall

Constructed between 1817 and 1833 overlooking Otsego Lake near Cooperstown, Hyde Hall was designed by the noted Albany archi-

tect, Philip Hooker. When the estate was acquired for parks purposes by New York State in 1964, the main house and outbuildings were seriously deteriorated and in some cases in ruins. When jurisdiction of the main house and the immediate outbuildings and grounds was transferred to the Historic Trust, the original carriage house designed by Hooker in 1819 had collapsed. The roof of the main house was leaking badly and in many sections the floor joists had rotted and collapsed.

Because of the extent of the preservation problems and the urgency in stabilizing the complex, Hyde Hall was selected as the location of the first State restoration crew. Once the crew was established, under the leadership of Douglas Clinton, the first project was to rebuild the collapsed carriage house using a modern framing system and retaining as much of the original exterior building fabric as possible. Inside the carriage house, a workshop was set up for use as the base of operations for the restoration of the rest of the complex. Other outbuildings have been restored to house additional shop facilities and the buildings presently house a large woodworking shop, sheet metal shop, building materials and fragments storage, drafting room and offices. These shops are now used to fabricate restoration components for other sites in the system.

The first stage in the restoration of the main house was the installation of a new roof (Fig. 20.7). Research indicated that the front sections of the original roof were tinplate while other areas were wood shingle. The crew began by replacing the wood shingles. Altogether there were sixteen sections restored in wood shingle and eleven in tinplate. Each roof section had a different series of problems which became apparent only when the roof was opened up (Fig. 20.8). These problems ranged from the rotting of heavy timber plates and rafters, to sections of the roof where the framing system had shifted causing structural instability. Fires during the nine-

teenth century had destroyed sections of the framing. Because of the complicated configuration of the roof, the lack of adequate flashing had caused major deterioration over the years. The bases of the limestone block chimneys had eroded away below the roof line causing a serious lack of stability. The lower sections of the chimneys had to be rebuilt and new flashings were designed and fabricated in such a way as to be visually compatible with the design of the building as well as function adequately in keeping water out.

Inside the house, the leaking roofs and moisture rising from the damp ground had combined to rot out much of the first floor framing and flooring (Fig. 20.9). In some areas fungus had completely consumed moulded wood baseboards and door entablatures. Because the house had never been adequately heated, quantities of soil from beneath the first floor had to be excavated to provide crawl spaces to accommodate heating and ventilating ducts.

Mantle pieces had to be disassembled and fireplaces rebuilt because moisture had penetrated through the chimneys, causing dis-

Fig. 20.7 Hyde Hall, from the southwest. One of the roof sections, originally covered with tinplate sheets, has been restored.

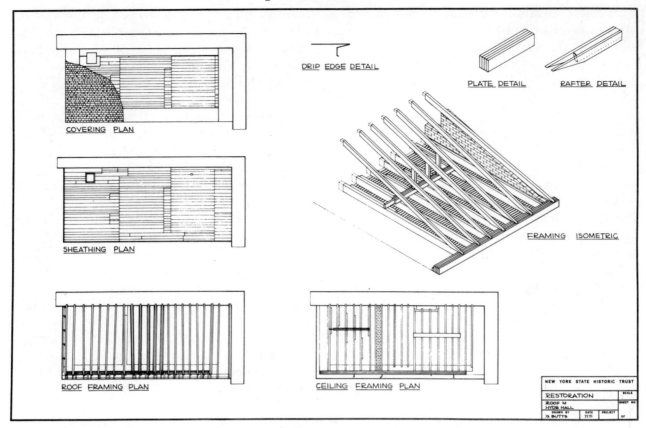

COVERING PLAN

SHEATHING PLAN

ROOF FRAMING PLAN

CEILING FRAMING PLAN

DRIP EDGE DETAIL

PLATE DETAIL

RAFTER DETAIL

FRAMING ISOMETRIC

NEW YORK STATE HISTORIC TRUST
RESTORATION
ROOF M
HYDE HALL
DRAWN BY G. BUTTS DATE 7/71 PROJECT

Fig. 20.8   Hyde Hall: isometric and details of a portion of the roof indicating conditions after restoration work had been carried out. The drawings include details of modern materials and techniques used in the restoration, and also show where and to what extent new materials have been added to complement the historic fabric of the building.
(Drawing by George Butts)

placement and eventual failure of parts of the fireplaces. Rotting had occurred, necessitating the replacement of damaged framing members with new pressure-treated lumber. The basic stabilization of the house is continuing. New mechanical systems will be installed in the next phase of work on the building. After the mechanical systems are installed, the restoration of the interior finishes will be completed.

Sackets Harbor

During the War of 1812, the small village of Sackets Harbor became a center of American naval and military activity for the upper St. Lawrence River Valley and Lake Ontario. A major shipyard and naval station was established and in 1813 Sackets Harbor was the site of a decisive battle which ended as an Ameri-

can victory. Sackets Harbor remained an important naval station until after the Civil War, when hostility between the United States and Great Britain began to calm down. During the 1840s the naval station experienced a major building program when a variety of structures were constructed, including impressive houses for the Commandant and Lieutenant, and several outbuildings. It remained a naval station until 1955. Part was acquired by the Historic Trust and combined with the battlefield to form an historic site.

The first structure to be restored at Sackets Harbor was the Lieutenant's House, rehabilitated as site offices and staff housing. The restoration of the Commandant's House as an exhibited building was begun by replacing the roof with wood shingles to match the appearance of the original. The chimneys, which

Fig. 20.9   Hyde Hall: floor framing system in the kitchen. Floor joists which had failed due to rotting were removed and a crawl space dug below the framing so that air circulation would prevent similar damage in the future.

Fig. 20.11   After restoration, the Commandant's House again boasts a porch which had been added about 5 years after construction.

were either missing or deteriorated were rebuilt or stabilized to their original condition (Fig. 20.10). The exterior brick walls were stabilized and repainted with their original colors.

Within five years of the construction of the

Fig. 29.10   Sackets Harbor, Commandant's House, constructed in 1845, is shown before restoration.

Commandant's House in 1845, splendid gothic revival porches were added to the front and back of the building (Fig. 20.11). As these wood porches deteriorated, sections were gradually removed until they completely disappeared sometime during the mid-twentieth century. However, the site staff found an 1860s photograph of the house which clearly showed the front porch. In addition, the archaeologists found a section of a quadrifoil buried in a light well, as well as evidence of the porch foundations. This evidence, both photographic and physical, made possible the accurate reconstruction of the porches on paper using reverse perspective techniques (Fig. 20.12). Once working drawings were prepared, the porches were rebuilt by the restoration crew to exactly match their historic appearances. An interesting aspect of the porch rebuilding was the finding of a small piece of painted canvas which had been the original porch roof covering. The appearance of the painted canvas was duplicated by using neoprene sheets.

Future work scheduled for the Commandant's House includes the restoration of the interior finishes and the installation of new mechanical systems.

Recently the Union Hotel, a large stone

383

Fig. 20.12 A fragment of the original quadrifoil detail and a photograph of the building from the 1860s are used in preparation of working drawings for the reconstruction of the Commandant's House porches.

building, was purchased by the State for use as a visitor orientation center (Fig. 20.13). The hotel was constructed as a hostelry in 1817 and today it contains some of the finest non-residential Federal period interiors in the entire state. The restoration crew has begun replacing the modern sheet metal roof with one of wood shingles and the rebuilding of missing or deteriorated chimneys. Work has already begun on the preservation and restoration of the very fine interior woodwork and the replacement of original walls, which were moved during the twentieth century. Again, future work includes the complete restoration of the interior finishes and the installation of new heating, plumbing and electrical systems.

Olana

One of the most spectacular sites in the state system, Olana is a complete environment—buildings, grounds and collections—shaped by Frederic E. Church, the noted Hudson River School artist. The house and 250 acres

of grounds were as carefully designed by Church as his paintings. He happily referred to his estate as "About one hour this side of Albany is the center of the world—I own it."

The center of Olana is the castle designed by Church and his architect, Calvert Vaux, on top of Church Hill (Fig. 20.14). Built between 1870 and 1874, the castle contains thirty-seven rooms. A studio wing was constructed in 1888–90. The entire exterior of the house was conceived as a single, unified design. The decorative slate roof, polychromed woodwork and cornices, and multicolored brick and tile masonry walls united to form a composition of great visual impact.

The site was acquired by the State in 1966 and, not unexpectedly, required major restoration work. The immediate problem was the sheet metal and slate roof which contained over seventy-five leaks. The existing roof was very carefully replaced by one of modern materials which reproduced as closely as possibly the appearance of the original roof. Research revealed the original slate patterns as well as the quarries which supplied the slate. It was possible to return to the areas where the original slate was quarried and to purchase new material that exactly matched the color of the old (Fig. 20.15). It was even possible to

Fig. 20.13 Sackets Harbor: exterior of Union Hotel undergoing restoration. The hotel, which dates from 1817, contains some of the finest nonresidential Federal period interiors in New York State. It will be used as a visitor center.

Fig. 20.14 Olana, south elevation before restoration of roof and exterior woodwork.
Built between 1870 and 1874, the "castle" was designed by Frederic E. Church, the noted
Hudson River School artist, and Calvert Vaux, his architect.

replace the gilded tinplate shingles which formed the centers of the decorative patterns of the roof of the belltower.

Many of the roof sections were flat and had been covered with tinplate pans; after structural and sheathing repairs were made, these sections were covered with lead-coated copper pans. Great care was exercised to match exactly the original profiles of the roof and to retain the configuration of original details, such as gutters, scuppers and downspouts.

Much of the polychromed exterior woodwork was deteriorated and had to be replaced or restored (Figs. 20.16 and 20.17). This work was done by the restoration crew assigned to a workshop on the site. As much as possible of the original woodwork was retained. Where it was not possible to retain original elements, new components were made to exactly match the originals, which were in turn retired to the study collection.

Although the wood cornices which visually connected the slate roofs and patterned brick walls were originally polychromed, they all had been painted brown during the twentieth century. Physical examination and documentary research has indicated the original colors and patterns which will be restored.

The exterior masonry walls were badly in need of repointing, as the design of the roofs

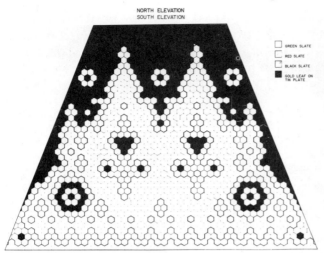

NORTH ELEVATION
SOUTH ELEVATION

GREEN SLATE
RED SLATE
BLACK SLATE
GOLD LEAF ON
TIN PLATE

Fig. 20.15 Olana: working drawing for restoration of
the slate roof of the main tower. Original quarries produced new slate which exactly matched the old green,
red and black slates. Gilded tinplate centers of the decorative patterns were also reproduced. *(Drawing by John
Shank)*

and towers created many pockets which
trapped moisture, causing the building fabric
to deteriorate. An extensive program has been
started to repair and rebuild the chimneys and

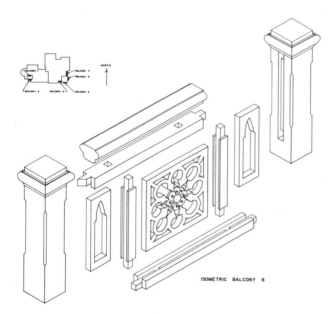

ISOMETRIC BALCONY 6

Fig. 20.16 Olana: isometric drawing of the original
polychromed wood balustrade of exterior balconies.
While some original woodwork could be retained, reproductions were required to replace heavily damaged
sections. *(Drawing by James L. Johnson)*

Fig. 20.17 A section of the original wood balustrade
being replaced in position after restoration in workshop.

repoint the walls where deterioration had
occurred (Fig. 20.18). In the future, a maintenance program will be implemented which
will retard further deterioration.

Another major problem was the deterioration and general inadequacy of the heating,
electrical and plumbing systems. It was decided to replace all of these mechanical systems
and to locate the furnace and tank for the
heating oil outside the main house. A short
distance from the castle, the archaeologists
found evidence of a small carriage house dating from the 1890s. Documentary research
revealed photographs of the structure, which
was located between the castle and main carriage house. The building had been sold about
30 years previously and moved away from the
site. The site staff was able to locate the building, which was in use as a garage on a local

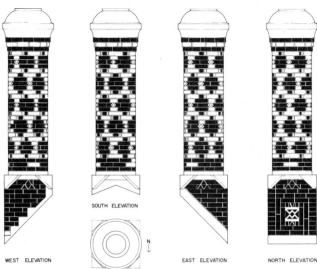

Fig. 20.18 Existing-conditions drawing of the poly-chromed round chimney guided its repair. Pattern is composed of red and yellow bricks, with red base. *(Drawing by James L. Johnson and Clayton R. Pauley)*

farm. After negotiations, the original building was reacquired and moved back to Olana.

There it was placed over a new concrete building constructed on the original site to house the furnace and the main electrical service panels. The furnace supplies steam, which is piped to the main house where it is converted by exchange units into hot air, which in turn is distributed throughout the building by the original heating ducts. This system permits the introduction of modern temperature and humidity control systems into a historic building using the original duct system. It also removed the danger of the furnace from the house to a modern, fireproof concrete building.

The main carriage house is now being restored for use as site offices and exhibit areas. Further work scheduled for the site includes the continuation of the interior and exterior restoration of the castle; the continued preservation and rehabilitation of the carriage house and other outbuildings; and the stabilization and restoration of the grounds and landscaping.

Index

Page numbers in *italics* indicate material in illustrations; page numbers indicating material in notes are followed by *n* and the number of the specific note.

389